Science Challenging AIDS

Proceedings Based on the VIIth International Conference on AIDS,
Florence, June 16–21, 1991

Science Challenging AIDS

Managing Editor
Gaetano Giraldo, Naples

Volume Editors
Giovanni Battista Rossi, Rome
Elke Beth-Giraldo, Naples
Luigi Chieco-Bianchi, Padua
Ferdinando Dianzani, Rome
Gaetano Giraldo, Naples
Paola Verani, Rome

52 figures and 16 tables, 1992

KARGER

Basel · München · Paris · London · New York · New Delhi · Bangkok · Singapore · Tokyo · Sydney

Library of Congress Cataloging-in-Publication Data
International Conference on AIDS (VIIth: 1991: Florence, Italy)
Science challenging AIDS / managing editor, Gaetano Giraldo;
volume editor, Giovanni Battista Rossi ... [et al.].
Includes bibliographical references and index.
1. AIDS (Disease) – Congresses. I. Giraldo, G. (Gaetano) II. Rossi, G.B. (Giovanni B.)
III. Title.
[DNLM: 1. Acquired Immunodeficiency Syndrome – congresses.]
ISBN 3-8055-5485-0

Drug Dosage

The authors and the publisher have exerted every effort to ensure that drug selection and dosage set forth in this text are in accord with current recommendations and practice at the time of publication. However, in view of ongoing research, changes in government regulations, and the constant flow of information relating to drug therapy and drug reactions, the reader is urged to check the package insert for each drug for any change in indications and dosage and for added warnings and precautions. This is particularly important when the recommended agent is a new and/or infrequently employed drug.

Printed in Switzerland on acid-free paper by Thür AG Offsetdruck, Pratteln
ISBN 3-8055-5485-0

Contents

The Best Papers Selected

Summary Reports

Preface

The Italian Experience in the Global Challenge against AIDS

Hon. *Francesco de Lorenzo*

Minister of Public Health, Italy

It is my pleasure to introduce this book containing the master lectures, state-of-the-art plenary lectures, the best papers selected and the summary reports from the VIIth International Conference on AIDS that was held in Florence in June 1991. The conference represented the major international forum on AIDS, and constituted a meeting place for initiatives on a problem that in a short time has had a dramatic global impact.

In addition to the scientific observations presented during the conference, discussion focused also on important questions relating to the war against AIDS – not only as a war against the disease, but also as a concrete prospect of human solidarity towards the persons suffering.

We obtained maximum benefits from these presentations in terms of exchange and collaboration between the different people, professional and nonprofessional, fighting against HIV infection. This is an extraordinary opportunity to unite efforts at all levels – whether they be researchers, clinicians, health care workers, NGOs or representatives of the mass media.

In Italy, the fight against AIDS is very long and became fully operational in March 1987, when the Ministry of Health, faced with the evolution of the AIDS epidemic in this country and the rest of the world, set up a national commission of experts. This group helped to provide guidelines and expertise to the Minister, who is also the chairman of the commission, in order to coordinate the efforts already underway in some regions which were hard hit by the HIV infection. This initiative was well integrated with the development of the AIDS policy at the European level: in November 1987, the European committee of Health Ministers declared the fight against AIDS to have priority at a national level.

The Italian National Commission has always worked in conjunction with the WHO document 'Guidelines for the Development of National AIDS Prevention and Control Program' in order to fully implement the global fight against AIDS. This policy has focused since the beginning on international collaboration and the exchange of information at a national level, in both AIDS research and the struggle against discrimination. As regards the Italian situation, the activity of the national commission has recently been enriched by the participation of NGOs aiming to harmonize legislative and social decisions.

I wish to emphasize this serious and important point. The resolution adopted on November 13th 1989 by the Council of the European Community and by the Ministers of Health of the member states, stressed the concept that 'the free movement of people ... in the community and the equity of treatment defined by the treaties are guaranteed and will continue to be so'. This concept was developed and ultimately elaborated in 1990, during the semester of the Italian presidency of the Community. In that context, several measures were adopted for an effective fight against AIDS and drug addiction, the most serious and rapidly increasing risk behavior.

A resolution adopted on December 3, 1990 underlined the need to improve home care and study of prevention interventions to counteract the increasing use of needles and syringes.

On the 4th of June 1991 in Luxembourg, the Council of the Ministers of Health of the member states approved a 3-year plan of action called *Europe against AIDS.*

As already stated, the main purpose of the Italian activities against AIDS is the tenacious pursuit of the most extensive international collaboration, especially with EEC countries and WHO. In fact, rejecting all discriminatory aspects, the Ministers of Health of the EEC countries unanimously agreed in the spring of 1990 to take a united stand against the restrictions regarding the entry of HIV-positive persons at national borders, to the point of not participating at the San Francisco Conference.

Today, there is no scientific basis to justify such a restriction on the free movement of HIV-positive individuals from country to country. That resolution was intended to convey a message to the US government to review its policy that is in serious conflict with the great democratic traditions of that country, with which we are closely associated. At the time of the Florence Conference, taking place in a country of free movement, to our regret the US had not yet found a solution based on the principle of freedom, as well as scientific knowledge. Therefore, we hope the US government will change its policy. This will allow everyone to meet in Boston in 1992. It is in a spirit of cooperation and brotherhood that we express this desire.

In Italy, a very comprehensive strategy plan for the fight against AIDS has been elaborated. These choices were incorporated in the act approved in June 1990 by Parliament, 'Program of Urgent Intervention for the Fight against AIDS' with a budget of US$ 1.7 billion.

The total strategy of this law, together with the law on drug addiction, favors prevention through information campaigns aimed at the general public and groups at risk.

Other initiatives aim to favor 'solidarity' against discrimination and ignorance. This commitment has proven that the groups at risk have indeed acquired a better knowledge of AIDS and its prevention. AIDS is no longer and must not be a taboo. The production of self-blocking syringes has been started and the production of condoms is now subject to quality control. This last objective has to also overcome cultural, psychological and political difficulties and resistance.

In addition, updated training has been organized for health care workers in order to make their activity more efficient and gratifying. Since 1988, research programs have contributed to the growth of international scientific knowledge and to detailed studies of the specific features affecting particularly the Italian population.

In this context, the Italian strategy of intervention is also based on international collaboration. Thus, we are already involved in European experiments on vaccines and clinical trials for the in vivo testing of new drugs. Now we intend to expand our collaboration with developing countries.

In the field of care services for people with HIV/AIDS, we want to provide treatment in hospitals as in-patient services during periods of acute crises and follow-up care mainly in day hospitals or out-patient clinics. In addition, an extensive program of home care treatment has been started under the supervision of the hospital units. In this way, we shall ensure to all citizens the free public health care they so rightfully deserve, including the distribution of necessary drugs, such as AZT.

The act against AIDS of June 1990 includes a whole chapter concerning the ethical problems relating to AIDS: from confidentiality to the request of informed consent to the elimination of discrimination toward the patient.

We are aware that laws and financial measures are not enough to rapidly change the course of events. At times we witness episodes and events that demonstrate how difficult it is to create a climate of collaboration and solidarity which encourage the prevention of the spread of HIV infection, and to provide the necessary help – both moral and economic. For this reason our commitment will be steadfast and constantly growing.

It would be unfair to understimate how much has been done and how much we are doing, in particular the efforts of health care workers who, with much diligence and responsibility, work to improve the conditions of the AIDS patients. In the same way, we are grateful to the NGOs, those groups of people who freely, silently and lovingly give of themselves, dedicating time and efforts to those suffering and sometimes outcast from society. They help to transform good intentions into concrete, and efficient interventions.

In conclusion, my hope is in presenting this book that all these important scientific contributions, which gave such prestige to the Florence Conference, will be useful for increasing the knowledge and interest of all those involved around the world in the fight against AIDS.

This demonstrates the irreplaceable role of these conferences as a forum and focal point for the cultural, but also humanistic, exchange that contributes so much to the development of a global strategy for the fight against AIDS.

Foreword

AIDS and Its Impact on the Health, Social and Economic Infrastructure in Developing Countries

His Excellency *Yoweri Kaguta Museveni*

President of the Republic of Uganda

The AIDS/HIV epidemic has reached catastrophic proportions. Globally, WHO estimates that 8–10 million people are infected by HIV. Over 1.2 million men, women and children have already developed AIDS and, by the year 2000, there may be 25–30 million people infected by the virus. There are millions of people around the world whose lives have been affected by the consequences of this epidemic in one way or another. AIDS has not only become the most important public health challenge of our time, but it has sweeping global consequences, social, economic, cultural and even political.

At the same time, social, economic and cultural conditions have encouraged and are fueling the epidemic especially in the developing world. According to WHO estimates, the global balance of HIV is rapidly tipping towards the developing world. In 1985, about 50% of the world total infections estimated were in developing countries. But now, it is estimated that by the year 2000, 75–80% will be in developing countries and be the year 2010 as much as 90%. Some think that this is but one of the first of many developmentally linked infectious diseases. If that is true, then, unless there is a concerted effort to redress the economic imbalance between the rich and poor nations, we may see untold millions of people die.

While AIDS is the pre-eminent public health threat of our time, socio-economic factors, crucial in the transmission of AIDS and other sexually transmitted diseases, have deep historical roots. In Africa, sexually transmitted diseases, such as gonorrhoea and syphilis, were a big health hazard before the advent of modern Western medicine. To discourage the spread of these in society, Africans had evolved cultural taboos against premarital sex and strict sanctions had been established against premarital sex or sex out of wedlock. With the advent of sulfas and penicillin, 'The Magic Bullet', in the early 1940s, the fear of sexually transmitted diseases subsided, ushering in the era of permissive sexuality. This was later encouraged by the era of the 'Sexual Revolution' of the 1960s, mainly inherited from events in Western Europe which in turn had been set in motion by the universal availability of the contraceptive pill. Traditional checks based on morality and self-control were thrown aside. This was accompanied by complete neglect of our traditional herbal and other medicines. Their further development was aborted. In the developing world, though, the promise of the wonder drugs in controlling sexually transmitted diseases was not to be fulfilled, because of the prevailing socio-economic circumstances.

One can appreciate why this was not to be if one considers the following. For sub-Saharan Africa, in particular, we find that, between 1960 and 1986, the average share of public health expenditure of gross national product (GNP) increased only from 0.7 to 0.8%. This average does not, however, tell us that in a number of African countries the share allotted to health actually declined. In some countries, for example, it fell from 1% in 1960 to 0.7% in 1986.

If we look at the per capita public health expenditure in real dollar terms, we note a clear downward trend. In 1987, the average per capita public health expenditure in the sub-Saharan countries was a mere US$ 3.50 per year as compared to more than US$ 1000.00 in the USA and Scandinavian countries. This trend is explained by the marked slow-down of economic growth in the 1980s. In 12 sub-Saharan countries, the per capita growth rate of the period 1965–1988 was negative. Starting in 1970, the average income per head of population fell to the level of the 1960s. It is easy to understand that, in such a depressed economic environment, expenditure on public health was severely constrained.

In Uganda, now, we have the following distorted ratios: (1) 1 doctor for every 23,000 people compared to 1 doctor for about 1,000 people in industrial countries; (2) 1 hospital for every 200,000 people; (3) 1 health unit for every 150,000 people; (4) 1 nurse for every 2,332 people, and (5) 1 health bed facility for every 800 people.

It is, therefore, not surprising that by the early 1970s, sexually transmitted diseases had reached epidemic proportions. Studies done at the time showed an incidence of up to 14% among mothers attending antenatal clinic in some of our cities. Where modern medical facilities are inadequate, sexually transmitted diseases are usually not diagnosed at all, or when they are diagnosed, proper or adequate treatment is not given.

We are told by scientists that heterosexual transmission appears to be much more important in the epidemiology of HIV infection in Africa than in Europe and North America, and sexually transmitted diseases which are hyperendemic in tropical Africa have been considered possible co-factors in heterosexual transmission of HIV in this region. Some studies have shown that in uncircumcized men who acquired genital ulcer disease, such as syphilis or chancroid, the cumulative transmission rate for HIV after a single sexual exposure was 43%. I understand that in developed countries the heterosexual transmission is less than 0.1% per single sexual exposure.

In view of the socio-economic factors enumerated above, it is, therefore, not surprising that when HIV arrived in East and Central Africa in the early 1980s it landed on fertile ground. The AIDS epidemic has demonstrated the historical continuity of social, political and medical reaction of epidemics among socially deprived populations. AIDS was heralded by epidemics of sexually transmitted diseases. A historical appreciation of the resilience of social conditions that promote diseases like AIDS may help society in rationally responding to protect the public's health.

Epidemiological studies have traditionally focused on the biological dynamics of disease such as incubation period, infectiousness of the disease organism and its ability to cause death. AIDS has, more than any other disease, brought about the recognition that the spread of disease may be driven more by sociological and economic dynamics. AIDS cannot be understood in biological terms alone. Sex is not a simple manifestation of a biological drive, it is socially dictated. Sexual opportunities available to an individual and the type of partners deemed appropriate will vary from one social group to another. It is in view of this background that I have been emphasizing a return to our time-tested cultural practices which emphasized fidelity and condemnation of premarital or extramarital sex. I believe that the best response to the threat posed by AIDS and other sexually transmitted diseases is to reaffirm publicly and forthrightly the reverence, respect and responsibility every person owes to his or her neighbour. Young people must be taught the virtues of abstinence, self-control and postponement of pleasure and sometimes sacrifice. Just as we were offered the magic bullet of penicillin from the early 1940s, our public health figures are now offering us the condom and 'safe sex'. In countries like ours, where a mother often has to walk 20 miles to get an aspirin for her sick child or 5 miles to get any water at all, the practical questions of getting a constant supply of condoms or using them properly may never be resolved.

Meantime, we are being told that only a thin piece of rubber stands between us and the death of our continent. I feel condoms have a role to play as a means of contraception especially in couples who are HIV-positive, but condoms cannot be the main means of stemming the tide of AIDS. The initial response of AIDS in Africa, similar to that in the rest of the world, was denial followed by accusation of an 'us' being victimized by a foreign 'them'.

Early reports on AIDS in Africa were clouded by sensationalism, panic, politics and narrowly focused academic research that ignored socio-cultural issues.

AIDS was first recognized in Uganda in 1981, but the negative feelings mentioned above prevailed and the government of the time decided to bury its head in the sand like the proverbial ostrich. A lot of time was, therefore, lost from 1981 up to 1986, when we got into government.

Our government has not had any qualms about being frank to our people on issues of a national catastrophe such as the AIDS epidemic. When we got into power in 1986, the problem had already spread to most parts of the country. We opened gates to national and international efforts to the control of the epidemic. Unfortunately, despite my government's effort, the AIDS situation in Uganda is becoming more and more serious, despite the high level of awareness among the population. However, this awareness has over the last few years started paying off. There has been a singular decline in the incidence of other sexually transmitted diseases. The disease has hit hardest those who are not only in their most sexually active years, but also in their most productive years. Loss of labour force is already being experienced both at the national and household levels. A number of professionals working in government and other institutions have died. In our countries, where education and technical expertise are at a premium, this will offset our economic and social gains.

For many rural households, current levels of agricultural production may be threatened, especially as the agricultural activities are labour-intensive. This will affect coffee production by small holders, which account for over 90% of the country's export earnings. This reduction will cause more economic strain, especially when the coffee are not favourable.

With regard to the social services, AIDS is already affecting our overstretched medical services. Apart from looking after AIDS patients, secondary infections such as tuberculosis (TB) have increased because of AIDS. We understand that already a 2- to 3-fold increase has been observed in our region where dormant TB is common, yet treatment for a single case of TB is US$ 126.00. In Uganda, over the last 6 years, with the help of UNICEF, we have managed to achieve over 90% immunization coverage for the 6 immunizable diseases including TB, thus dramatically bringing down the infant mortality rate. AIDS might reverse these achievements.

UNICEF predicts that, by the year 2000, there will be 590,000 maternal AIDS deaths in Africa leaving behind 5.5 million AIDS orphans. Traditionally, in an African society, when parents die, the children go to live with another member of the extended family. These families' resources, already overstretched by their own children's needs and the orphaned children, are going to wind up at the end of the family line. If the orphaned children survive, they will find it increasingly difficult to occupy a niche in society as they will have missed education opportunities and, later, have no hereditary rights in their adopted families. Government and other bodies have, therefore, to come in and fill the gap.

Apocalyptic visions of the virtual decimation of much of Africa may be unwarranted, but the growing devastation of a range of national aspirations is very real unless something is done quickly.

In Uganda we have realized that the AIDS problem goes beyond the mere health of the people. We have, therefore, adopted a multisectoral strategy for the control of the epidemic. Where control measures were centered in the health sector, we are now establishing full-fledged control programmes in other key sectors of communication, rehabilitation, education, community services, defence and economic planning. An independent body, the Uganda AIDS Commission, has been established to guide, direct, co-ordinate and monitor the strategy. The Commission, which I am chairing myself, so as to give the necessary political support, will establish policies and guidelines on HIV and AIDS-related matters. The Commission will monitor the control measures of the various sectors by listening to the voices of the people through the Commission's field officers and the grass-root leaders. This strategy, we hope, will be able to harness our efforts in combating AIDS.

I understand that in Romania and the Soviet Union hundreds of children got infected through use of unsterile injections in medical facilities. If this is true, then, I wonder what is happening in our countries where sterility is worse. I am surprised that our scientists have not looked at this exhaustively with a view to eliminating this possibility and protecting our people.

Is it not possible that people with sexually transmitted diseases are more likely to get unsterilized injections in unlicensed clinics, in addition to their genital ulcers being a good port of entry of the HIV virus?

Uganda and Africa's forests and savannahs still conserve the largest variety of plant and animal life. These are a potential source of natural chemotherapeutics for AIDS and other diseases.

In my country, we have embarked on research on these. We welcome international collaboration.

Our emphasis, though, has been in health education and change of sexual behaviour, having realized that HIV infection is mostly dependent on voluntary behaviour unlike most other transmittable diseases. Even so, we shall ultimately depend on the concerted efforts of you scientists in your various disciplines to lead us out of this dark abyss.

It is in view of this that I salute you for the achievements of the last 10 years in the pursuit of an affordable drug and/or vaccine.

Introduction

The volume is based on the *VIIth International Conference on AIDS,* held in Florence, June 16–21, 1991, and sponsored by the Istituto Superiore di Sanita', Rome, Italy with the co-sponsorship of the Italian Ministry of Health, the World Health Organization, International AIDS Society, European Economic Community and the Regione Toscana, Italy as well as also sponsored in association with: Associazione Nazionale per la Lotta contro l'AIDS (ANLAIDS), Associazione per la Salute della Donna, Comune di Firenze, Comitato Firenze '91 and Fondazione Italiana per la Ricerca sul Cancro (FIRC).

These proceedings contain the Master- and State-of-the-Art Lectures, followed by the four best papers selected from an impressive total of 3,500 accepted abstracts, and finally five Conference Summary papers concerning *basic science, clinical science and trials, epidemiology and prevention, social and behavioral science,* and *communities challenging AIDS.*

The book opens with three main topics: (a) the *Impact of Biomedical Research on the AIDS Epidemic* identifying also the gaps in knowledge and some of the goals for the coming years: (b) *Progress and Challenges in Therapy:* progress is possible if continuous attention is paid to the basic causes of the disease; and (c) the *Implications of AIDS in Developing Countries;* WHO estimates a total of 40 million HIV-infected persons worldwide by the end of this decade with more than 90% of infections taking place in developing countries.

The following state-of-the-art overviews deal with present and future dimensions of the HIV/AIDS pandemic: the *virology of AIDS* – biological aspects including co-factors; the *molecular biology of HIV,* looking backwards to 10 years of discovery and hope for the future; as well as the *cellular pathogenesis of both viruses, HIV-1 and HIV-2;* as well as the *etiopathogenesis of AIDS-related malignancies.*

A critical outlook for *clinical trials* urges that if progress is to be made in 1991–1992, science must continue to be questioned and challenged in order to move faster. Advances in *AIDS vaccine research* are reviewed thereafter. Significant progress has been made and the prospects for efficacious vaccines by the year 2000 are not outside the realm of possibility.

The *psychosocial aspects and impact of AIDS on social, economic, health and welfare systems* in developed and developing countries is the subject of the following two contributions which demand that if mankind is to withstand the impact of AIDS and eliminate it from our midst, then a family perspective which heightens global solidarity and global learning is needed with an imperative global ethic.

The following four papers selected from the best abstracts of the four major areas of this International Conference report novel results on (a) genes or domains within genes that control cellular tropism, cytopathology and replicative properties of HIV-1; (b) the development of opportunistic non-Hodgkin's lymphomas in AIDS patients receiving long-term antiretroviral therapy; (c) risk reduction and stabilization of HIV seroprevalence among drug injectors in New York City and Bangkok; and (d) change in AIDS risk behaviors from adolescence to adulthood.

The book ending with the Conference commentaries on: (a) *Basic Science Challenging AIDS;* (b) *Clinical Science and Trials;* (c) *Epidemiology and Prevention;* (d) *Social and Behavioral Science,* and (e) *Communities Challenging AIDS* is highlighting the most pertinent new data, overviews and reflections on AIDS mechanisms, novel designs and strategies, enhanced international cooperations, and call for global solidarity.

Lastly, the editors wish to thank Francesco A. Manzoli, Director General, Istituto Superiore di Sanità, Rome, Italy, the Members of the International Steering Committee, and of the Program, Communications, Developing Countries Sponsored Delegates and Communities Committees. Furthermore, we thank Cristina D'Addazio, Secretary General, and all other Conference Personnel as well as Daniela Rosselli, Rita Pestellini and their staff for their contributions to this International Conference.

Editors

Giovanni Battista Rossi, Rome
Elke Beth-Giraldo, Naples
Luigi Chieco-Bianchi, Padua
Ferdinando Dianzani, Rome
Gaetano Giraldo, Naples
Paola Verani, Rome

Acknowledgments

The Executive Committee for the VIIth International Conference on AIDS was constituted of Giovanni B. Rossi, Conference Chair; Luigi Chieco-Bianchi, Conference Co-Chair; Gaetano Giraldo; Paola Verani; Elke Beth-Giraldo, Chair, Developing Countries Delegates Committee. The Steering Committee included the Members of the Executive Committee and Michael Merson, Lars Olof Kallings, Manuel Carballo, World Health Organization; Paul A. Volberding, Meinrad Koch, Friedrich Deinhardt, International AIDS Society; Alexandre Berlin, European Economic Community (EEC); Ferdinando Dianzani, Chair, Program Committee; Elio Guzzanti, Chair, Communities Committee, and Antonio Siccardi, Chair, Communication Committee. The Program Committee included Luigi Chieco-Bianchi, Coordinator, Basic Science; Carlo Zanussi, Coordinator, Clinical Science and Trials; Gaetano M. Fava, Coordinator, Epidemiology and Prevention, and Elio Guzzanti, Coordinator, Social and Behavioral Science.

Communities Challenging AIDS Sessions were organized by: Manuel Carballo, World Health Organization; Giovanni Rezza and Laura Thomas in collaboration with the World Federation of Hemophilia and the World Health Organization.

The core staff for the conference included Cristina D'Addazio, Secretary General; Lucy Felicissimo, Coordinator and Ulla Johansson, Assistant Coordinator, Scientific Program; Catherina Ramel, Cooordinator, Registration; Benjamin Junge, Coordinator, Developing Countries Sponsored Delegates Program; Laura Thomas, Coordinator, Community Relations.

International Conferences on AIDS

Ist International Conference on AIDS, Atlanta, Ga., April 14–17, 1985
Chairman: Gary Noble

IInd International Conference on AIDS, Paris, June 23–25, 1986
Chairman: Jean-Claude Gluckman

IIIrd International Conference on AIDS, Washington, D.C., June 1–5, 1987
Chairman: George J. Galasso

IVth International Conference on AIDS, Stockholm, June 12–16, 1988
Chairman: Lars Olof Kallings

Vth International Conference on AIDS, Montreal, June 4–9, 1989
Chairman: Ivan L. Head
Leitmotiv/Slogan of the Conference: The Scientific and Social Challenge

VIth International Conference on AIDS, San Francisco, Calif., June 20–24, 1990
Chairman: John L. Ziegler
Co-Chairman: Paul A. Volberding
Leitmotiv/Slogan of the Conference: AIDS in the Nineties: From Science to Policy

VIIth International Conference on AIDS, Florence, June 16–21, 1991
Chairman: Giovanni B. Rossi
Co-Chairman: Luigi Chieco-Bianchi
Leitmotiv/Slogan of the Conference: Science Challenging AIDS

The *International Conference on AIDS* is the most prestigious, mondial forum for the life-threatening pandemic of HIV infection which is taking its heaviest tolls now and will continue to do so in the years to come, particularly in all the developing countries. The conference is hosted each year by local/national organizers and cosponsored by the *International AIDS Society* and the *World Health Organization* in recognition of the global scientific and social challenge of HIV infection/AIDS. The conference promotes mondial cooperation, participation, solidarity, and respect for the human rights and dignity of HIV-infected people and people with AIDS.

The uniqueness of this conference series resides in the *blueprint* of program which is built around four balanced themes:

(1) Basis research.
(2) Clinical research and trials.
(3) Epidemiology and prevention.
(4) Social and behavioral science.

Thus ensuring joint participation by scientists, clinicians, public health workers and volunteers – any and everyone working in AIDS research or involved in the care of AIDS patients – in solidarity to enrich knowledge and understanding, i.e. the impact of new research data and scientific observations on treatment, preventive strategies, public health policy, and all other fields affected by the AIDS epidemic.

Master Lectures

Rossi GB, Beth-Giraldo E, Chieco-Bianchi L, Dianzani F, Giraldo G, Verani P (eds): Science Challenging AIDS. Basel, Karger, 1992, pp 1–14

Impact of Biomedical Research on the AIDS Epidemic

Anthony S. Fauci

National Institute of Allergy and Infectious Diseases,
National Institutes of Health, Bethesda, Md., USA

In the present paper, I would like to discuss my perspectives on the impact of biomedical research on the AIDS epidemic, and outline some of the broad scientific bases for advances in AIDS research, a number of AIDS-specific research accomplishments which have been realized, as well as certain critical gaps in our knowledge which will require further intensive investigation.

Fundamental to our understanding of AIDS is the identification of the human immunodeficiency virus (HIV) as the etiologic agent [1, 2]. It should not go unnoted that decades of basic research in animal retrovirology was critical to the ability of the biomedical research community to identify so quickly the etiologic agent of AIDS [3]. However, many mysteries remain, since this particular retrovirus is unique in its complexity, particularly with regard to its regulatory genes whose precise functions still elude us.

The *tat* and *rev* genes are clearly the most well studied and arguably the best understood [4]. Not only have such studies provided insight into the regulation of HIV expression, but invariably they will heighten our understanding of the entire area of viral regulatory genes as well as their relationship to cellular factors which impact on viral gene expression.

The uncertainty of least year concerning the precise role of the *nef* gene still remains amid new insights as well as questions concerning the roles of *vpu, vpr* and *vif* [5]. We will certainly hear more about this. At best we will gain insight into HIV regulation; at worst we will come closer to an understanding of the complex, if not inscrutable broader area of gene regulation.

The discipline of structural biology utilizing such tools as X-ray crystallography and nuclear magnetic resonance imaging has found a welcome home in AIDS research, for precise delineation of the molecular structure of viral proteins is essential for the optimization of truly targeted drug development. This past year has seen the realization of the crystal structure of the HIV protease [6] as well as the ribonuclease H domain of HIV-1 reverse transcriptase [7]; and this is only the beginning. Clearly in the 1990s virtually all of the important viral encoded proteins will have their molecular structure clarified.

Perhaps the most complex and perplexing area of AIDS research is that of pathogenesis (fig. 1). Despite enormous insights over the past decade, our lack of complete understanding of the pathogenic mechanisms of HIV infection stands out as a sobering balance to the excitement of so many important discoveries in this and in other areas of AIDS research. Pathogenesis encompasses the spectrum from initial infection; through the mechanisms of cytopathicity; the acute HIV syndrome; the mechanisms and implications of latency; the regulation of virus expression involving the induction of virus from the latent state, the role of co-factors in initial infection, induction of HIV expression and cytopathicity; the puzzling question of the relationship between viral burden and disease progression; the uncertainties of tissue distribution of HIV; the role of autoimmunity in the pathogenic process as well as the potential deleterious effects of immunity against HIV; the spectrum of infectable cells as well as the range of syndromes directly or indirectly related to HIV infection; the impact of infection of precursor cells on the course of disease; the enigmatic mechanisms of neuropathogenesis, and finally the mystery of the efficacy or lack thereof of the immune response to HIV. As we learn more about each of these facets of the process, additional questions emerge ... not an unusual experience in biomedical research.

Starting with the initial event of the primary infection, we are immediately faced with numerous questions. We have made enormous progress in recreating events in vitro and thus hopefully understanding by extrapolation the in vivo events of initial virus binding to $CD4^+$ target cells in which the precise domains of binding of virus envelope protein to CD4 molecule have been delineated. This is followed by the process of envelope fusion to cell membrane, internalization, reverse transcription and ultimately integration followed by virus replication [8]. However, we really do

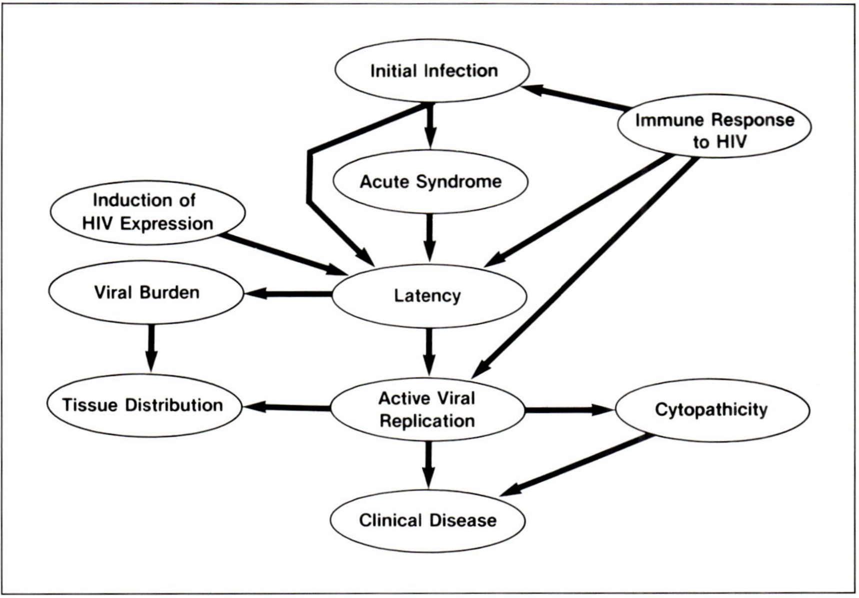

Fig. 1. Pathogenesis of HIV infection. Schematic diagram of the pathogenic events that occur from initial infection with HIV to the development of clinical disease.

not know what the very first infective event is and where it takes place. When the infection is transmitted grossly via blood, there is greater clarity. But what about sexual transmission which is the most prevalent mode of spread of HIV worldwide? Must virus bind to receptor on a cell in the bloodstream or can infection occur at or just beneath mucosal surfaces? Must there be breakdown of mucosal integrity? Once the virus enters the blood, does infection take place within the bloodstream or must the virus first be cleared into and reside in the peripheral lymphoid tissues? What are the relative contributions of cell-free virus versus cell-associated virus in the initial infective event? Appropriate animal models, particularly primates, are essential to answering these critical questions.

We have now come to the realization that the acute HIV syndrome occurs in as many as 70% of infected individuals (fig. 2) [9]. The mechanisms of this complex of signs and symptoms are still unclear. There is a burst of viremia several weeks after initial infection accompanied by marked disturbances in the distribution of immunocompetent cells of mul-

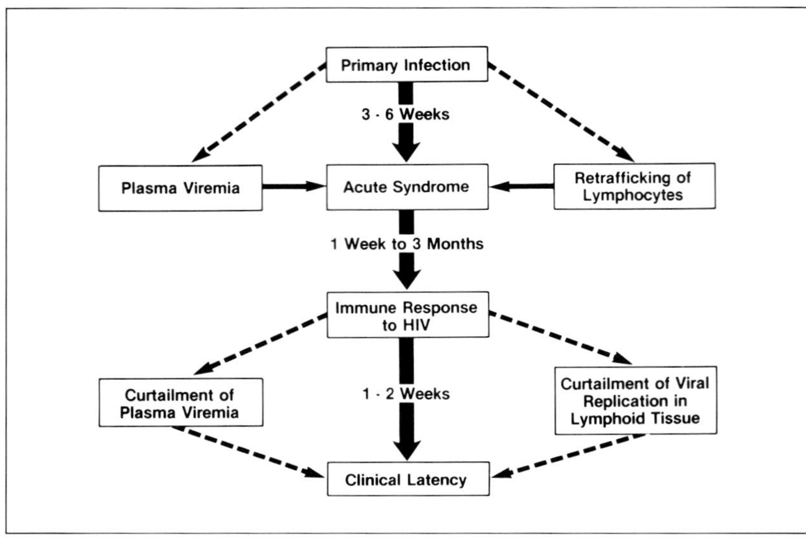

Fig. 2. Acute HIV syndrome. Schematic diagram of the progression of HIV infection from an initial infective event to the establishment of clinical latency.

tiple classes in the peripheral blood and the lymph nodes. However, it is uncertain what role the trafficking of immunocompetent cells out of the circulation into the lymphoid tissue has in the clinical syndrome itself and what effect this trafficking has on the propagation of fulminant viral replication. No less a mystery is the role of the immune response in the curtailment of virus replication that is presumed to occur within weeks after infection. In fact, it may even be an incorrect assumption that virus replication is curtailed at all since detectable p24 antigenemia may decrease only as a result of complexing with antibody and virus replication may proceed steadily in the relatively inaccessible lymphoid organs. Since many of the most important disease-determining events may occur during this initial phase of infection, it is essential that the acute syndrome be intensively studied, since this may well be the most opportune time to intervene therapeutically.

One of the most fascinating areas of research in HIV infection is that of viral latency and the regulation of HIV expression. In this regard, the word 'latency' has unfortunately engendered a good deal of confusion. One of the reasons for this is the inappropriate usage of the word to describe

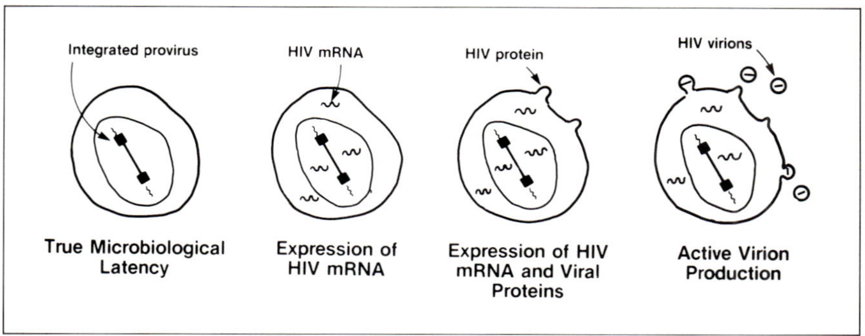

Fig. 3. Stages of HIV expression. Schematic diagram depicting the potential phases of HIV expression in HIV-infected cells.

both a clinical phenomenon as well as a microbiological phenomenon. As most of you know, HIV-infected individuals experience long periods of clinical latency characterized by lack of symptoms [10]. Microbiologically, even during the stage of clinical latency, it is almost certain that complete viral latency never exists in all infected cells in a given individual (fig. 3). Although at any given point in time there are cells in the body that are truly latent in that there is only integrated provirus and no trace of viral mRNA or protein, cells that are experiencing variable degrees of viral replication are also present at all stages of infection.

A number of endogenous and exogenous mechanisms have been shown to be involved in the induction of virus expression in an infected cell. Here again investigators have built upon the broad base of scientific knowledge accumulated over decades in the study of expression of a wide array of viruses as well as endogenous cellular regulatory mechanisms, particularly involving the immune system, the primary target of HIV expression. A number of heterologous viruses including HTLV-I likely operating via the inductive capability of its *tax* gene, and HHV-6 can up-regulate HIV expression in a system of co-infection [11, 12]. In addition, endogenous cellular factors have been shown to directly or indirectly induce HIV expression [13–16]. My own laboratory and others have been intensively involved in the delineation of the role of the endogenous immunoregulatory cytokine network in the regulation of HIV expression (fig. 4) [17, 18]. It seems that the virus has adapted itself quite comfortably to this extraordinarily complex network that immunologists have been dis-

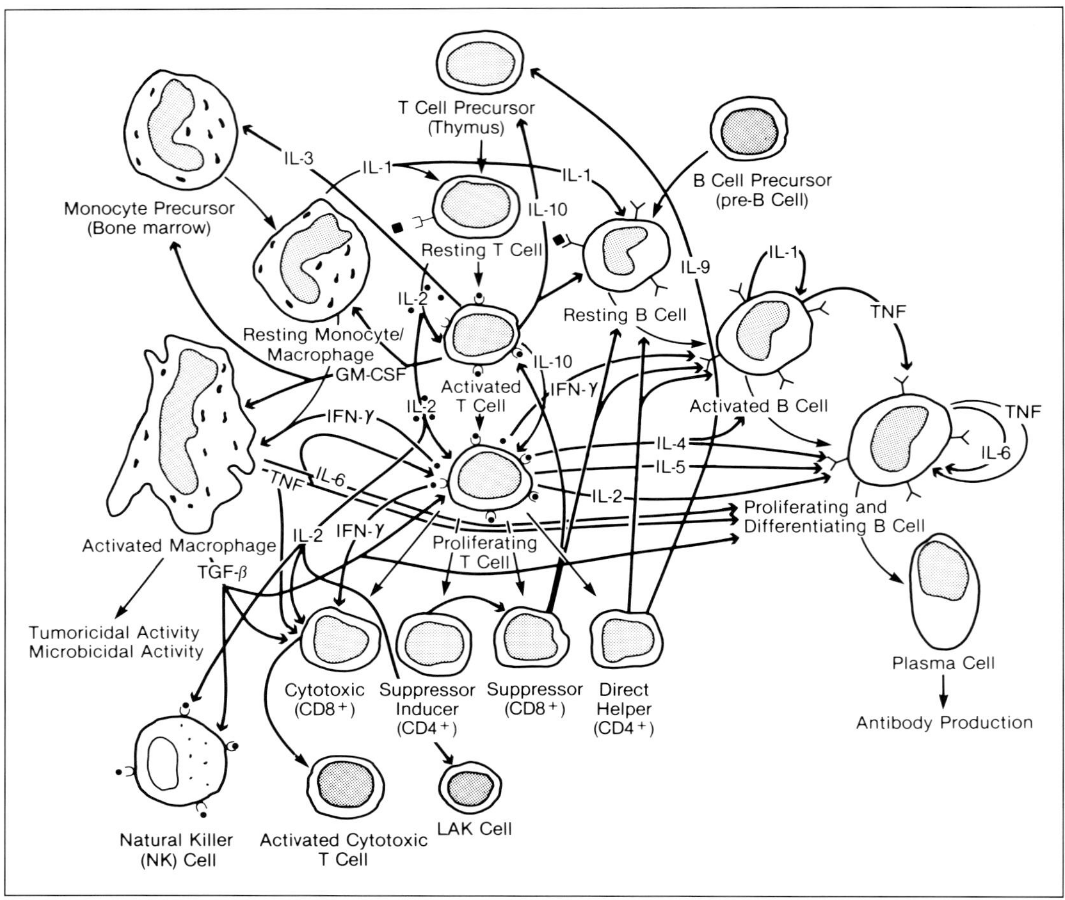

Fig. 4. Human immunoregulatory network – 1991. (Updated from fig. 1 of Fauci et al., Ann Intern Med 1987;106:421.)

covering and redefining for years. Not only can these factors whose physiologic role is the regulation of immunologic homeostasis precisely regulate virus expression, but they do so by molecular mechanisms that normal cells have adopted in the responses to activation by mitogens, antigens and cytokines [14, 19, 20]. These molecular mechanisms range from transcriptional activation iva NF-kB binding to consensus sequences in the HIF long terminal repeat as seen with mitogen or tumor necrosis factor-α stimulation of infected cells to posttranscriptional modulation as seen with cytokines such as interleukin-6 and granulocyte/macrophage-colony-stimulating factor. Other cytokines such as transforming growth factor-β can up-regulate or down-regulate HIV expression by both transcriptional and posttranscriptional mechanisms [17, 21].

Intimately intertwined with the concept of induction of HIV expression is the area of co-factors. The cytokines and co-infecting viruses just mentioned strictly speaking are co-factors as are the genital ulcerations that allow entry of virus into the host. No doubt co-factors can influence infection and virus expression [8]. However, there is not evidence at this time that any given co-factor is absolutely essential either for initial infection, induction of HIV expression or cytopathicity.

Recent data have strongly supported the concept that precursor cells either in the bone marrow or the thymus can be infected with HIV. Of particular interest is the finding that so-called 'triple-negative' thymic precursors are in fact not triple-negative at all, but actually express a low level of CD4 [22]. We have recently demonstrated that these cells can in fact become infected with HIV in vitro. This has important implications in understanding the perplexing lack of regenerative capability of the $CD4^+$ T cell pool in HIV-infected individuals even during apparently effective antiretroviral therapy (fig. 5). If infection of these T cell precursors does indeed occur in vivo, then this argues strongly for treatment of HIV infection at the earliest possible stage in order to avoid depletion of the very pool of precursor cells that would be essential for adequate regeneration of the mature $CD4^+$ T cell repertoire. Years of fundamental research on thymic ontogeny made possible our ability to dissect out potential intrathymic events associated with HIV infection.

The entire question of the level and distribution of viral burden deserves intensive study. We had thought early on that the frequency of infected cells was so low that it could not explain the immunosuppression that was clinically observed. However, it is now clear that the peripheral lymphoid tissues maintain a much greater viral burden per given number of $CD4^+$ T cells than does the peripheral blood [G. Pantaleo et al. and C. Fox et al., unpubl. observations]. In fact, as with many human diseases that have an immunologic component, the peripheral blood may not accurately reflect the total body immunologic status [23]. It is very likely that from the very first infective event throughout the entire process of immunopathogenesis the peripheral lymphoid tissues are the most significant sites of HIV replication, propagation and cytopathicity.

Another puzzling area of HIV research is that of neuropathogenesis. Neuronal damage is severe in certain HIV-infected individuals despite the fact that there is no convincing evidence that neurons themselves are infected with HIV [5, 24]. Substantial data, however, point to the fact that cells of the central nervous system, particularly those of the monocyte/macrophage lineage such as microglial cells, are readily infectable and likely secrete along with other infected cells soluble factors which exert considerable toxicity on neurons [5, 24]. Certainly neuropathogenesis is one of the most important areas of research on HIV that deserves intensive focus.

If the study of pathogenesis of HIV infection has been a humbling experience for investigators, then attempts to dissect out the correlates of immunity have been equally as frustrating. Multiple components of immunity have been described and immunodominant epitopes of several viral proteins have been identified [5]. However, we still do not understand why and how disease progresses in the face of an apparently adequate immune response. Is the response never totally adequate to control infection? The demonstration of active viral replication even early during the apparently quiescent stage of infection argues against an effective immune response at any time during infection. Is the inadequacy due to rapid mutation of the virus? This may be the case; however, cytolytic $CD8^+$ T cells that are present early in infection are no longer detectable as the disease progresses. Have they themselves gotten infected in the lymph nodes by antigen-specific recognition of and direct contact with infected $CD4^+$ T cells and monocytes [25]? We have recently demonstrated that HIV-specific $CD8^+$ cytotoxic T lymphocytes can in fact be infected in vitro in the process of killing HIV-infected $CD4^+$ T cells [26]. Even more perplexing is the possibility that an immune response against HIV may actually propagate spread of HIV. During the initial stage of infection, there is an exuberant recruitment of $CD4^+$ T cells, $CD8^+$ T cells, as well as B cells, to lymphoid organs presumably as part of an appropriate immune response. However, recruitment of susceptible $CD4^+$ T cells to the site of infection and the state of activation that would normally accompany such a process may potentially enhance spread at the same time that an immune response is being initiated.

The natural offshoots of an understanding of the virus itself and its immunopathogenic mechanisms are attempts at developing safe and effective therapies for HIV and for the opportunistic diseases that result from the profound immunosuppression that accompanies HIV infection. With regard to drugs for HIV itself, as we have just discussed, it is critical to

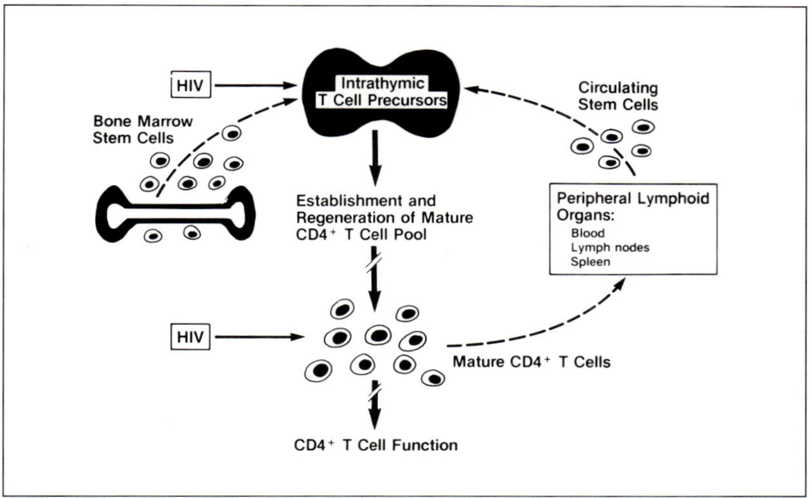

Fig. 5. HIV infection of intrathymic T cell precursors. Schematic diagram that illustrates one possible explanation for the lack of regeneration of mature $CD4^+$ T lymphocytes during HIV infection.

understand the mechanisms of viral spread, latency, and endogenous and exogenous regulation of viral gene expression in order to effectively target therapy. Structural biological delineation of the viral proteins will pave the way in the 1990s for safe, specific and effective inhibition of HIV. Some concrete examples of this approach which are in the process of being realized include specific inhibitors of HIV proteins such as *tat* and protease [27–32].

The targeted structural biological approach to HIV itself has a natural clarity to it by virtue of the well-defined nature of the virus. However, the situation is infinitely more complex in the area of the multitudinous opportunistic infections associated with HIV infection. Directed approaches will require a commitment to years of basic research on these diverse microorganisms which will ultimately have positive impact not only on the HIV-infected individual, but on future generations of patients who become immunosuppressed for whatever reason notwithstanding HIV. It is an opportunity to benefit HIV-infected individuals by prophylaxis and treatment at the same time that a whole new field of research is embarked upon.

Table 1. HIV vaccine research agenda for the 1990s

Expand vaccine challenge studies in animal models focusing on:
Heterologous challenge
Duration of immunity
Immune correlates of protection
Protection against cell-associated virus challenge
Protection against mucosal challenge
Determine if virus components, alone or in combination, are as effective immunogens as whole killed virus
Develop improved adjuvants
Move products currently in pilot clinical trials through efficacy studies if/when appropriate

Much less well studied than drug development for HIV and opportunistic infections is the entire area of immunologic enhancement and/or reconstitution [33, 34]. All of the many problems associated with bone marrow and thymic transplantation that have been faced in attempts at reconstituting defective immunity in other diseases are still present with AIDS patients and are compounded by the added problem of persistent virus potentially infecting the transplanted tissue. HIV-specific immunologic enhancement may serve as a temporizing measure and deserves further study. Nonspecific immune enhancement, while predictably advantageous, carries the added potentially harmful effects of globally activating the immune system and propagating HIV spread.

Drug development flows naturally into the clinical trials process and both have yielded impressive, but at this time still inadequate results. Over the past few years accomplishments have included the demonstration of effective but temporary anti-HIV activity for several compounds, the benefit of early intervention and the striking advantage of prophylaxis against *Pneumocystis carinii* pneumonia [35, 36]. Realistic goals for the next decade include a combination of drugs which are relatively nontoxic, which can be administered early in the course of infection and which can significantly prolong the disease-free state, orally administered nontoxic prophylaxes against a panel of opportunistic infections commonly afflicting HIV-infected individuals, and safe and effective regimens of specific immunologic enhancement and/or global immunologic reconstitution. Commitment to a rigorous course of basic biomedical research together

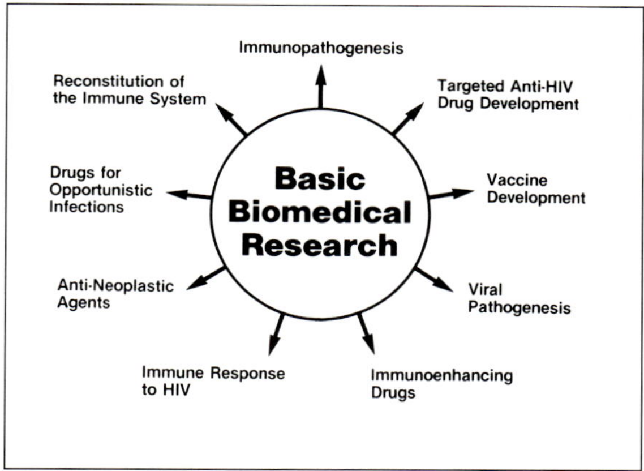

Fig. 6. Importance of basic biomedical research in AIDS.

with a sense of urgency to make such therapies widely available once they are proven effective is essential.

Finally, vaccine development creates an imposing challenge to the biomedical research community. A number of important questions should dictate our research agenda for the 1990s (table 1). The lack of clearcut correlates of immunity has been a major stumbling block in the rational approach to vaccine development. Here again the linkage between delineation of the basic mechanisms of pathogenesis and a directed approach to vaccine therapy is clear. The recent results from live virus challenge studies in the macaque with SIV [37–40] and the chimp with HIV [41, 42] have proven extremely encouraging in firmly establishing the feasibility of a vaccine for HIV in humans. Research on vaccine adjuvants which has lagged over the past few decades will be critical not only to an HIV vaccine, but to vaccinology in general. Over the past year we have seen several potential vaccine candidates in phase I clinical trials, some of which are being considered for progression to phase II clinical trials. Hopefully, and I am optimistic in this regard, we will witness the availability of a safe and effective vaccine against HIV in the decade of the 1990s.

I have tried to highlight the impact of biomedical research on the AIDS epidemic, the gaps in our knowledge and some of the goals for the coming years. In conclusion, let me underscore the importance of the basic biomedical research effort in AIDS (fig. 6). Urgency and swiftness in so many areas are essential because of the worldwide scope and impact of this problem. Yet, we must also continue to support the somewhat undifferentiated approach to our research efforts in AIDS, an approach which in other diseases has historically led to discoveries and advances with the greatest long-term positive impact. I and others that many of the issues that I have raised will meet further clarification and insight.

References

1 Barre-Sinoussi F, Chermann JC, Rey F, Nugeyre MT, Chamaret S, Gruest J, Dauguet C, Axler-Blin C, Vezinet-Brun F, Rouzioux C, Rozenbaum W, Montagnier L: Isolation of a T-lymphotropic retrovirus from a patient at risk for acquired immune deficiency syndrome (AIDS). Science 1983;220:868–871.

2 Gallo RC, Salahuddin SZ, Popovic M, Shearer GM, Kaplan M, Haynes BF, Palker TJ, Redfield R, Oleske J, Safai B, White G, Foster P, Markham PD: Frequent detection and isolation of cytopathic retroviruses (HTLV-III) from patients with AIDS and at risk for AIDS. Science 1984;224:500–503.

3 Fauci AS: Basic immunology: The path to the delineation of the immunopathogenic mechanisms of HIV infection. Trans Assoc Am Phys 1988;101:160–173.

4 Rosen CA, Pavlakis GN: Tat and Rev: Positive regulators of HIV gene expression. AIDS 1990;4:499–509.

5 Rosenberg ZF, Fauci AS: The immunopathogenesis of HIV infection. Adv Immunol 1989;47:377–431.

6 Miller M, Schneider J, Sathyanarayana BK, Toth MV, Marshall GR, Clawson L, Selk L, Kent SB, Wlodawer A: Structure of complex of synthetic HIV-1 protease with a substrate-based inhibitor at 2.3 A resolution. Science 1989;246:1149–1152.

7 Davies JF II, Hostomska Z, Hostomsky Z, Jordan SR, Matthews DA: Crystal structure of the ribonuclease H domain of HIV-1 reverse transcriptase. Science 1991;252: 88–95.

8 Fauci AS: The human immunodeficiency virus: Infectivity and mechanisms of pathogenesis. Science 1988;239:617–622.

9 Tindall B, Cooper DA: Primary HIV infection: Host responses and intervention strategies. AIDS 1991;5:1–14.

10 Fauci AS, Schnittman SM, Poli G, Koenig S, Pantaleo G: NIH Conference. Immunopathogenic mechanisms in human immunodeficiency virus (HIV) infection. Ann Intern Med 1991;114:678–693.

11 Zack JA, Cann AJ, Lugo JP, Chen IS: HIV-1 production from infected peripheral blood T cells after HTLV-I induced mitogenic stimulation. Science 1988;240:1026–1029.

12 Lusso P, Ensoli B, Markham PD, Ablashi DV, Salahuddin SZ, Tschachler E, Wong-Staal F, Gallo RC: Productive dual infection of human $CD4^+$ T lymphocytes by HIV-1 and HHV-6. Nature 1989;337:370–373.

13 Jones KA, Kadonaga JT, Luciw PA, Tjian R: Activation of the AIDS retrovirus promoter by the cellular transcription factor, Sp1. Science 1986;232:755–759.

14 Nabel G, Baltimore D: An inducible transcription factor activates expression of human immunodeficiency virus in T cells. Nature 1987;326:711–713.

15 Gatignol A, Buckler-White A, Berkhout B, Jeang KT: Characterization of a human TAR RNA-binding protein that activates the HIV-1 LTR. Science 1991;251:1597–1600.

16 Kato H, Horikoshi M, Roeder RG: Repression of HIV-1 transcription by a cellular protein. Science 1991;251:1476–1479.

17 Rosenberg ZF, Fauci AS: Immunopathogenic mechanisms of HIV infection: Cytokine induction of HIV expression. Immunol Today 1990;11:176–180.

18 Rosenberg ZF, Fauci AS: Activation of latent HIV infection. J NIH Res 1990;2:41–45.

19 Bohnlein E, Lowenthal JW, Siekevitz M, Ballard DW, Franza BR, Greene WC: The same inducible nuclear proteins regulate mitogen activation of both the interleukin-2 receptor-alpha gene and type 1 HIV. Cell 1988;53:827–836.

20 Greene WC, Bohnlein E, Ballard DW: HIV-1, HTLV-1 and normal T-cell growth: Transcriptional strategies and surprises. Immunol Today 1989;10:272–278.

21 Poli G, Kinter AL, Justement JS, Bressler P, Kehrl JH, Fauci AS: Transforming growth factor beta suppresses human immunodeficiency virus expression and replication in infected cells of the monocyte/macrophage lineage. J Exp Med 1991;173:589–597.

22 Schnittman SM, Denning SM, Greenhouse JJ, Justement JS, Baseler M, Kurtzberg J, Haynes BF, Fauci AS: Evidence for susceptibility of intrathymic T-cell precursors and their progency carrying T-cell antigen receptor phenotypes TGR alpha beta + and TCR gamma delta + to human immunodeficiency virus infection: A mechanism for $CD4^{+}$ (T4) lymphocyte depletion. Proc Natl Acad Sci USA 1990;87:7727–7731.

23 Westermann J, Pabst R: Lymphocyte subsets in the blood: A diagnostic window on the lymphoid system? Immunol Today 1990;11:406–410.

24 Price RW, Brew B, Sidtis J, Rosenblum M, Scheck AG, Cleary P: The brain in AIDS: Central nervous system HIV-1 infection and AIDS dementia complex. Science 1988;239:586–592.

25 De Maria A, Pantaleo G, Schnittman SM, Greenhouse JJ, Baseler M, Orenstein JM, Fauci AS: Infection of CD8+ T lymphocytes with HIV: Requirement for interaction with infected CD4+ cells and induction of infectious virus from chronically infected CD8+ cells. Proc Natl Acad Sci USA 1991;146:2220–2226.

26 De Maria A, Colombini S, Pantaleo G, Schnittman SM, Greenhouse JJ, Koenig S, Moretta L, Fauci AS: Lysis of HIV-1 infected target cells by HIV-1 specific CD8+ cytotoxic T-lymphocytes results in infection of the effector cells. VII Int Conf on AIDS, Florence, 1991, abstract 59.

27 Hsu M-C: Searching for inhibitors of HIV tat. Presented at Advances in Molecular Biology and Targeted Treatments for AIDS, Washington, May 15–18, 1990.

28 Erickson J, Neidhart DJ, VanDrie J, Kempf DJ, Wang XC, Norbeck DW, Plattner JJ, Rittenhouse JW, Turon M, Wideburg N, Kohlbrenner WE, Simmer R, Helfrich R, Paul DA, Knigge M: Design, activity, and 2.8 A crystal structure of a C2 symmetric complexed to HIV-1 protease. Science 1990;249:527–533.

29 Kempf DJ, Norbeck DW, Codacovi L, Wang XC, Kohlbrenner WE, Wideburg NE, Paul DA, Knigge MF, Vasavanonda S, Craig-Kennard A, et al: Structure-based, C2 symmetric inhibitors of HIV protease. J Med Chem 1990;33:2687–2689.

30 Roberts NA, Martin JA, Kinchington D, Broadhurst AV, Craig JC, Duncan IB, Galpin SA, Handa BK, Kay J, Krohn A, et al: Rational design of peptide-based HIV proteinase inhibitors. Science 1990;248:358–361.
31 DesJarlais RL, Seibel GL, Kuntz ID, Furth PS, Alvarez JC, Ortiz de Montellano PR, DeCamp DL, Babe LM, Craik CS: Structure-based design of nonpeptide inhibitors specific for the human immunodeficiency virus 1 protease. Proc Natl Acad Sci USA 1990;87:6644–6648.
32 Ashorn P, McQuade TJ, Thaisrivongs S, Tomasselli AG, Tarpley WG, Moss B: An inhibitor of the protease blocks maturation of human and simian immunodeficiency viruses and spread of infection. Proc Natl Acad Sci USA 1990;87:7472–7476.
33 Lane HC, Fauci AS: Immunologic reconstitution in the acquired immunodeficiency syndrome. Ann Intern Med 1985;103:714–718.
34 Lane HC, Zunich KM, Wilson W, Cefali F, Easter M, Kovacs JA, Masur H, Leitman SF, Klein HG, Steis RG, Longo DL, Fauci AS: Syngeneic bone marrow transplantation and adoptive transfer of peripheral blood lymphocytes combined with zidovudine in human immunodeficiency virus (HIV) infection. Ann Intern Med 1990; 113:512–519.
35 Broder S, Mitsuya H, Yarchoan R, Pavlakis GN: NIH Conference. Antiretroviral therapy in AIDS. Ann Intern Med 1990;113:604–618.
36 Kovacs JA, Masur H: Prophylaxis of Pneumocystis carinii pneumonia: An update. J Infect Dis 1989;160:882–886.
37 Desrosiers RC, Wyand MS, Kodama T, Ringler DJ, Arthur LO, Sehgal PK, Letvin NL, King NW, Daniel MD: Vaccine protection against simian immunodeficiency virus infection. Proc Natl Acad Sci USA 1989;86:6353–6357.
38 Murphey-Corb M, Martin LN, Davison-Fairburn B, Montelaro RC, Miller M, West M, Ohkawa S, Baskin GB, Zhang JY, Putney SD, Allison AG, Eppstein DA: A formalin-inactivated whole in SIV vaccine confers protection in macaques. Science 1989;246:1293–1297.
39 Carlson JR, McGraw TP, Keddie E, Yee JL, Rosenthal A, Langlois AJ, Dickover R, Donovan R, Luciw PA, Jennings MB, Gardner MB: Vaccine protection of rhesus macaques against simian immunodeficiency virus infection. AIDS Res Hum Retroviruses 1990;6:1239–1246.
40 Stott EJ, Chan WL, Mills KH, Page M, Taffs F, Cranage M, Greenaway P, Kitchin P: Preliminary report: Protection of cynomolgus macaques against simian immunodeficiency virus by fixed infected-cell vaccine. Lancet 1990;336:1538–1541.
41 Berman PW, Gregory TJ, Riddle L, Nakamura GR, Champe MA, Porter JP, Wurm FM, Hershberg RD, Cobb EK, Eichberg JW: Protection of chimpanzees from infection by HIV-1 after vaccination with recombinant glycoprotein gp120 but not gp160. Nature 1990;345:622–625.
42 Girard M, Kieny MP, Pinter A, Barre-Sinoussi F, Nara P, Kolbe H, Kusumi K, Chaput A, Reinhart T, Muchmore E, Ronco J, Kaczorek M, Gomard E, Gluckman JC, Fultz PN: Immunization of chimpanzees confers protection against challenge with human immunodeficiency virus. Proc Natl Acad Sci USA 1991;88:542–546.

Anthony S. Fauci, MD, National Institute of Allergy and Infectious Diseases, National Institutes of Health, Bethesda, MD 20892 (USA)

Rossi GB, Beth-Giraldo E, Chieco-Bianchi L, Dianzani F, Giraldo G, Verani P (eds): Science Challenging AIDS. Basel, Karger, 1992, pp 15–23

AIDS: Progress and Challenges in Therapy

Samuel Broder

National Cancer Institute, Bethesda, Md., USA

AIDS, caused by infection with the T cell lymphotropic retrovirus known as human immunodeficiency virus (HIV), was recognized more than a decade ago and has become a global pandemic since that time [1]. In 1981, there were isolated reports of a devastating illness that was eventually recognized as a constellation of findings – a syndrome – obligatorily linked to profound cellular immunodeficiency [2]. The immunodeficiency was often accompanied by a then-rare tumor, Kaposi's sarcoma (KS), and punctuated by overwhelming infections with opportunistic organisms such as *Pneumocystis carinii* or cytomegalovirus or *Toxoplasma gondii* that, as a rule, threaten the lives of only the most deeply immunocompromised hosts. The syndrome mimicked many of the manifestations of known congenital immunodeficiency diseases but was clearly acquired in a fashion suggestive of transmission of an infectious agent – hence the name acquired immunodeficiency syndrome (AIDS). Other than these clinical observations, virtually nothing was known and there were few meaningful interventions.

Then, in 1983, Montagnier and his colleagues first reported the association of this clinical syndrome with a candidate virus [3]. In 1984, Montagnier and Gallo provided conclusive evidence that this virus was indeed the etiologic factor for AIDS [4]. Since then, major strides in the basic understanding and therapeutic management of all stages and manifestations of HIV infection have emerged from intensive research efforts in the broad areas of biomedical research, natural history, vaccine development and treatment. Yet as AIDS and AIDS-targeted scientific endeavors enter the second decade, the spectrum of HIV infection has broadened in new directions: specifically, the increasing occurrence of disparate and aggressive AIDS-related cancers, and the growing occurrence of HIV infection in women and children. These areas of emerging challenge are of paramount biologic and societal importance, and continue to command intensive commitment.

The areas of progress can be measured in the translation of basic virologic and immunologic discoveries into practical therapies. The decoding of viral targets that can be attacked therapeutically has not been straightforward, as HIV is a genetically complex retrovirus [5]. Like all retroviruses, HIV has a gene encoding reverse transcriptase (RT), the HIV enzyme responsible for transcription of HIV RNA into DNA. RT is the focus of inhibition by the dideoxynucleoside drugs and many other agents now being tested. But HIV is now known to contain at least 9 genes, making it much more complex than the previously identified retroviruses. Antigenic variation may complicate attempts to develop vaccines capable of inducing cytotoxic T cell and humoral responses that target HIV for a broad spectrum of individuals. The panel of HIV regulatory genes, such as *tat* and its Tat protein product, subvert host cell genetic machinery in ways that modulate host cell gene expression, T cell antigenic responses, and the production of immune-active cytokines that in turn enhance HIV replication [5, 6]. New targets for therapy are emerging rapidly. For example, recent data in simian models suggest that a poorly understood gene (*nef*) is required for maintaining high viral loads during the course of persistent infection, and so this gene could serve as a novel site for antiviral therapy [7].

The dissection of the complexity of HIV and its interactions with host cell transcriptional processes has provided a number of mechanisms that can be targeted for viral destruction or inhibition by practical antiretroviral therapies. The dideoxynucleosides, of which azidothymidine (AZT) is the prototype, act through RT inhibition [1, 8]. The availability of additional effective drugs of this class (ddI, ddC, d4T) affords the opportunity to create multidrug combinations aimed at ameliorating cumulative drug toxicities and preventing the emergence of drug resistance. Drug resistance seems to depend on several clinical parameters including stage of disease [9]. The HIV protease, an enzyme that appears to have a pivotal role in both the early stages of viral replication and the late stages of viral maturation and packaging, is another promising therapeutic target [1, 8]. The three-dimensional structure of this complex enzyme has been deciphered and used to develop effective inhibitors that are coming rapidly to clinical testing. In another area of molecular virology, the Tat protein has an autocrine function with regard to HIV replication and, thus, may have a central role in disease perpetuation and progression [5]. Elucidation of the specific RNA sequence within the long terminal repeat (LTR) of the HIV genome to which the Tat protein binds (known as TAR) offers a novel target for therapy, namely inhibitors of TAR-Tat binding such as TAR decoy molecules or antisense DNA constructs directed against the *tat* genomic RNA or mRNA [1].

Knowledge of the many steps in the HIV life cycle has opened other avenues of attack. For example, initial attachment of the HIV envelope glycoprotein gp 120 to its molecular target CD4 can be blocked by recombinantly produced soluble CD4 (r-CD4) or a recombinant toxin conjugate coupled to CD4 (specifically, CD4 coupled to *Pseudomonas* exotoxin, CD4-PE 40) [1]. The recent development of an experimental vaccine that stimulates both a humoral and cellular response against gp 160 (encompassing both the gp 120 binding component and the gp 41 fusion component of the HIV envelope) in humans represents a major advance for therapy and prevention [10]. While it is too early to conclude that any given vaccination strategy will be safe and effective, it is quite likely that a functional vaccine will be available in the decade of the nineties.

The ability to abrogate drug toxicity, in particular toxicity that affects the ability to administer effective drug doses, is an important adjunct to anti-HIV therapy. In this regard, genetically engineered hematopoietic stimulators such as erythropoietin (EPO), granulocyte-macrophage colony-stimulating factor (GM-CSF) or granulocyte (G)-CSF may ameliorate both HIV-related and dideoxynucleoside-related bone marrow suppression [1, 11].

The effects of GM-CSF are especially provocative, as this biomodulator may actually augment both HIV replication and the anti-HIV effects of AZT in some systems [11]. This apparent paradox can be explained by the enhanced effects of AZT against an actively replicating viral genome and by GM-CSF-induced perturbations in AZT intracellular biochemical activation [11]. In another critical area of adjunctive therapies, the armamentarium directed against opportunistic organisms continues to expand, for instance, trimetrexate for resistant *P. carinii* or the new erythromycin-related macrolide antibiotic clarithromycin for *Myobacterium avium* complex infections [12].

Future innovations in anti-HIV therapy and prevention will likely come from several promising directions The discovery of various natural products with unique mechanisms of action through high-capacity drug screening mechanisms, the use of sophisticated biophysical tools (crystallography and the supercomputer, as examples) to define the structures of key viral components and design drugs to specifically target those components, and the rational use of immunomodulating growth factors such as a newly discovered thymic hormone or multilineage interleukins (IL) such as

IL-1 or IL-3 are but a few examples of novel approaches that are already under active investigation [1].

The momentum of progress on the molecular level, both in the laboratory and in the arena of practical clinical application, is unfortunately matched by provocative challenges in at least two major areas mentioned above: the escalating numbers of women in whom HIV infection and its consequences are now being detected, and the increasing numbers of aggressive HIV-related malignancies – ironically a consequence of lengthening survival for AIDS patients in the deeply immunocompromised state [13, 14].

HIV infection in the United States, initially thought to be a disease virtually unique to men, is occurring in growing numbers in women [15]. The dramatic increase in AIDS noted for women since 1987 is occurring at a time when a stabilization or decrease in AIDS is being noted for other segments of the population, notably for gay men and for recipients of blood product transfusions. The increase in women is, sadly, accompanied by a similar increase in perinatally acquired AIDS, although the trend in perinatal infection may be starting to plateau over the last 1–2 years. Of the almost 160,000 AIDS cases reported in adults as of December, 1990, approximately 10% are women; however, while approximately 3,600 women (10% of all adult cases) were diagnosed with AIDS in 1989, there was a 35% increase in the numbers of women diagnosed with AIDS in 1990 (about 4,900 women, comprising 11.5% of all adult cases). This escalating prevalence is a surpassingly important public health issue that demands attention to the multifaceted biological, public health, and societal manifestations of HIV infection that uniquely affect women.

There are other demographic considerations worth noting. In the USA and in many other countries, AIDS is now inextricably linked to poverty and the societal consequences of poverty. Moreover, AIDS is a cause of economic devastation at the individual and geopolitical levels in many parts of the world. Since its recognition, AIDS has had a close association with a broad spectrum of cancers, includin KS, lymphomas, basal cell carcinoma, anogenital carcinomas (including carcinoma of the cervix), and hepatitis B-related hepatocellular carcinoma [13, 16–18]. This association is not unexpected, and, indeed, was presaged by malignancies linked to other severe immunodeficiency disorders such as ataxia-telangiectasia, Wiskott-Aldrich syndrome, and the immunocompromised state that results from organ transplantation (and the iatrogenic immunosuppression used in transplantation) [19]. In the setting of AIDS, there are multiple factors that can promote or permit the emergence of cancers – namely, the chronic and profound defects in both cellular and humoral mechanisms of immune surveillance, B cell stimulation, and possibly the presence of additional viruses that may act directly or indirectly as cofactors to induce malignant transformation. Further, the types of malignancies and their incidence rates are increasing as the development of effective antiretroviral therapies and prophylaxis against opportunistic infections leads to prolonged survival in the immunodeficient state for AIDS patients.

The increasing incidence of AIDS-related malignancies for all HIV-infected persons is an emerging challenge, and women are not an exception. This is perhaps not surprising, in view of the association between certain strains of human papillomavirus (HPV) and cervical cancer. In this regard, cervical cancer is an opportunistic cancer, much like KS, non-Hodgkin's lymphomas (NHL) and HPV-associated anal squamous cell cancers in HIV-infected men [19]. Like other malignancies arising in the setting of HIV infection and immunosuppression, HIV-associated cervical cancers are distinctly aggressive [18]. HPV co-infection is common in women with HIV; in fact, 60% of HIV-infected women have cervical dysplasia noted on Pap tests (similar to anal epithelial dysplasia in men) [17, 18]. HPV and HIV act synergistically to increase each other's gene expressions [16]. The enhanced expression of viral proteins that increase viral replication, abrogate host tumor suppressor functions and further exacerbate cellular imunodeficiency contribute to the high incidence of advanced stage disease that is rapidly progressive and refractory to therapy. Studies to define the true incidence and natural history of concomitant HPV and HIV infections, cervical intraepithelial neoplasia (an early or 'premalignant' lesion) and cervical cancer will lay the groundwork for effective strategies aimed at early detection and prevention. There is a critical need to develop multi-center networks for the prospective epidemiologic study of HIV-associated malignancies in women (aggressive HPV-related cervical cancer) and children (for example, rhabdomyosarcoma). These networks can be built upon the Multicenter AIDS Cohort Study (MACS) for men which has provided extensive data regarding HIV-associated cancers, in particular specific cancer risks as related to infections with potentially oncogenic viruses.

KS and NHL occur with striking frequencies in AIDS [13, 16, 19]. While KS has been a notorious accoutrement of AIDS since the beginning of the epidemic (in the early years of the AIDS epidemic, often an index marker of AIDS), NHL has perhaps emerged as a major sequela of HIV infection only recently and now occurs predominantly in patients who survive other consequences of AIDS for protracted periods of time. Profound cellular immunodeficiency plays a central role in lymphomagenesis, as demonstrated by the striking relationship between the depletion of CD4 lymphocytes and the development of NHL, particularly when the CD4 count falls below 50/mm^3 [20].

Although KS and NHL arise from different primordial target cells, they share distinctive features that likely reflect similar pathogenic mechanisms as a consequence of the underlying HIV infection. In the setting of AIDS, both malignancies display aggressive clinical behavior and unusual patterns of organ involvement with a penchant for monocyte-laden tissues, including the gastrointestinal tract for both tumor types, the lungs for KS, and the central nervous system and bone marrow for NHL. Both are associated with other viruses which are likely to have at least a cofactor role in tumorigenesis – cytomegalovirus (CMV) is frequently present in KS patients and Epstein-Barr virus (EBV) often accompanies NHL, with genomic sequences detectable in about 50% of AIDS-related lymphomas which in some cases precedes NHL development [21, 22].

Both tumors may arise via polyclonal proliferation of multiple interactive cell types, for example monocytes and vascular endothelium in KS or monocytes and B cells in NHL. In the case of KS, HIV-infected T cells, monocytes, and the transformed cells themselves produce multiple growth factors and viral regulatory proteins (e.g. Tat) that can result in both autocrine and paracrine stimulation of host cell(s) proliferation and viral expression [24]. In particular, the monocyte, as both a cellular reservoir for infective viral particles and a production site for cytokine synthesis and secretion, may be central to the growth dysregulation and eventual malignant transformation in both KS and NHL. The monocyte and its cytokine products such as IL-6 may be important to growth abnormalities and eventual malignant transformation in both KS and NHL. Definition of the roles of monocytes, AIDS-KS and/or B-stimulating growth factors and potential T-suppressing lymphokines (for example, the newly discovered IL-10) in the genesis and clonal expansion of transformed B cells may lead to identification of lymphokine inhibitors for eventual clinical application [23]. For example, IL-6 inhibitors might have potential therapeutic value for both AIDS NHL and AIDS-KS (where IL-6-mediated autocrine stimulation drives KS cell proliferation). Recent studies of multiple myeloma cells by Sidell et al. [25] demonstrate that retinoic acid can down-regulate IL-6 receptors and thus inhibit IL-6-mediated autocrine growth stimulation. These findings may be pertinent to therapeutic considerations in AIDS-KS and possibly AIDS NHL as well. The potential contributions of specific viruses (EBV, CMV and human herpesvirus-6, for instance) or antiviral therapies to the pathogenesis of AIDS-NHL by inducing genetic instability and DNA damage (DNA breaks, aberrant gene rearrangements) are also areas for investigation. The ability to define multiple molecular factors that confer heightened susceptibility to malignant transformation in the AIDS patient will translate into the development of both therapeutic and preventive strategies with broad implications for the entire spectrum of AIDS-related cancers.

As with all malignancies, the most effective strategies are those targeted to prevention. The ability to interfere early in the course of active HIV infection with mechanisms that promulgate transformed cell hyperproliferation and clonal expansion – especially growth factors (for example, fibroblast growth factors or IL-6) and concomitant viral infections (mainly EBV but also CMV and human herpesvirus-6) – might decrease the occurrence or prolong the time to development of AIDS-related KS or lymphomas. Ultimately, however, the key strategies will be those directed toward maintaining the CD4 cell count at a level that prevents the establishment and perpetuation of transformed clones. At least for AIDS lymphomas, one critical level appears to be a CD4 count of approximately 50/mm^3. Indeed, a CD4 count of 50 mm^3 seems to be a significant threshold as a mortality risk indicator related not only to lymphomas but to death from any cause [20]. If further work supports this concept, a critical goal of stabilizing the immune system may be practical and within reach. The identification of factors that portend a significant risk of tumorigenesis for an HIV-infected individual and the development of antiretroviral strategies that confer long-term suppression of HIV activity and relative preservation of immune function are essential to the ultimate prevention of all malignancies that arise as a consequence of HIV-induced immunosuppression.

Conclusion

In the decade since AIDS was first noted, much has changed. There are definite areas of progress, but many challenges and unresolved issues remain. Continued progress is possible if we pay attention to the root causes of the disease and if we continue to adhere to science and the scientific method.

References

1 Yarchoan R, Pluda JM, Perno C-F, Mitsuya H, Broder S: Antiretroviral therapy of HIV infection: Current strategies and challenges for the future. Blood 1991; in press.

2 Gottlieb MS, Groopman JE, Weinstein WM, Fahey JL, Detels R: The acquired immunodeficiency syndrome. Ann Intern Med 1983;99:208–220.

3 Barre-Sinoussi F, Chermann JC, Rey F, Nugeyre MT, Chamaret S, Gruest J, Dauguet C, Axler-Blin C, Vezinet-Brun F, Rouzioux C, Rozenbaum W, Montagnier L: Isolation of a T-lymphotropic retrovirus from a patient at risk for acquired immune deficiency syndrome (AIDS). Science 1983;220:868–871.

4 Gallo RC, Salahuddin SZ, Popovic M, Shearer GM, Kaplan M, Haynes BF, Palker TJ, Redfield R, Oleske J, Safai B, White G, Foster P, Markham PD: Frequent detection and isolation of cytopathic retroviruses (HTLV-III) from patients with AIDS and at risk of AIDS. Science 1984;224:500–503.

5 Greene WC: The molecular biology of human immunodeficiency virus type I infection. N Engl J Med 1991;324:308–316.

6 Fauci AS, Schnittman SM, Poli G, Koenig S, Pantaleo G: Immunopathogenic mechanisms in human immunodeficiency virus (HIV) infection. Ann Intern Med 1991; 114:678–693.

7 Kestler HW III, Ringler DJ, Mori K, Panicali DL, Sehgal PK, Daniel MD, Desrosiers RC: Importance of the *nef* gene for maintenance of high virus loads and for development of AIDS. Cell 1991;65:651–662.

8 Broder S, Mitsuya H, Yarchoan R, Pavlakis GN: Antiretroviral therapy in AIDS. Ann Intern Med 1990;113:604–618.

9 Larder BA, Darby G, Richman DD: HIV with reduced sensitivity to Zidovudine (AZT) isolated during prolonged therapy. Science 1989;243:1731–1734.

10 Redfield RR, Birx DL, Ketter N, Tramont E, Polonis V, Davis C, Brundage JF, Smith G, Johnson S, Fowler A, Wierzba T, Shafferman A, Volvovitz F, Oster C, Burke DS, and the Military Medical Consortium for Applied Retroviral Research: A phase I evaluation of the safety and immunogenicity of vaccination with recombinant gp160 in patients with early human immunodeficiency virus infection. N Engl J Med 1991;324:1677–1684.

11 Pluda JM, Yarchoan R, Smith PD, McAtee N, Shay LE, Oette D, Maha M, Wahl SM, Myers CE, Broder S: Subcutaneous recombinant granulocyte-macrophage colony-stimulating factor usd as a single agent and in alternating regimen with azidothymidine in leukopenic patients with severe human immunodeficiency virus infection. Blood 1990;76:463–472.

12 Rastogi N, Labrousse V: Extracellular and intracellular activities of clarithromycin used alone and in association with ethambutol and rifampin against *Mycobacterium avium* complex. Antimicrob Agents Chemother 1991;35:462–470.

13 Pluda JM, Yarchoan R, Jaffe ES, Feuerstein IM, Solomon D, Steinberg SM, Wyvill KM, Raubitschek A, Katz D, Broder S: Development of non-Hodgkin's lymphoma in a cohort of patients with severe immunodeficiency virus (HIV) infection on long-term antiretroviral therapy. Ann Intern Med 1990;113:276–282.

14 Gail MH, Pluda JM, Rabkin CS, Biggar RJ, Goedert JJ, Horm JW, Sondik EJ, Yarchoan R, Broder S: Projections of the incidence of non-Hodgkin's lymphoma

related to acquired immunodeficiency syndrome. J Natl Cancer Inst 1991;83:695–700.

15 Ellerbrock TV, Bush TJ, Chamberland ME, Oxtoby MJ: Epidemiology of women with AIDS in the United States, 1981 through 1990. A comparison with heterosexual men with AIDS. JAMA 1991;265:2971–2975.

16 Cremer KJ, Spring SB, Gruber J: Role of human immunodeficiency virus type 1 and other viruses in malignancies associated with acquired immunodeficiency disease syndrome. J Natl Cancer Inst 1990;82:1016–1024.

17 Kiviat N, Rompalo A, Bowden R, Galloway D, Holmes KK, Corey L, Roberts PL, Stamm WE: Anal human papillomavirus infection among human immunodeficiency virus-seropositive and virus-seronegative men. J Infect Dis 1990;162:358–361.

18 Maiman M, Fruchter RG, Serur E, Remy JC, Feuer G, Boyce J: Human immunodeficiency virus infection and cervical neoplasia. Gynecol Oncol 1990;38:377–382.

19 Groopman JE, Broder S: Cancer in AIDS and other immunodeficiency states; in DeVita VT (ed): Cancer. Principles and Practice of Oncology. Philadelphia, Lipincott, 1989, pp 1953–1970.

20 Yarchoan R, Venzon DJ, Pluda JM, Lietzau J, Wyvill KM, Tsiatis AA, Steinberg SM, Broder S: Cd4 count as a mortality risk indicator in human immunodeficiency virus (HIV)-infected patients receiving antiretroviral therapy: Experience in a research hospital. Ann Intern Med 1991;in press.

21 Neri A, Barriga F, Inghirami G, Knowles DM, Neequaye J, Magrath IT, Dalla-Favera R: Epstein-Barr virus infection precedes clonal expansion in Burkitt's and acquired immunodeficiency syndrome-associated lymphomas. Blood 1991;77:1092–1095.

22 Shibata D, Weiss LM, Nathwani BN, Brynes RK, Levine AM: Epstein-Barr virus in benign lymph node biopsies from individuals infected with the human immunodeficiency virus is associated with concurrent or subsequent development of non-Hodgkin's lymphoma. Blood 1991;77:1527–1533.

23 Birx DL, Redfield RR, Tencer K, Fowler A, Burke DS, Tosato G: Induction of interleukin-6 during human immunodeficiency virus infection. Blood 1990;76:2303–2310.

24 Ensoli B, Nakamura S, Salahuddin SZ, Biberfeld P, Larsson L, Beaver B, Wong-Staal F, Gallo RC: AIDS-Kaposi's sarcoma-derived cells express cytokines with autocrine and paracrine growth effects. Science 1989;243:223–226.

25 Sidell N, Taga T, Hirano T, Kishimoto T, Saxon A: Retinoic acid-induced growth inhibition of a human myeloma cell line via down-regulation of IL-6 receptors. J Immunol 1991;146:3809–3814.

Samuel Broder, MD, National Cancer Institute, Bethesda, MD 20892 (USA)

Rossi GB, Beth-Giraldo E, Chieco-Bianchi L, Dianzani F, Giraldo G, Verani P (eds): Science Challenging AIDS. Basel, Karger, 1992, pp 24–32

The Implications of AIDS in Developing Countries

V. Ramalingaswami

All India Institute of Medical Sciences, New Delhi, India

The Developing Scenario in Developing Countries

AIDS is increasingly becoming a problem of developing countries; the gap between developing and developed countries is widening. While HIV infection rates appear to be slowing down in some developed countries, in itself a welcome feature, it is increasing markedly in many developing countries, especially in sub-Saharan Africa but also in Asia, Latin America and the Caribbean, where it is poised for dramatic spread in the future both in the short and long term [1, 2]. In this developing country setting, heterosexual transmission is the dominant mode, rising new infections and rapid spread are the rule, and men, women and children are affected. WHO estimates that the cumulative total of 10 million HIV infected worldwide now will increase to about 40 million infected by the end of the decade with more than 90% of infections taking place in developing countries; that the current cumulative total of about one million adult cases of AIDS worldwide will increase to 10 milion, again with almost 90% of cases occurring in developing countries [1].

The Asian drama of AIDS is unfolding dark horizons and the barometer is dropping. With more than 50% of the world population but with less than 0.2% of AIDS cases reported, Asia was for the few initial years a huge continent with invisible AIDS but this alas is no longer so. Danger signals of aggressive spread of HIV have begun to appear in parts of India and Thailand in the past 3 years, transmission being largely heterosexual but AIDS infection had also been spreading with lightning speed among groups of intravenous drug users in parts of both of these countries as well as the southern part of China and Myanmar [1, 3]. For example, in the north eastern state of Manipur in India, there were no incidences of HIV-positive infection recorded up to September 1989; in October 1989 only one blood sample was positive but after vigorous search, by June 1990 54% of 1,564 blood samples drawn from intravenous drug users were positive [3].

The Moment of Truth

The inequities of health care resources in caring for AIDS patients reveal a stark reality, i.e. the total funding for the average national AIDS programme in the developing world today is less than the cost of caring for only 40 persons with AIDS in the USA and prompts one to ask: Are there going to be two worlds of AIDS? Developing countries are characterised by severely limited resources to match large unresolved and competing priorities. AIDS struck at a time when several of the developing countries, particularly in sub-Saharan Africa, were reeling under severe economic recession, cyclical droughts and cyclones and unending wars, with more than 20 African countries now needing emergency food help this year and parts of Africa now passing through the worst famine in living memory. The hard-earned improvements in child survival in these countries over the past two to three decades are likely to be wiped away in the 1990s in the wake of the AIDS pandemic [5]. The situation of women, already desperate, will worsen. Indeed, the impact of AIDS is so wide-ranging on overall socio-economic development in sub-Saharan Africa that AIDS is not only a health issue but also a societal, a developmental, and an ethical one. Interventions that strengthen coping responses are needed at a time when coping capabilities are stretched to the limits with fragmentation of attention and action, i.e. [6]: What type of community and family coping mechanisms could respond to adverse economic impact and demographic change of the type being witnessed today in sub-Saharan Africa? What are the policy options to labour losses, to lack of caring facilities for the terminally ill and the like?

AIDS should open our eyes to the need for long-range vision in dealing with development by building the capabilities of the countries to resolve their problems, for knowing what help is needed and when to seek it from outside, and for seeking international partnerships out of their own enlightenment in addressing common problems. The capability building needs of developing countries relate to infrastructure and human resources, information systems, policy and management research, biomedical and epidemiologic research, and social and behavioural research. In the field of health, this process, called Essential National Health Research is what is advocated by the Commission on Health Research for Development and by the World Health Assembly at its meeting held last year [7, 8]. Developing countries need to be helped to come out of ad hoc, staccato responses and move towards sustainable human-centered development.

The apparent low prevalence or no prevalence or only a few foreign imported cases of AIDS in large parts of the developing world as the Asia-Pacific should no longer lull us into complacency and inaction. Low prevalence of AIDS had so far proved to be transient while the virus had been spreading clandestinely and had only been a prelude to pandemic transformation. What price the long incubation period of AIDS! Osborn [9] called AIDS a snapshot picture which took 10 years or more to develop. Low prevalence is an opportunity to grasp as a stimulus for vigorous research and preventive action for we know that AIDS is preventable. Where the numbers are still small, lies the geatest opportunity to prevent. To foresee is to govern. No country is immune, no population has innate biological resistance, no country is free from chinks in its cultural and behavioural armour to drive AIDS away, no other disease is so closely interwoven with human values and attitudes [4].

Nor can we in the developing countries sit back in the hope that vaccines and miracle drugs discovered in the laboratories of developed countries will come to the developing countries at some propitious moment. Developed and developing countries have to join forces together in the fight against AIDS and developing countries have to participate in the whole range of endeavour against AIDS from identifying and characterising the local strains of HIV, defining how they differ from those in the developed countries, how the backdrop triad of malnutrition, infectious disease and excessive population growth influences the entire natural history of the disease; how local social, economic, cultural and behavioural factors influence transmission dynamics; how on the basis of such knowledge each country can proceed to fashion its own approach to AIDS, an approach based on accurate information and intensive education that ensures medical and social support in an atmosphere that is humane, caring and non-discriminatory using community structures and locally trusted people. In particular, we need in developing countries rapid and practical research into social and behavioural domains with feedback to intervention strategies [10]. The road to HIV infection is not irreversible; there are several escape exits to be explored in each culture and context. With or without effective technologies for prevention and cure of AIDS and at all times, education holds the key to prevention. As H.G. Wells once said. 'The choice before mankind is between education and catastrophe.' Every man and women is a secret teacher.

Women and Children – The Gathering Storm

Three million women worldwide have already been infected with HIV, 0.5 million have developed AIDS, and 2 million will die of the disease in the 1990s [11]. One million children are already infected with HIV, 0.5 million have developed AIDS, many more millions will have been infected by the end of the 1990s, and at least 10 million children will be orphaned. The bulk of these HIV infections, diseases and deaths will take place in the developing countries, mostly sub-Saharan Africa but also in some countries of Latin America and the Caribbean attributable to predominance of heterosexual transmission, rising HIV infection rates in women of reproductive age, high birth rates and the perinatal transmission potential including transplacental transmission of HIV in most, if not all, perinatal transmissions. Such is the inexorable march of AIDS.

As mother-to-child transmissions increase, AIDS embeds itself in the future generations and starts an inter-generational trajectory, posing a dramatic shift in the dynamics of health care and revealing the true universality of HIV [9]. I have described on another occasion how women in South Asia are born into inequity characterised by historical negligences and ecological handicaps. Lack of educational opportunity and low societal status, early marriages, and excessive fertility [12]. Women in development is a profound issue stirring our times. Half a million women lose their lives every year throughout the world from causes related to pregnancy and child birth, in the very natural process of giving life as it were. And over 99% of these deaths take place in developing countries [12]. It is a tragic irony that a new equality is evolving among men and women with regard to HIV infections as this decade of the 1990s moves forward with parity in AIDS being expressed between men and women [13]. This decade will also see AIDS emerging as the leading cause of death in women between 20 and 40 years of age in many regions. When HIV infection reaches 20% in pregnant women, there may be a substantial increase in under 5 mortality due to perinatal transmission [14].

Women have little bargaining power to negotiate the practice of safe sex with their partners and few women have any control over the sexual behaviour of their men folk. 'To ask for the use of condoms is to risk conflict; not to ask is to risk infection and death' [13]. Fortunately, a fruitful approach lies in integrating AIDS control activities into the intensified maternal and child health and Safe Motherhood initiatives currently under way providing access for infected and at-risk mothers to services including family planning to avoid pregnancy, and to counselling and psycho-social support. Antenatal and inter-pregnancy visits by trained personnel provide a setting where mothers could be educated and counselled regarding HIV infection and where high-risk pregnancies could be detected for referral preparedness for severe anaemia, toxaemia and anticipated obstructed labour. Integrated maternal and child care services and AIDS control strategies constitute a stimulus to health services triggered by AIDS control thrusts [15].

AIDS and Health Care

HIV may be new but sexually transmitted infections (STIs) are as old as the hills. An estimated 250 million or more sexually transmitted infections including syphilis, gonorrhoea, genital herpes, chancroid, hepatitis B, HIV and others occur annually around the world and their number is increasing. It is the young who are most affected [16]. It is well known that STIs, in particular, ulcerative genital lesions caused by sexually transmitted infections, may increase the risk of contracting HIV by more than 300% [16–18]. An opportunity exists here to reduce HIV infectivity by revamping the sagging STI control programmes in developing countries.

That endemic tropical parasitic diseases through their continued stimulation of the immune system may have an influence on the outcome of the host: HIV interaction has been suggested. Malaria alone is believed to be infecting 270 million people with an estimated 110 million clinical cases of malaria a year, affecting 100 countries and alone leading to a mortality of one million annually from this single disease. The tuberculosis situation in many developing contries remains serious with 2–3 million people dying annually from this disease. This situation is made worse by AIDS. In Africa, a number of countries have documented dramatic increases in tuberculosis, some with rates which doubled in 3 years and similar events can be expected in other parts of the developing world, especially the Asia-Pacific [19]. It is clear that the health connection of AIDS provides a window of opportunity which must be seized forthwith. Furthermore, the routine testing and elimination of blood if infected before transfusion for markers of infection with HIV, hepatitis B virus and syphilis will eliminate this route of infection for these diseases. AIDS provides an opportunity to renew the lagging efforts of many developing countries to improve the safety of their blood transfusion services and to eliminate this route of infection not only for HIV but also for hepatitis B virus and syphilis as well. The same can be said of contaminated needles, syringes, skin piercing instruments and blood products. Integration of AIDS control strategies into the broad framework of health services using common entry points and common channels of service delivery and deriving strength from its close epidemiological relationships with a number of diseases constitutes practical, viable and immediate response. Reproductive health, maternal and child care, family planning, sexually transmitted diseases, tropical disease control are all targets for integration. There is also a wonderful opportunity to streamline hospital infection control procedures which will not only have an impact on AIDS but also on the whole range of nosocomial infections which plague many hospitals in developing countries. The moral is to ramify AIDS concerns and control approaches into a wide range of health services, at the same time taking care to see that the basic services themselves, fragile as they may be, are not jeopardised but strengthened [20].

Lessons Learned

Despite the difficulties and the need for speed, much has been accomplished to mobilise forces nationally and internationally to cope with AIDS, and the WHO through its special programme, the GPA, has been playing a leading role in this regard, Many valuable lessons have been learned. Among these are: the need for protection of human rights, countering of discrimination against AIDS everywhere, the futility of the cordon sanitaire approach, the overriding need for care and counselling and for a deeper understanding of the diversity of cultural and economic conditions within and between countries and a sense of social realism. Testing is after all a basic instrument in public health, but a cautious approach to testing for HIV infection is being advocated because of the discrimination and its consequences – the segregation, the incarceration, the criminalisation, the abandonment of care, the rejection, the consecration of female prostitutes to remand homes – the string of inhumanities which are still, sadly, features of the AIDS landscape and reminiscent of societal attitudes towards diseases such as leprosy not yet fully eliminated even today [4].

The crucial importance of raising the level of understanding and cooperation on the part of men to provide greater autonomy and social control for women should not be ignored [10]. There is also a need to recognise that policies are readily made but determined action often lags behind to implement those policies in many developing countries with low HIV prevalence rates. Also notable is the widening chasm between testing for HIV infection and building medical and social support services.

The tendency for performing seroprevalence surveys and then reacting must yield to anticipatory proactive approaches. Success in football lies in players positioning themselves where the ball is likely to come rather than running after the ball all the time.

Is the New Biology a match for the challenge of AIDS [21]? Science had never been in a stronger position than today and the world awaits with optimistic expectancy the fruits of research and development currently under way for effective vaccines and drugs against AIDS. Developing countries must be deeply involved in all stages of the fight against AIDS; they have a crucial role to play not only in promoting a better understanding of HIV transmission dynamics at the local level and in developing local strategies for intervention, but also in collaborating in a global effort in clinical and field trials of new tools of AIDS control ensuring the highest scientific and ethical standards.

Conclusion

AIDS is increasingly becoming a disease of developing countries, sub-Saharan Africa is currently the worst affected with predominant heterosexual transmission, rising fresh infections, rapid spread and an adverse impact on socio-economic development. Increasing prevalence rates in women and children is the most sinister feature of AIDS in developing countries. Hard-won improvements in child health are in danger of being wiped away and adult mortalities will increase.

Perinatal transmission brings AIDS into the inter-generational arena making urgent the need for supreme efforts to interrupt this transmission through the integration of AIDS control strategies into maternal and child care services including family planning. There is an opportunity for the containment of AIDS by integrating its control strategies into the broader framework of health services including control of sexually transmitted diseases, endemic tropical diseases, reproductive health and family planning.

At the other extreme is the vast Asia-Pacific area with more than 50% of the world's population but with an apparent low overall prevalence of HIV/AIDS till now. Low prevalence induces a false sense of security. The rapidity with which HIV infection had been spreading in recent years both heterosexually and amongst intravenous drug users in India and Thailand is a cause for deep concern. The teeming millions of this region are exposed to the danger of the spread of AIDS on an unimagined scale. It is imperative that AIDS should be placed high on the political agenda. Precious time is being lost.

AIDS can become a force for human centered development including improvement of women's status and health. It can also usher in a new era of emphasis on human life-styles and human behaviour as a key to the future health of man. There need no longer be a conflict between technological solutions and socio-cultural, behavioural and educational approaches.

AIDS heightens the imperative need of looking at development from a long-term point of view in a process involving local capacity building, infrastructure development, information, policy and management, sustained national commitment and international resolve.

Men, women and children all threatened by AIDS are asking for a response from humanity as a whole. They ask for tapping the fountain springs of humanism, revealing every man and woman as maximally human. They ask for elevation of humankind to new levels of feeling and consciousness.

To care is a duty; to prevent is a responsibility. Prevention and care are the twin engines that should drive our effort for the containment of AIDS. Mahatma Gandhi, and countless others like him, unhonoured and unsung, have a degree of relevance today as never before. Many years ago Mahatma Gandhi said: 'I am hard hearted enough to let the sick die if you can tell me how to prevent others from falling sick.' At the same time he nursed with his bare hands the weeping sores of a patient with lepromatous leprosy. It is this balanced mix of *caring* and *prevention* that the world needs today.

References

1 WHO: In Point of Fact. The Global HIV/AIDS Situation: WHO Projects 40 Million HIV Infections by the Year 2000. May 1991, No 74.

2 Chin J: Epidemiology of AIDS: Keynote Address, VIIth Int Conf AIDS, Florence, July 16–21, 1991.

3 Pal SC, Sarkar S, Naik TN, Singh KP, Ao SLT, Lal S, Tripathy SP: Explosive epidemic of HIV infection in north eastern states of India, Manipur and Nagaland. CARC Calling 1990;3:2–6.
4 Siverman MF: AIDS and Ethics. AIDS in Asia and the Pacific Conference, Canberra, Aug 5–8, 1990, pp 67–70.
5 Ramachandran P: Impact of AIDS on women and children. AIDS in Asia and the Pacific Conference, Canberra, 1990, pp 122–130.
6 Nabarro D, McConnell C: The impact of AIDS on socioeconomic development. AIDS 1990;3(suppl 1):S265–S272.
7 Commission on Health Research for Development: Health Research Essential Link to Equity in Development. London, Oxford University Press, 1990, p 136.
8 World Health Assembly Resolution, WHA43.19: The Role of Health Research, May 1990.
9 Osborn JE: Cultural, social and political aspects of AIDS: 2001; in Dupuy JM, Lemaire JF, Valetta L (eds): Sida 2001 AIDS 2001, 22–23 Apr, 1989. Compte-rendu de la réunion organisée aux Pensières à Veyrier-du-Lac, Annecy, Marcel Merieux Foundation, 1989, pp 75–81.
10 Carballo M: Psychosocial aspects of AIDS: Policy Implications of AIDS: In: Asia and the Pacific Conference, Canberra, 1990, pp 59–63.
11 Preble EA: Impact of HIV/AIDS on African children. Soc Sci Med 1990;31:671–680.
12 Ramalingaswami V: Into Inequity Born: The state of health of the women in South Asia. Sci Publ Affairs 1990;6:67–76.
13 Huston P: Opinion, International Herald Tribune, Dec 1–2, 1990.
14 Kallings L: The global situation on AIDS in the nineties. Third Panhellenic Congr on AIDS, Athens, March 1991, pp 9–10.
15 Mann J: AIDS Guidelines for the Second Decade. Opinion, International Herald Tribune, Dec 1–2, 1990.
16 WHO Features. Sexually Transmitted Infections Increasing – 250 Million New Infections Annually, No 152, Dec 1990.
17 Holmberg SD, Horsburgh GR Jr, Ward JW, Jaffe HW: Biological factors in the sexual transmission of human immunodeficiency virus. J Infect Dis 1989;160:933–937.
18 Peprin J, Plummer FA, Branham RC, Piot P, Cameron DW, Ronald RR: The interaction of HIV infection and other sexually transmitted diseases, an opportunity for intervention (Editorial review). AIDS 1989;3:3–9.
19 UNDP/World Bank/WHO: Tropical Diseases, Progress in Research 1989–1990. Tenth Programme Report of Special Programme for Research and Training in Tropical Diseases.
20 Ko Ko U: Paper presented in AIDS in Asia and hte Pacific Conference, Canberra, Aug 1990, pp 25–27.
21 Krause RM: Is the biological revolution a match for the trinity of despair? In: Biotechnology in Society, Pergamon Press, 1986, pp 31–46.

Prof. Vulimiri Ramalingaswami, All India Institute of Medical Sciences,
Department of Pathology, Ansari Nagar, New Delhi 110-029 (India)

Rossi GB, Beth-Giraldo E, Chieco-Bianchi L, Dianzani F, Giraldo G, Verani P (eds): Science Challenging AIDS. Basel, Karger, 1992, pp 33–50

Present and Future Dimensions of the HIV/AIDS Pandemic

James Chin

World Health Organization, Geneva, Switzerland

Introduction

Uncertainty about the future dimensions of the pandemic of HIV infections and AIDS (HIV/AIDS) has existed since initial recognition of AIDS in the early 1980s. Uncertainty in formulating even short-term projections persists because of the very substantial difficulties associated with obtaining reasonably precise measurements of the prevalence and incidence of HIV infections and AIDS cases in any given population. These uncertainties remain in large part due to our limited knowledge of such major variables as: the spectrum of behaviours and the number of persons who engage in those behaviours associated with HIV transmission – unprotected sexual intercourse (without a condom), especially with multiple partners, and shared injection equipment for drug use; the proportion of individuals infected with HIV who will ultimately progress to AIDS; and progression rates from HIV infection to AIDS.

Despite the fact that much vital information is not available, estimates and projections of the HIV/AIDS pandemic have been made. This paper first reviews the transmission and natural history of HIV-1 infection and progression from HIV infection to AIDS and subsequent death. Then current estimates of global and regional HIV infections and AIDS are presented along with major trends and projections of the pandemic's impact that have been developed by the World Health Organization (WHO).

Transmission and Natural History of HIV Infection

HIV Transmission

Only three modes of HIV transmission have been documented: (1) via sexual intercourse (vaginal or anal); (2) by injection or transfusion of HIV-infected blood, blood products, semen, tissues, or organs; sharing drug-injecting equipment; (3) from a mother to her fetus/infant (perinatal transmission). Each mode involves exposure to infected body fluids such as semen, vaginal excretions, and blood, tissues, or organs [1].

Table 1 summarizes WHO's estimate of global HIV transmission by type and efficiency of exposure, and percent of worldwide infections attributed to each as of 1991.

Transfusion of infected blood or blood products almost invariably leads to HIV infection: the likelihood of infection after such exposure is estimated to be greater than 90%. Despite the very high probability that a large volume of infected blood will transmit infection to recipients, only an estimated 3–5% of infections worldwide are due to such exposure because the number of persons at such potential risk is relatively small and blood in many countries since 1985 has increasingly been subjected to routine screening for HIV antibody.

Perinatal exposure ranks second to blood transfusion in efficiency of HIV transmission: transmission rates have ranged from a low of 11% to a high of almost 70% – the higher value is probably related to later clinical stages of maternal infection. The average probability is about 30% that an infected woman will transmit HIV to her fetus or infant. Up to 10% of global HIV infections, as of 1991, are estimated to be due to perinatal transmission.

HIV transmission through sexual intercourse has a markedly lower efficiency compared with blood transfusion or perinatal exposure. The risk of contracting HIV infection from sexual intercourse with an infected person depends in part on the type and number of sexual contacts with those partners. Studies have consistently indicated that there is a greater probability that infected males will transmit infection to susceptible female partners as compared with the probability that infected females will infect susceptible male partners [2, 3]. Accumulating evidence suggests that several contributory factors ('co-factors') may increase HIV transmission via sexual intercourse. Several studies have implicated concurrent infection with other sexually transmitted diseases (STDs), especially those associated with genital ulcers such as syphilis and chancroid as significant 'co-

Table 1. HIV transmission, global summary – 1991

Type of exposure	Efficiency per single exposure, %	Global total %
Blood transfusion	>90	3–5
Perinatal	30	5–10
Sexual intercourse	0.1–1.0	70–80
(Vaginal)		(60–70)
(Anal)		(5–10)
Injecting drug use: sharing needles, etc.	0.5–1.0	5–10
Health care: needle sticks, etc.	<0.5	<0.01

factors' in the spread of HIV [4, 5]. The probability of transmitting infection from a single sexual act may be low – the reported range is 1/100 (1%) to 1/1,000 (0.1%). However, as a result of very large numbers of sexual exposures which currently occur between HIV-infected and uninfected persons, sexual transmission is estimated, as of mid-1991, to account for up to 80% of global HIV infections; up to 70% of total infections have been due to heterosexual exposure and up to 10% to homosexual exposure.

On average, a single exposure through sharing injecting drug equipment poses a somewhat greater risk of HIV transmission than a single exposure via sexual intercourse – the probability is estimated to vary from 1/100 (1%) to 1/200 (0.5%), but is closely correlated with the amount of blood which may be exchanged. As of 1991, up to 10% of infections worldwide are estimated to be due to sharing drug injection equipment.

Needle-stick exposure in a medical or non-medical setting has the lowest estimated efficiency among known modes of HIV transmission – chances are less than 1/200 (about 0.31%) that accidental puncture with an HIV-contaminated needle will cause infection. This type of exposure is estimated to account for less than 1/10th of 1% of global infections as of 1991.

Progression from HIV Infection to AIDS

Figure 1 presents results from studies that followed HIV seropositive men from the time they were infected until they developed AIDS; data for 4 cohorts of white males are shown [6–8]. The San Francisco cohort of white homosexual men has been followed longer than any other group. In the first 3 years few men progressed to AIDS – less than 3% but by the 4th year close to

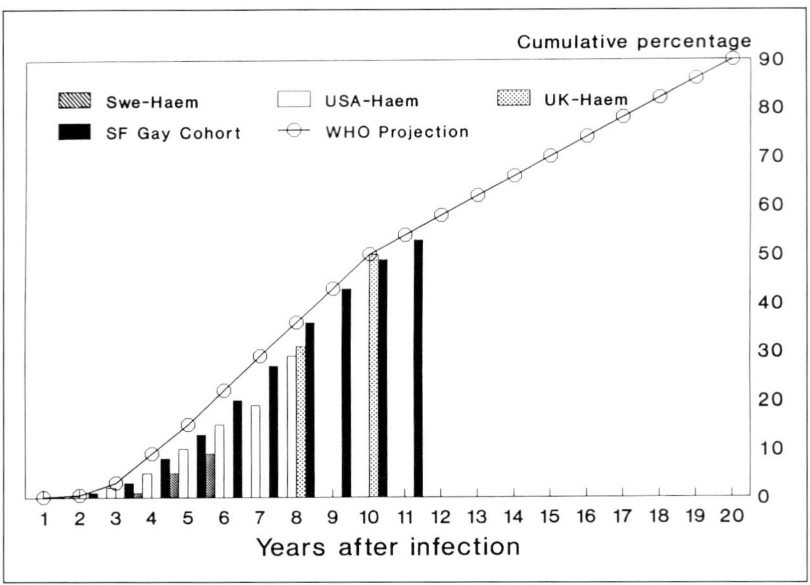

Fig. 1. Progression from HIV infection to AIDS.

10% had developed AIDS; thereafter, a steady progression rate of 6–7% per year was seen. By the 10th year, about half had developed AIDS.

While observed progression rates in the haemophilia cohorts are similar to those for the San Francisco cohort, they are generally a bit lower for each time period. However, the most recent observations in the UK haemophilia cohort [8] show that half of the cohort developed AIDS by the 10th year – almost identical to the San Francisco cohort. A detailed analysis of annual progression rates to AIDS by age at HIV acquisition for the USA haemophilia cohort showed that the younger members of this cohort progressed to AIDS more slowly than the older men. The average time to AIDS for men with haemophilia aged 35 and over is about 7–8 years, while that for men aged under 35 is approximately 12 years [9].

Scant data are available for determination of progression rates in women and other ethnic or racial populations. In addition, prospects of collecting reliable data on progression from infection to AIDS in other populations are not good since this requires long-term, detailed follow-up of large numbers of HIV-infected persons for whom the year of infection (seroconversion) is known with some degree of confidence. In the absence of reliable data that indicate otherwise, WHO has assumed that progression rates do not differ significantly for adults by sex, race, or geographic area.

HIV/AIDS Models

Many HIV/AIDS models have been developed in an attempt to understand the dynamics and interrelationships of the determinants of HIV spread and/or to develop reliable estimates of its eventual extent. These models range from simple to very complex. In general, simple models are designed to produce short-term projections; the more complex models to evaluate determinants of the HIV/AIDS pandemic and to test hypotheses. WHO has developed a simple model for short-term projections of AIDS [10].

The WHO model requires estimates of the following three variables: (1) HIV prevalence for a given year; (2) the year in which extensive epidemic spread of HIV began, and (3) the general shape of the infection curve during the epidemic period. These data and assumptions are used to calculate annual cohorts of HIV-infected adults. Short-term (up to 5 years) projections of AIDS cases are then calculated by multiplying each annual cohort of HIV-infected persons by annual progression rates from infection to AIDS. For modelling purposes WHO has assumed that 90% of all HIV-infected adults will develop AIDS within 20 years (fig. 1). For paediatric AIDS, the limited data suggest a much more rapid progression rate, so for modelling purposes, WHO has assumed that 80% of perinatally infected infants will develop AIDS by age 5 years [11].

Once diagnosed, survival with AIDS is limited. The case fatality rate for AIDS is very high and may approach 100% [12]. Where current antiviral treatment and effective medication for opportunistic infections are available, survival now averages about 1.5–2 years; while in most developing countries, almost all AIDS patients die within a year of diagnosis. The WHO model used average survival times from diagnosis (or onset of AIDS symptoms) to death to derive estimates and projections of AIDS deaths.

Estimates of Persons at Increased Risk of HIV Infection

Those at increased risk of infection with HIV, are predominantly homosexual or heterosexual persons with multiple sex partners, especially male or female prostitutes, and drug users who share injection equipment.

It should be emphasized that it is the behaviour(s) which places a person at risk for HIV and not just membership in a particular population subgroup – such as homosexual men or injecting drug users.

It is extremely difficult to estimate the number of individuals defined by behaviours that are illegal; of a private, personal nature; or relatively uncommon. These populations are essentially socially invisible or hidden in the sense that they engage in activities concealed from mainstream society or agencies of social control. Members of hidden populations become visible when they enter official governmental or institutional settings. Studies of captive institutional and clinical populations abound: however, data obtained from such selected subgroups if generalized to the larger noninstitutional population, may be seriously biased.

Random sample surveys of various design have been used to estimate the size and behaviours of subpopulations at greater risk of HIV infection. Surveys may provide reasonably accurate data on sexual activity in heterosexuals (adolescents and adults) and in homosexual men, but they are not useful for estimating injecting drug users or prostitutes – groups that may be systematically underrepresented because they more often do not live in households, are school dropouts, or are in prison or other institutions.

To overcome the problems associated with random sample surveys, other approaches such as multiplier and projection techniques, and capture-recapture methods have been utilized [13, 14]. Recent efforts to estimate numbers of homosexual men, injecting drug users, and prostitutes in several developed countries using these methods suggest that these populations may be smaller than previously believed. However, even if these estimates are seriously in error, it is clear that heterosexuals at risk of HIV greatly outnumber homosexual men and intravenous drug users. WHO estimates that throughout the world there may be from a quarter to a half billion heterosexuals who are at moderate-to-high risk of exposure to HIV through multiple sexual partners; about 10 million homosexual men who have multiple sex partners; and up to 5 million drug users who routinely share injection equipment.

Global HIV Prevalence and Trends

Since the first AIDS cases were described in 1981, HIV transmission patterns have shifted to involve population groups not initially affected. In the industrialized countries of North America, Europe, Australia, and New

Zealand the highest incidence first occurred in homosexual men. Studies of homosexual male cohorts have noted decreasing incidence rates since the mid-1980s, possibly due to both saturation of infection in those with the highest risk behaviours as well as significant reduction in risk behaviours for most homosexual men. Large differences in transmission rates among injecting drug users have been observed in various regions of the world. Some countries have experienced explosive increases in HIV among injecting drug users, while in others transmission rates have remained relatively low. In some areas, notably Latin America, including the Caribbean, a marked shift from homosexual to heterosexual transmission has been noted during the last half of the 1980s. In addition, there has been a slow but steady increase in heterosexual transmission in industrialized countries – a trend which is expected to continue. The expectation is that HIV will spread more widely through populations where STD rates are high, and increasingly heterosexual transmission will become the predominant mode of HIV transmission throughout the world [15].

WHO's estimates of adult HIV infections, based on available HIV serological data, indicate that from the late 1970s when HIV spread became extensive through 1990, incidence has been greatest in sub-Saharan Africa (fig. 2). The cumulative total of infected adults in this region as of 1991 is close to 6 million. In the 'western' industrialized countries, HIV prevalence increased more slowly and national incidence in most of these countries may have reached a peak by the mid-1980s: the cumulative total of HIV infections as of 1991 is about 1.5 million. HIV incidence in Latin America may be increasing still, but as of 1990–1991, the cumulative total for Latin America was estimated to be about 1 million [16]. Several countries in south and southeast Asia reported large increases in HIV infections beginning in 1988: as of mid-1991, over one million infections may have occurred in this region of the world [17].

HIV prevalence projections based on estimated current incidence rates are shown on the right side of figure 2. A continued increase in HIV transmission is expected for sub-Saharan African with a cumulative total of close to 10 million infections by 1995. Transmission in industrialized countries is projected to remain relatively low so that cumulative HIV infections may reach only about 2 million by 1995. While HIV transmission is expected to increase in Latin American countries, increases may not be as rapid as in other developing countries so that the cumulative total by 1995 may also reach only 2 million. South and southeast Asian countries, on the other hand, because of the very large populations involved may experience greater

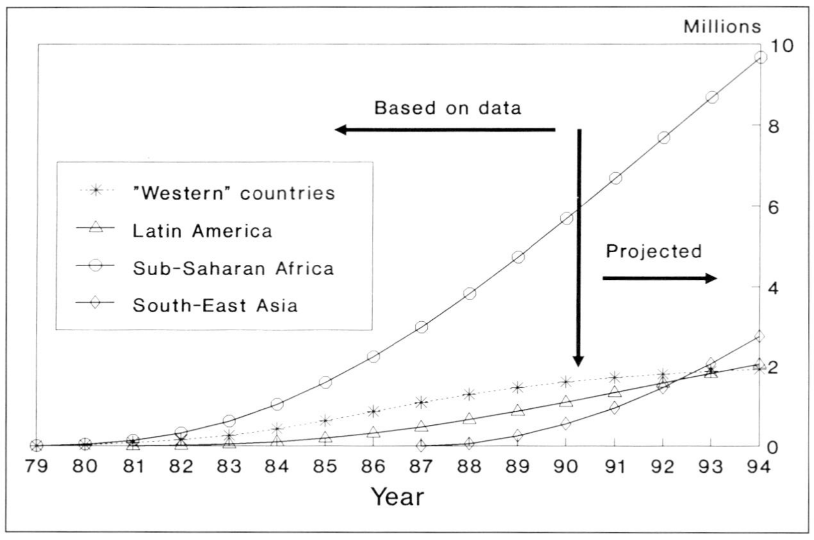

Fig. 2. Estimates/projections of cumulative adult HIV infections.

numerical increments in HIV: infections are expected to reach a cumulative total of over 2.5 million by the mid-1990s. Thus, the projected global cumulative total is over 15 million HIV-infected adults by 1995.

Estimates and Projections of AIDS

The estimated and projected HIV seroprevalence values described above were used in the WHO model to estimate and project annual adult AIDS cases in the major geographical regions through the year 2000 (fig. 3). Until 1990, the WHO model's estimates of AIDS were based on public health surveillance and HIV seroprevalence data. Projections of annual AIDS cases through the mid-1990s are based primarily on HIV seroprevalence estimated as of 1990/1991; projected AIDS cases to the year 2000 were based on projected HIV trends and prevalence levels up to the mid-1990s. Annual adult AIDS cases can be projected with some confidence through the mid-1990s since such projections are based on current estimates of HIV seroprevalence, and close to 90% of the projected AIDS cases can be expected to occur even if new HIV infections were to decline

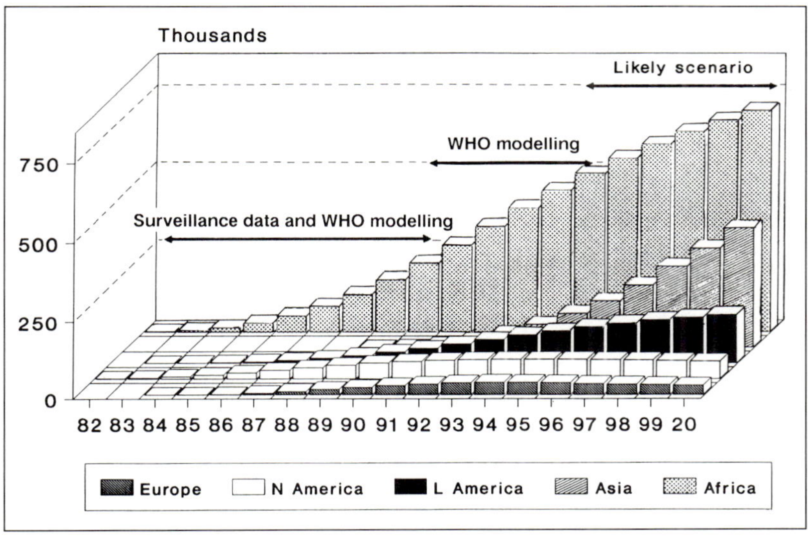

Fig. 3. Distribution of estimated and projected annual adult AIDS cases.

sharply during the first half of this decade. From the mid-1990s to the year 2000 projections of annual adult AIDS cases are less certain since they are based on projected HIV trends. AIDS projections need constant review and revision to assure that they are in line with data on HIV prevalence and trends.

During the 1980s, it was estimated that the global cumulative total of adult AIDS cases was about one million. In constrast, during the 1990s an additional 9 million adult AIDS cases are projected. The temporal pattern of projected annual AIDS cases reflects the observed trends and prevalence of HIV infections previously described after an average delay of 10 years. During the 1980s, the largest number of annual AIDS cases occurred in Africa and North America. In 1990, an estimated 200,000 AIDS cases occurred in African adults and the cumulative case total was close to 700,000, about 2/3rds of the global total at that time. The corresponding numbers for North America were about 50,000 cases in 1990 and a cumulative total of 200,000 (about 20% of the global total).

Between 1990 and the mid-1990s, still greater annual totals of AIDS cases are projected for developing regions; while in developed countries annual AIDS cases are projected to reach a peak before the middle of the decade. From the middle to the end of this decade, these trends in annual AIDS cases are expected to continue. By the year 2000, worldwide annual adult AIDS cases are expected to be more than one million; over 90% of the cases are projected to occur in developing countries – about half in Africa, over a quarter in Asia, and over 10% in Latin America.

Estimated and Projected AIDS Mortality

Analyses of mortality data in the USA provide a striking picture of the increasing impact of HIV/AIDS. During the last decade AIDS has caused a reversal in the long-term decline in death rates for men aged 25–44 [18]. Between 1970 and 1983, death rates for this group of men fell from about 300 per 100,000 to 212 per 100,000 (fig. 4a). After 1983, death rates started to increase and by 1987 the rate was 236 per 100,000. It was estimated that the death rate in 1987 in the absence of AIDS would have been 201–209 per 100,000. AIDS deaths for men aged 25–44 increased from 583 or 1.9 per 100,000 in 1980 to 10,240 or 26.6 per 100,000 in 1987. These deaths represented 0.8% of all deaths to men 25–44 in 1980, but 11.3% in 1987. The WHO model projects that the total death rate among US men in this age group will increase to about 300 per 100,000 by the mid-1990s (fig. 4a).

Even though HIV/AIDS in women in the USA constitutes only 10% of reported AIDS cases, an analysis of 1980–1988 deaths for women aged 15–44 found that mortality trends paralleled those for men [19]. The death rate for AIDS quadrupled between 1985 and 1988 from 0.6 per 100,000 to 2.5 per 100,000; by 1987, it had become the eighth leading cause of death in women of reproductive age (fig. 4b). In contrast, other causes of death among women in this age group have remained relatively stable during the past decade. Because of the large number of women already infected, and the continued increase in new AIDS cases in women 15–44, mortality due to AIDS will undoubtedly rise, particularly in black and Hispanic populations. Rates for the leading causes of death in US women aged 15–44 are expected to remain essentially unchanged during the 1990s, but the AIDS death rate in these women is expected to reach about 10 per 100,000 according to the WHO model. An increase of that magnitude would place AIDS among the top three or four causes of death in this group of women by the mid-1990s (fig. 4b). In cities such as New York, with documented high HIV prevalence levels during the 1980s, AIDS became the leading cause of death in men and

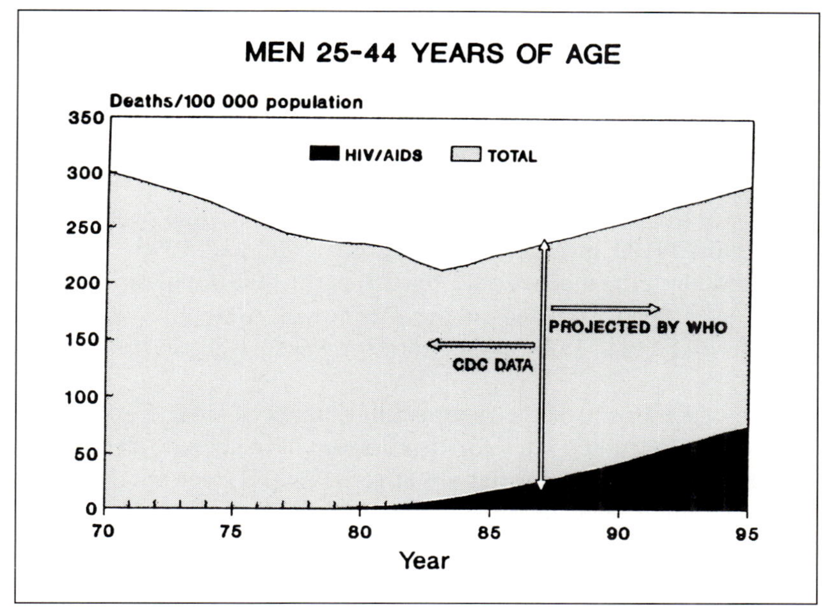

a

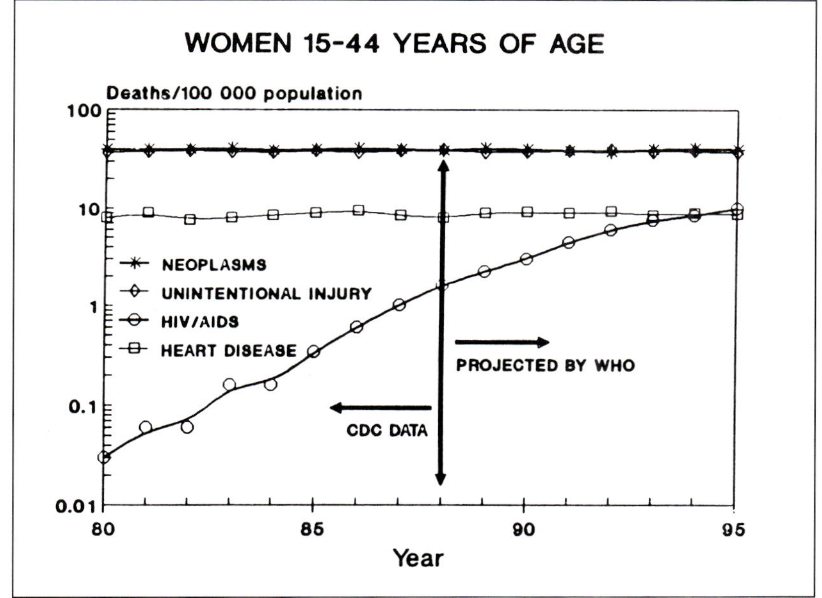

b

Fig. 4a, b. Projected death rates – USA.

women aged 25–34 in the late 1980s [20]. These mortality trends among young and middle-aged adults in the USA can be expected to occur in many other industrialized countries during the 1990s.

Sub-Saharan Africa

For a hypothetical, sub-Saharan African country, the WHO projection model was used to estimate and project AIDS cases and deaths up to the mid-1990s. WHO estimated that between 1985 and 1990, AIDS deaths increased the crude death rate by 0.3 per 1,000 population over what would have been expected in the absence of AIDS [21, 22]: a further increase of 1.1 per 1,000 population (or 8.7%) is expected by the mid-1990s.

Any assessment of the demographic impact of AIDS must consider age-specific mortality rates for two reasons. First, crude death rates for sub-Saharan African countries are affected by the great number of deaths among children. For all sub-Saharan Africa, it has been estimated that deaths among children under five account for more than 45% of total deaths [23]. Second, AIDS in sub-Saharan African countries affects some age groups more severely – particularly children under age five and sexually active adults. WHO estimated that between 1985 and 1990, AIDS added an average of 10% to the annual death rate for those aged 15–49; by the mid-1990s, AIDS is projected to add more than 40% to that rate. Figure 5a shows that for those 15–49, AIDS is expected to reverse the declining trend in adult mortality rates. The data presented are projections for the entire country: however, in many urban areas the death rate for this age group is projected to double or triple by the mid-1990s.

At the start of the epidemic, almost no effect on infant or child mortality was noticeable. Yet by the mid-1990s, infant mortality is expected to be 4% higher than it would be in the absence of AIDS, and the probability of dying before age five is expected to increase by 7.6% as a consequence of AIDS. By the mid-1990s the projected increase in AIDS deaths among children will begin to negate the reductions in mortality achieved by child survival programmes over the past two decades (fig. 5b).

Demographic Impact of AIDS in Sub-Saharan Africa

Any modelling of the potential demographic impact due to AIDS in sub-Saharan Africa has to be interpreted with caution because it is extremely difficult to precisely simulate with mathematical equations such a complex, dynamic, and interactive human problem. Thus, the results of

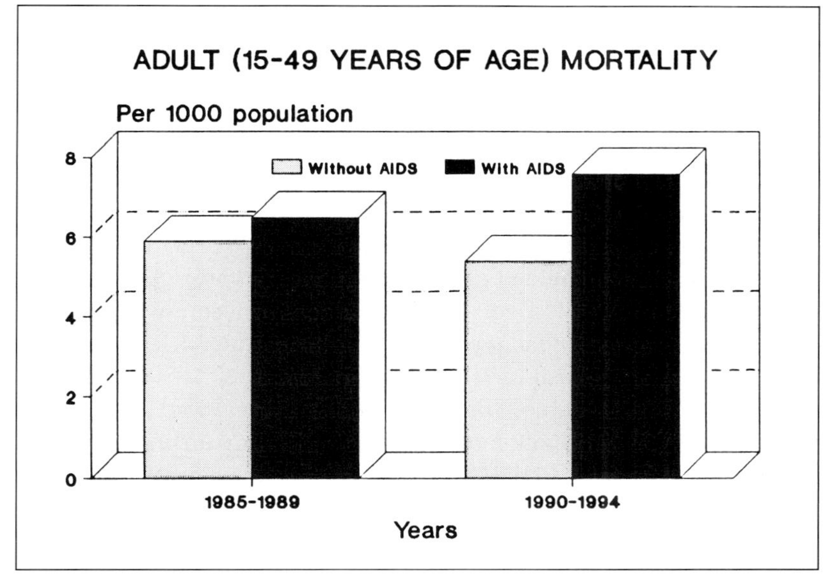

a

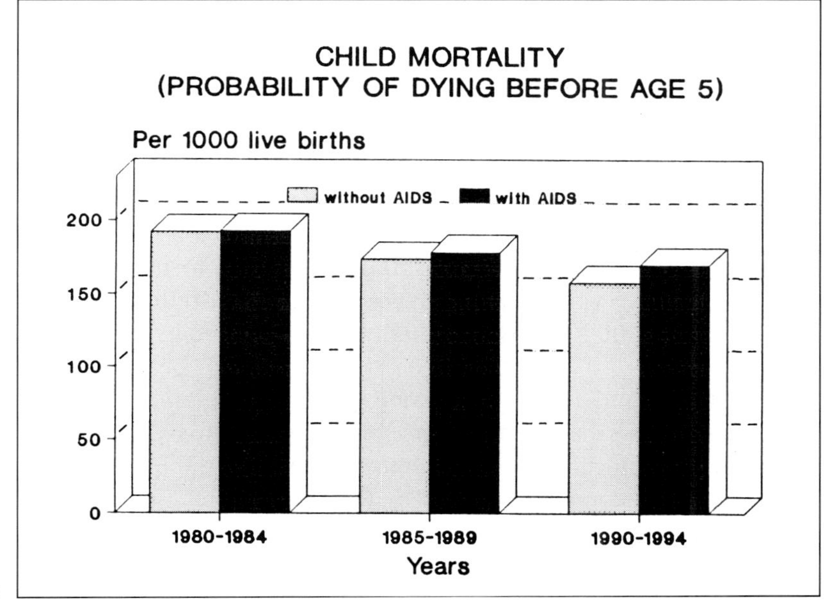

b

Fig. 5a, b. Estimated/projected adult and child mortality in a sub-Saharan African country.

any HIV/AIDS demographic model on specific demographic indicators for a sub-Saharan African country several decades into the future cannot be accepted as precise projections but should rather be looked upon as initial attempts to project how AIDS may grossly affect these indicators over the next few decades.

The potential effects of AIDS on selected demographic indicators were derived from a demographic projection model which includes HIV/AIDS developed by the World Bank [24]. The World Bank model was selected because it provided detailed demographic projections which were in the general range of results of several other deterministic models. The World Bank model used a hypothetical sub-Saharan African country population where the spread of HIV started in the early 1980s, and also used standardized input values for the major biological and epidemiological variables. The World Bank model provided some disturbing projections of the potential impact of HIV/AIDS on several major demographic indicators in a hypothetical sub-Saharan African country by the year 2010.

(A) The total crude death rate in this sub-Saharan African country was projected in the absence of HIV/AIDS to fall from 14.8 per 1,000 population in 1985 to 8.4 per 1,000 population in 2010. However, as a result of AIDS, the crude death rate is projected to rise to 16.4 by 2010 – about double the rate without AIDS (fig. 6a).

(B) Infant mortality rates in the absence of AIDS would have been expected to have dropped from 101.2 in 1985 to 61.3 in 2010, but instead remains almost unchanged by 2010 when it is projected to be 104.7 per 1,000 live births. Similarly, when considering AIDS, no decline is projected for the probability of dying by age five; it is expected to increase from 180 per 1,000 live births to 220 per 1,000 live births between 1985 and 2010, whereas without AIDS, it was expected to decline by about 50% from 170 per 1,000 live births in 1985 to 90 per 1,000 live births in 2010 (fig. 6b).

(C) By the mid-1990s, the United Nations Population Division estimated that a sub-Saharan African person is expected to live about 24 years less than a person in a more developed country. In the hypothetical sub-Saharan African country modelled by the World Bank, life expectancy at birth in the absence of AIDS is expected to increase from 51.5 years in 1985 to 61.4 years by 2010, an increase of about 10 additional years of life. AIDS is expected to decrease life expectancy from about 50 in 1985 to 47 in 2010 and thus decrease life expectancy by more than 12 years compared to what was projected in the absence of AIDS (fig. 6c).

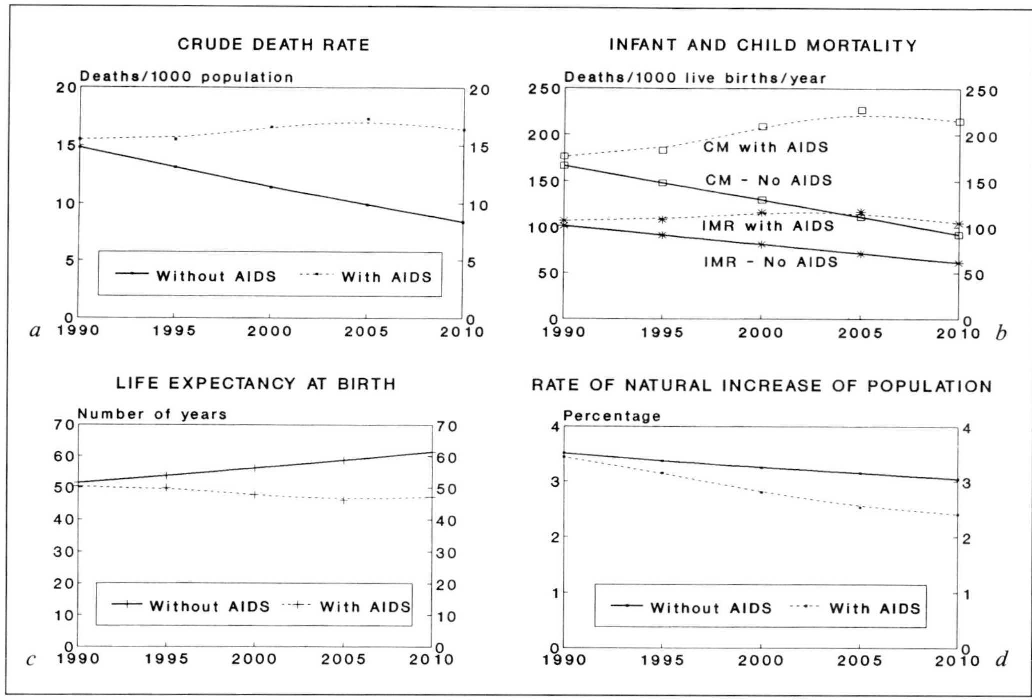

Fig. 6. Projected impact of AIDS on demographic indicators in a sub-Saharan African country.

(D) The rate of natural increase for the hypothetical sub-Saharan African country is projected to decrease more rapidly due to AIDS. As a result of AIDS, the growth rate would decline from 3.4 to 2.4% in 2010; without AIDS the growth rate is projected to decline more slowly to about 3.0% in 2010 (fig. 6d).

Conclusions

What appeared only a decade ago to be an epidemic that primarily involved homosexual men and injecting drug users in industrialized countries is evolving into a pandemic that predominantly affects heterosexuals in developing countries. As the HIV/AIDS pandemic enters its second decade, its general dimensions have been more clearly delineated though precise estimates and long-term projections cannot be made with great certainty.

Epidemiological data indicate that in industrialized countries, where extensive spread of HIV began in the late 1970s or early 1980s, the majority of HIV infections occurred during the first half of the 1980s. As a result, peak incidence for AIDS cases and deaths is expected to occur around the mid-1990s. AIDS will, nevertheless, continue to be a major disease problem in industrialized countries for the foreseeable future, and will exert far-reaching effects on many other sectors. Heterosexual HIV transmission increased slowly but steadily in almost all developed countries during the latter half of the 1980s, and the future demographic impact of HIV/AIDS in these countries will depend on the extent of HIV spread during the rest of this decade.

For the developing countries of Africa, Latin America and Asia, all epidemiological data collected so far point to continuing large increases in HIV seroprevalence. Thus, during the 1990s and the following decade, large increases in AIDS cases and deaths can be expected. Although all projections must be interpreted cautiously, there can be no doubt that during the next several decades, AIDS in most developing countries will become the leading cause of death among adults in their most productive years, and will also be one of the leading causes of infant and child mortality in many regions. In addition, AIDS will have a profound impact not only on health and social welfare systems, but on social, economic and other developmental aspects of life in Africa, Latin America and Asia.

During the last 10 years, HIV/AIDS has emerged as a human disease problem of staggering proportions. At the present time, a large proportion of new infections can be prevented by effective educational programmes that lead to modification or elimination of behaviours that place a person at increased risk of acquiring HIV infection. This is a difficult and challenging task, but given the appropriate motivation and resources, public health programmes can be successful. However, in the absence of effective and inexpensive treatment the many millions of persons infected with HIV during the 1980s cannot be prevented from developing AIDS before the end of this century.

Since initial recognition of AIDS in 1981, knowledge about this syndrome and its causative agent, the human immunodeficiency virus has grown enormously. Still, as the VIIth International Conference on AIDS will be informed, the development of effective and inexpensive drugs and biologics for the treatment and prevention of HIV infection and HIV-related conditions continues to be an elusive goal.

The HIV/AIDS pandemic clearly poses formidable problems for the future. All of the methods used to project HIV infections and AIDS cases indicate, for both the short and long term, substantial increases in total men, women, and children who will become infected, progress to AIDS, and subsequently die. Improved surveillance and estimation/projection of global HIV infections and AIDS cases will, therefore, be needed to direct AIDS programme activities, plan for support of health care and social welfare systems, and assess the growing impact of this pandemic.

References

1 Lifson AR: Do alternate modes for transmission of human immunodeficiency virus exist? A review. JAMA 1988;259:1353–1356.

2 Holmberg SD, Horsburgh CR, Ward JW, Jaffe HW: Biologic factors in the sexual transmission of human immunodeficiency virus. J Infect Dis 1989;160:116–125.

3 Peterman TA, Stoneburner RL, Allen JR, et al: Risk of human immunodeficiency virus transmission from heterosexual adults with transfusion-associated infections. JAMA 1988;259:55–58.

4 Simonsen JN, Cameron DW, Gakinya MN, et al: Human immunodeficiency virus among African men with sexually transmitted diseases. N Engl J Med 1988;319: 274–278.

5 Handsfield HH, et al: Association of anogenital ulcer disease with human immunodeficiency virus in homosexual men. IIIrd Int Conf AIDS, Washington, June 1987.

6 Moss AR, Bacchetti P: Natural history of HIV infection. AIDS 1989;3:55–61.

7 Rutherford GW, Lifson AR, Hessol NA, et al: Course of HIV-1 infection in a cohort of homosexual and bisexual men: An 11 year followup study. Br Med J 1990;301: 1183–1188.

8 Lee CA, Phillips AN, Elford J, et al: Ten year follow-up of a cohort of 111 anti-HIV seropositive haemophiliacs (Abstract Th.C.37). VIth Int Conf AIDS, San Francisco, June 1990.

9 Goedert JJ, Kessler CM, Aledort LM, et al: A prospective study of human immunodeficiency virus type 1 infection and the development of AIDS in subjects with haemophilia. N Engl J Med 1989;321:1141–1148.

10 Chin J, Lwanga SK: Estimation and projection of adult AIDS cases: A simple epidemiological model. Bull WHO 1991;69:(4)399–406.

11 Chin J: Current and future dimensions of the HIV/AIDS pandemic in women and children. Lancet 1990;i:221–224.

12 Fauci AS: The human immunodeficiency virus: Infectivity and methods of pathogenesis. Science 1988;237:612–622.

13 Newmeyer JA: The prevalence of drug use in San Francisco in 1987. J Psychoact Drugs 1988;20:185–189.

14 Watters JK, Biernacki P: Targeted sampling: Options for the study of hidden populations. Soc Probl 1989;36:416–430.
15 Chin J: AIDS: 1990. Global estimates of AIDS cases and HIV infections. AIDS 1991;4(suppl 1):S277–S283.
16 Quinn TC, Narain JP, Zacarias FRK: AIDS in the Americas: A public health priority for the region. AIDS 1990;4:709–724.
17 Global Programme on AIDS, World Health Organization: Unpubl. data, 1991.
18 Buehler JW, Devine OJ, Berkelman RL, Chevarley FM: Impact of the human immunodeficiency virus epidemic on mortality trends in young men in the United States. Am J Publ Health 1990;80:1080–1086.
19 Chu SY, Buehler JW, Berkelman RL: Impact of the human inmmunodeficiency virus epidemic on mortality in women of reproductive age, United States. JAMA 1990;264:225–229.
20 State of New York, Department of Health: AIDS in New York State, 1988 edition.
21 Population Division of the United Nations: World Population Prospects 1990. New York, United Nations, 1991.
22 United Nations: Mortality of Children Under Age Five. Population Studies, No 105. New York, United Nations, 1988.
23 Population Division of the United Nations: Age-specific mortality estimates, unpubl data.
24 Bulatao R: Projecting the demographic impact of the HIV epidemic using standard parameters; in: Modelling the Demographic Impact of the AIDS Epidemic in Pattern II Countries: Progress to Date and Policies for the Future. Proc United Nations/World Health Organization Workshop, New York, December 13–15, 1989. New York, United Nations Population Division and the Global Programme on AIDS of the World Health Organization, 1991.

Dr. J. Chin, Global Programme on AIDS, World Health Organization,
CH–1211 Geneva 27 (Switzerland)

Rossi GB, Beth-Giraldo E, Chieco-Bianchi L, Dianzani F, Giraldo G, Verani P (eds): Science Challenging AIDS. Basel, Karger, 1992, pp 51–70

Factors and Mechanisms of AIDS Pathogenesis[1]

L. Montagnier[a], *M.L. Gougeon*[a], *R. Olivier*[a], *S. Garcia*[a], *C. Dauguet*[a], *M. Adams*[a], *J.M. Béchet*[a], *D. Guétard*[a], *A. Laurent*[a], *A. Hovanessian*[a], *M. Kirstetter*[a], *R. Roué*[a], *G. Pialoux*[b], *B. Dupont*[b]

[a]Department of AIDS and Retroviruses, and [b]Hôpital, Institut Pasteur, Paris, France

AIDS is characterized by a profound alteration of the immune system, bearing mostly but not exclusively on function and number of helper T4+ lymhocytes. Although the retroviral etiology of the disease is beyond doubt, much remains to be understood in the mechanism by which T4 lymphocytes progressively disappear, and also how the other alterations of the immune system are brought about. Particularly, there is evidence of early alterations bearing on the function of antigen-presenting cells, namely monocytes and dendritic cells, as well as some progressive changes in the CD8+ population. The latter changes are exemplified by the loss of functionality of CD8+ cytotoxic lymphocytes directed against viral gag proteins, and by the appearance of some markers normally present on immature thymocytes such as the CD38 surface antigen (leu 17) [1]. Since the number of circulating infected cells (T4 lymphocytes and monocytes) is small during the long silent period of the disease, the effect of the virus must be amplified by its interaction with the functional network of the immune system or with some other infectious agents. In this review, we will concentrate on recent data obtained from our laboratory and from

[1] This work was supported by grants from the ANRS, the CNRS and the Pasteur Institute. The authors wish to express their thanks to C. Axler, S. Chamaret, M.C. Cumont, H. Lecœur and V. Rame for technical assistance.

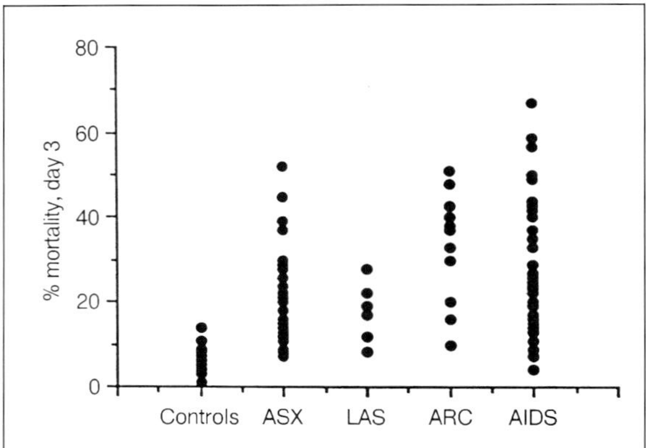

Fig. 1. In vitro loss of viability of lymphocytes from HIV-infected patients as compared to seronegative donors. Lymphocytes were isolated from heparinized blood by centrifugation over a Ficoll gradient (Lymphoprep) and cultured at 10^6 cells/ml in culture medium composed of RPMI-1640, glutamine, Hepes and 10% (v/v) fetal calf serum. After 3 days, dead cells were enumerated after blue-trypan coloration. Each point corresponds to a single donor lymphocyte culture. Same results were obtained if cultures were made in synthetic medium derived of fetal calf serum. ASX = Asymptomatic patients; LAS = lymphadenopathy syndrome (stage II of CDC); ARC = AIDS related complex (stage IV, A); AIDS = acquired immune deficiency syndrome (stages IV, C and IV, D) [reproduced from ref. 4, p. 532, by courtesy].

others shedding light on the mechanisms of lymphocyte cell death and its possible effectors, and on the possible role of some ubiquitous infectious agents, mycoplasmas.

Rapid in vitro Loss of Viability of Lymphocytes from AIDS Patients and HIV-Positive Individuals

Our early studies [2] indicated that when PBLs separated on Ficoll gradient were put in culture in a survival medium lacking lymphocyte growth factors, there was a significant difference between the daily loss of viability of cells from healthy HIV negative individuals and those of HIV-infected individuals: the former had generally a loss of viability (as measured by trypan blue) inferior or equal to 10% after 3 days, whereas in the latter the number of dead cells ranged from 10 to 60% (fig. 1).

The presence or absence of 10% fetal calf serum in the RPMI 16-40 culture medium did not change the difference, and the phenomenon could be observed even at the asymptomatic phase of the disease, although it was more marked and more frequent in its symptomatic phase. In some patients, an increased cell mortality was observed even at time zero of the culture, just after the separation of PBLs, suggesting that the in vitro phenomenon could be an accentuation of a process occurring in the patient. The phenomenon was partly or totally prevented by addition at the first day of culture of a crude preparation of T cell growth factors (TCGF, Biotest) containing mostly IL-2. In some, but not all cases, TCGF could be replaced by purified recombinant IL-2 alone.

To analyse more in depth the mechanism of this increased cell death, we set up a two-color flux cytofluorometry analysis using two dyes: acridine orange which will stain nucleic acids, essentially DNA, in living cells, and ethidium bromide, which will stain DNA of dead lymphocytes, and to some extent also living macrophages and monocytes. If the analysis is applied to cells of healthy HIV-negative donors, a picture as that shown in figure 2a, b is obtained. On the contrary, in cells of HIV-positive patients which show an increased mortality after 3 days of in vitro culture, we observed a different picture: the population of living lymphocytes is split into two subgroups, a group with normal high staining with acridine orange, and a group with lower staining (fig. 2c, d). We have quantitated the phenomenon, using the formula $\frac{D+L}{H}$, where D is the percentage of dead cells, L that of the low acridine orange stained cells, and H that of the highly stained (living) lymphocytes. Usually, this ratio is around 0.2 for the lymphocytes from healthy HIV-negative individuals. In HIV-positive subjects, there is a large spectrum of values, with an average of 0.7. There is a significant, inverse correlation between this ratio and the T4/T8 cell ratio, indicating a trend for increase with progression towards AIDS.

Evidence for an Apoptosis Process in This Population

The lower staining of some lymphocytes by acridine orange was possibly due to a decrease DNA content of these cells. This suggested to us that such cells, although still living, were losing DNA, perhaps by an apoptosis process. Apoptosis – programmed cell death – involves the induction or activation of an endogenous endonuclease which will destroy the chromatin structure, by cutting doubled-stranded DNA in exposed sites interspac-

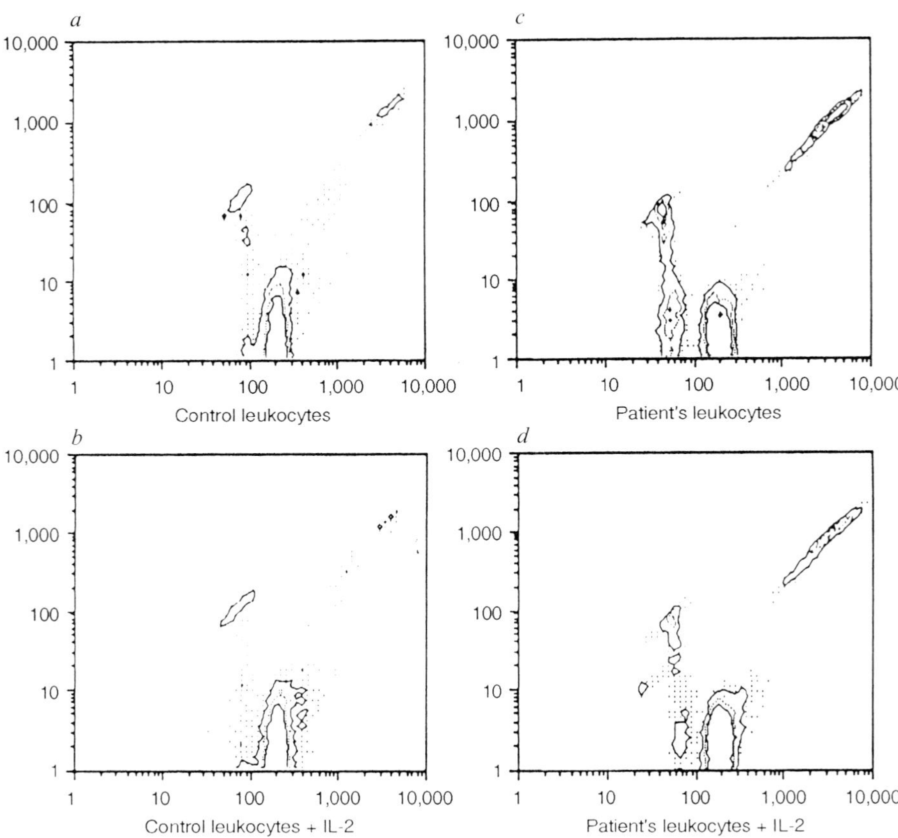

Fig. 2. Flow cytometry analysis of double-stained lymphocytes. Peripheral blood lymphocytes were prepared and cultured as described in figure 1 and stained with ethidium bromide which enters the dead cells and acridine orange which stains nucleic acids of living cells. Stained cells were then analyzed on a FASCAN (Becton Dickinson). Contour curves correspond, from the center to the periphery, to cellular densit levels. The cell population in the upper right angle of each plot is an aspecific labelled monocyte population. *a* Cells from a seronegative donor. The majority of cells are living and highly stained with acridine orange. *b* Same culture but in the presence of TCGF (Biotest, lymphocult, 10 U/ml). *c* Lymphocytes from a seropositive patient. A second living cell population with a lower amount of nucleic acid is observed. *d* Same culture but in the presence of TCGF. The population weakly stained by acridine orange has partly disappeared [reproduced from ref. 4, p. 533, by courtesy].

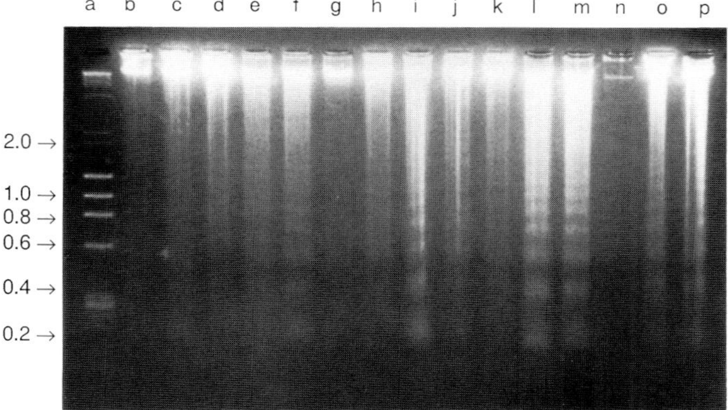

Fig. 3. Gel analysis of DNA from cultured lymphocytes. Lymphocytes from a seronegative donor (lanes b, c, d, and k) or from seropositive patients (lanes e–j and l–p) were cultured during 3 days in the absence of activating agents (lanes k–p) or during 24 h in the presence (lanes c, f, i) or the absence (lanes b, e, h) of ionomycin (Calbiochem, 1 μg/ml) or in the presence of ionomycin and TCGF (lanes d, g, j). At the end of the culture, cells were lysed in a buffer containing SDS (10%) and proteinase K (10 mg/ml) and DNA was extracted by phenol-chloroform. After overnight incubation at –80 °C, DNA samples were incubated with RNAse (0.5 mg/ml) and loaded on a 2% agarose gel. Electrophoresis was carried out during 6 h under a 50-mV voltage. Molecular sizes in kilobases are indicated in the left of the gel in lane (a). Oligonucleosomal fragments appear as ladders of diffuse bands whose molecular sizes are approximately multiples of 200 basepairs [reproduced from ref. 4, p. 537, by courtesy].

ing nucleosomes. As a consequence, when analyzed by electrophoresis on an agarose gel, the DNA is resolved in discrete bands, which have the size of nucleosomal DNA unit (200 bp) or a multiple of this size, if the digestion has not been complete. This is a physiological process, occurring for instance in the thymus to eliminate autoreactive T lymphocyte clones [3]. We have found indeed [4] that a significant fraction of lymphocytes from HIV-infected individuals undergoes apoptosis after a 3-day culture, so that the total DNA shows the characteristic oligonucleosomal bands (fig. 3).

Confirmation was obtained by observing cell sections at the electron microscope, displaying characteristic ultrastructural changes such as chromatin condensation and nuclear shrinkage (fig. 4). The phenomenon could be prevented by adding at the beginning of the culture, as for the appearance of the low staining population, a crude preparation of T cell growth

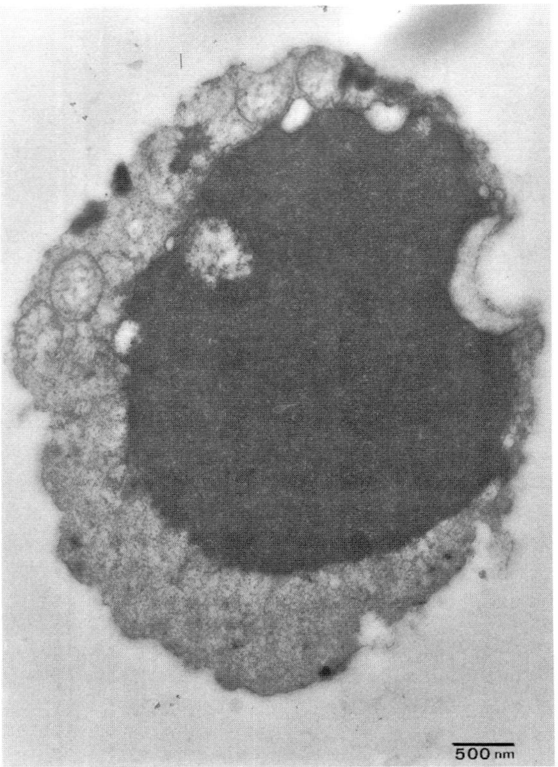

Fig. 4. Electron micrograph of an ultrathin section (uranyl acetate staining) of a lymphocyte from a HIV-positive patient cultured during 24 h in the presence of ionomycin: Note the condensation of chromatin characteristic of apoptosis.

factors (TCGF). However, purified IL-2 could not replace TCGF for the prevention of apoptosis, suggesting that some other cytokines are involved. The nature of those is under current analysis.

On the contrary, the apoptosis process could be accelerated by adding a calcium ionophore (ionomycin) to activate lymphocytes by increasing the calcium influx. Under these conditions, apoptosis could be observed in lymphocytes of HIV-positive individuals after 24 h of culture. In a study bearing on 70 HIV-negative individuals, apoptosis could be observed in 93%, under these conditions, and none in the case of HIV-negative healthy individuals.

Since, however, the phenomenon may reflect an abnormal level of T cell activation in vivo (see below), it may be not specific of HIV infection. Indeed, it was also observed in a case of infectious mononucleosis. But in the case of HIV infection, it could be observed very soon after primary infection. In a patient who was sequentially studied, apoptosis was apparent in his lymphocytes 3 weeks after seroconversion, and then disappeared. It is possible therefore that the 7% of HIV-positive, apoptosis-negative patients may reflect some rather quiet periods in T cell activation, interspacing some periods of high activation. Whatever the underlying mechanism be, the apoptosis test – or the two staining with acridine orange, which probably corresponds to a pre-apoptosis stage – could be added to the list of immunological markers useful for studying the progresion of the disease. Its early manifestation after HIV infection indicates that profound alterations in the immune system occur soon after this infection.

To our knowledge, two other groups have observed abnormal induction of apoptosis in lymphocytes from AIDS patients and HIV-infected individuals: Miedema et al. [personal commun.] have observed the phenomenon after activation of the T cell receptor of T4 and T8 lymphocytes by anti-CD3 antibodies. Groux et al. [5] have described an apoptosis phenomenon unique to the CD4+ population of HIV-infected patients, after stimulation by some mitogens (Pokeweed mitogen) and superantigens (SEB).

Since the conditions of stimulation were not identical in the three laboratories, it is premature to correlate these observations with each other. But they all point to abnormalities existing in a significant fraction of lymphocytes (T4, T8 and others) from HIV-infected patients, all leading to accelerate their decay by apoptosis.

We will now examine what signals could possibly trigger apoptosis: some are specific of T4 lymphocytes, some could be more general and be extended to any abnormally activated lymphocyte subset.

Mechanisms Specific of CD4+ Lymphocytes

Virus Infection

It has been shown by Terai et al. [6] and Laurent-Crawford et al. [7] that lymphocytes or lymphoid cells infected with cythopathic strains of HIV die, at least in part, of apoptosis.

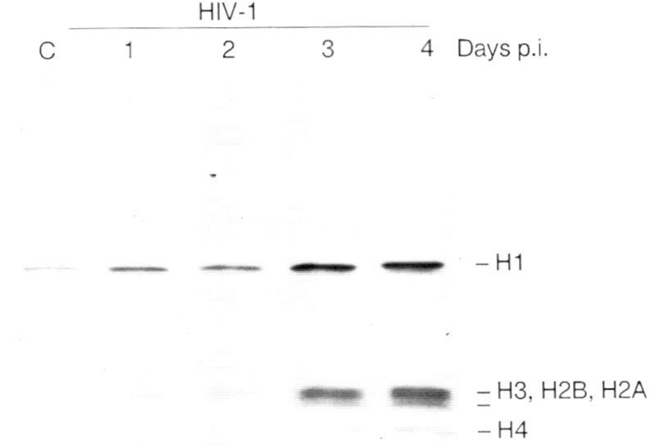

a

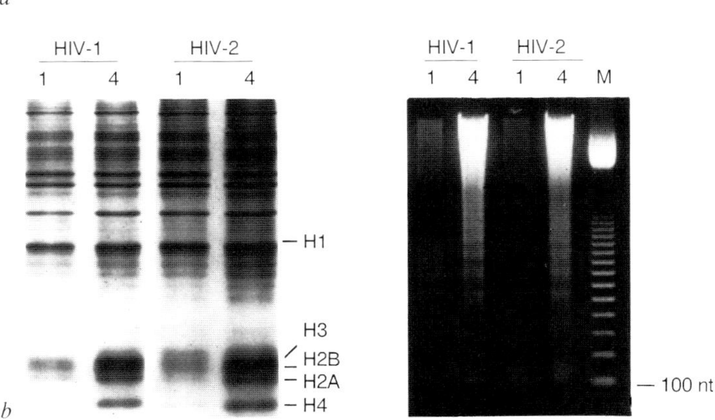

b

Fig. 5. Accumulation of histones in the nucleoplasm of HIV-1-infected cells. CEM cells were infected with HIV-1 and at different days postinfection (p.i.) nucleoplasm was prepared (lanes 1–4) Lane C refers to nucleoplasm from uninfected CEM cells. Samples corresponding to material from 1 × 10^6 cells were analyzed by SDS-PAGE (15%) electrophoresis. The figure shows a Coomassie blue-stained gel. On the right is the position of the different histone subtypes: H1, H2A, H2B, H3 and H4. *b* Accumulation of oligonucleosomes in HIV-infected cells. CEM cells infected with HIV-1 LAI or HIV-1 ROD were analyzed 1 and 4 days p.i. (shown in lanes 1 and 4). Proteins in total cellular extracts (left-hand side): Cells were disrupted in lysis buffer containing 10 m*M* Tris-HCl pH 7.6, 0.4 *M* NaCl, 1 m*M* EDTA and 1% Triton X-100. This suspension was then centrifuged at 12,000 *g* to pellet aggregated material and cellular chromatin and recover the supernatant comprising cytoplasm and nucleoplasm. Samples corresponding to material from 1 × 10^6 cells were analyzed by PAGE. The figure shows a Coomasssie blue-stained gel. DNA in the nucleoplasm (right-hand side): For the recovery of the cytoplasm separately from the

In the work of Laurent et al. [this conf.], the phenomenon was detected in an indirect way, by the release of histones. After infection of CEM cells with the prototype LAI strain of HIV-1, at the peak of virus production preceding the cytopathic effect, it was possible to detect the 4 histones which constitute the core of nucleosomes in the nucleoplasm (fig. 5). This was due to the breakdown of chromatin in nucleosomes as evidenced by the concomitant appearance of the characteristic apoptosis bands in DNA. The phenomenon could be blocked by adding to the cells at time zero of infection AZT or Zn^{2+} ions; the latter are known to inactivate the endonuclease involved in apoptosis. In collaboration with S. Muller (Strasbourg), the same authors have also detected antibodies against the four histones in the serum of HIV-infected patients.

This result tends to indicate that apoptosis also occurs in vivo to a large extent in these patients. It is also reminiscent of the earlier observations of histone antibodies by Stricker and Fultz in rhesus macaques infected with pathogenic SIV [8]. It is unlikely that in vivo, at least in the asymptomatic phase of the disease where the number of infected lymphocytes is small, cell apoptosis results only from the death of HIV-infected lymphocytes. Therefore, some indirect mechanisms have to be envisaged.

Abnormal Signalling due to Viral Proteins

Newell et al. [9] observed that when preceded by ligation of CD4, signalling through the T cell receptor results in unresponsiveness due to the induction of apoptosis in T helper mouse lymphocytes [9]. Following this observation, Ameisen and Capron [10] postulated that interaction of the HIV-1 glycoprotein with its receptor sequence on the CD4 molecule, could also trigger apoptosis when the lymphocyte meets its specific antigen. This mechanism will induce clonal deletion of lymphocytes activated by spe-

---

nuceloplasm, cells were disrupted in the cytoplasm extraction buffer containing 10 m*M* Tris-HCl pH 7.6, 0.15 *M* NaCl, 5 m*M* $MgCl_2$ and 0.5% Triton X-100. After centrifugation at 1,000 *g*, the supernatant contained the cytoplasm whereas the intact nuclei were pelleted. The nuclei were then extracted in the lysis buffer and centrifuged at 12,000 *g* to separate the nucleoplasm from the high-molecular-weight chromatin. The nucleoplasm was extracted with phenol/chloroform and DNA was precipitated with ethanol. The figure shows the profile of low-molecular-weight DNA analyzed on 1.4% agarose gels. On the right is the position of 100 bp DNA ladder.

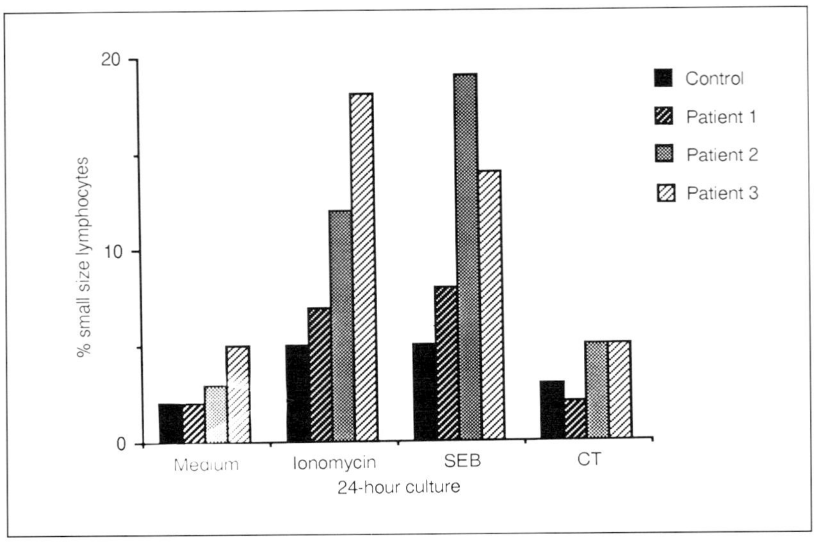

Fig. 6. Histogram showing the induction of lymphocytes with low acridine orange staining in 3 HIV-positive patients after treatment, by ionomycin (1 μ*M*), staphylococcal enterotoxin (SEB; 1 μg/ml) and cholera toxin (CT; 1 μg/ml). Control is with lymphocytes from a HIV seronegative donor.

cific antigens, but it can be operational for a large fraction of lymphocytes, via the pathway of superantigen activation. Exogeneous superantigens, generally some bacterial toxins, will interact with both the MHC of the antigen-presenting cell and the antigen receptor (its Vβ-chain) of the T4 lymphocyte presignalled by the gp120 of HIV. This could lead to the destruction by apoptosis of a significant fraction of T4 lymphocytes.

What is the evidence that such a mechanism is operating in vivo in patients? Groux et al. [5] reported apoptosis in T4 cells from asymptomatic patients induced by a superantigen, the Staphylococcal B toxin (SEB). We ourselves have observed the appearance of cells with low staining by acridine orange after in vitro treatment by SEB of lymphocytes from HIV-positive subjects. The phenomenon is not observed with PBLs from HIV-negative patients, or with cholera toxin, a bacterial toxin devoid of superantigen activity (fig. 6). Staphylococcal infections are not considered as opportunistic infections characteristic of AIDS; nor are streptococcal infections.

However, there are data indicating the expansion of some mycoplasma infections in AIDS (see below). A species of mouse mycoplasma, *M. arthritidis,* contains a superantigen active on murine and also on human T lymphocytes [11]. Concerning the mycoplasmas isolated from AIDS patients., *M. fermentans* and *M. pirum,* nothing is known about their superantigen activity. In vitro, we have preliminary indications that infection of PBLs from HIV-positive individuals by *M. fermentans* and *M. pirum* can induce the appearance of a lymphocyte population weakly stained with acridine orange, which we link to a pre-apoptosis state. These results are consistent with, but do not prove the idea that some mycoplasma proteins can act as superantigen in AIDS patients.

It will be interesting, along this line, to see if we can pre-program to apoptosis the lymphocytes of healthy, HIV-negative donors by some HIV proteins, such as gp120 interacting wit the CD4 receptor. Work along this direction has so far failed [Aiuti, personal commun.].

Factors Able to Induce Apoptosis in T8 and T4 Cells

It is known [12] that deprivation of IL-2 from the culture medium of cells which had been previously activated (for instance by mitogenic lectins), will induce apoptosis in these cells.

Therefore, our observations of in vitro apoptosis in medium deprived of IL-2 and the fact that we can prevent it by adding IL-2 or a cytokine mixture, could be interpreted as in vitro artefactual behavior of cells being in vivo in a normal activation state. There is evidence of general activation of T cells, as well as B cells and monocytes, in HIV-infected patients [13]. Moreover, there is appearance of CD8+ lymphocyte specific of HIV proteins, sugesting the occurrence of a cell-mediated immune response against HIV-infected cells [14, 15].

However, if all cells going to apoptosis were *normally* activated, we should be able to detect activation markers such as functional IL-2 receptor (CD25). This is not the case (data not shown). Therefore, a normal activation of a large number of lymphocytes could not explain the high level of apoptosis we observed in lymphocyte cultures from HIV-infected subjects.

As a corrolary, at least a significant fraction of these cells must be in an abnormal state of pre-activation. It is known that a double signalling is required to bring about normal T cell activation: besides the binding of antigen to the T cell receptor, a second signal coming from interleukins secreted by the antigen-presenting cell is required. If this second signal is not properly emitted or received, the cell will undergo apoptosis. Defects of monocyte function and disturbance in the array of circulating cytokines have been reported in HIV infection, making this hypothesis plausible.

It is also possible that soluble factors released by infected cells or by specific subsets of lymphocytes will have an inhibitory effect on the function of antigen-presenting cells or lymphocytes. Autran, Debré and colleagues [16] have isolated such a soluble heat-resistant factor released by CD57+ lymphocytes.

Finally, a default of intrathymic maturation can also generate immature T lymphocyte which will then die of apoptosis. Alterations of thymus architecture have been reported to occur in AIDS patients, but it is difficult to appreciate whether they are secondary to the immune depression or are an early event associated with the asymptomatic phase of the disease.

Table 1 summarizes the various hypotheses which could explain the programming to apoptosis in AIDS.

The last part of this presentation will review the circumstantial evidence that mycoplasma infection could be a co-factor of AIDS.

As already alluded to by us, some mycoplasmas may participate in the activation of T cells, through antigens or superantigens. But they might also amplify virus replication. Our interest in mycoplasma and AIDS started from our early observations [17] that antibiotics active in vitro on mycoplasmas can reduce the cytopathic effect of HIV in continuous T4+ cell lines. This result has been confirmed independently by Lo et al. [18]. The same group had also previously reported the isolation of a new infectious agent from Kaposi sarcoma material which turned out to be a particular strain of *M. fermentans* [19]. These findings led us and others to search for the presence of mycoplasma in AIDS patients, and to study the in vitro interaction between HIV and these microorganisms.

Isolation of Mycoplasma from AIDS Patients and HIV-Positive Individuals

Some mycoplasma species are ubiquitous, and are present on most mucosas without being harmful. Several species are pathogenic in animals. In man, only *M. pneumoniae* has been shown to be associtated with severe

Table 1. Possible signals triggering apoptosis in vivo

		Cells involved
1	Virus infection	T4
2	Default in double signalling of T cell activation	T4 and T8
3	Abnormal activation via CD4 by viral glycoprotein. Subsequent activation via T cell receptor (antigens, superantigens) will induce apoptosis	T4
4	Defaults of intrathymic maturation can generate immature T lymphocytes programmed for apoptosis	T4 and T8

pneumonia. If mycoplasmas have to play a role in AIDS, they should be associated also with cells or tissues of the immune system. We have therefore attempted to isolate mycoplasmas from circulating lymphocytes, either directly from blood or from short-term cultures. For direct isolation we used SP4, an enriched medium defined to isolate fastidious mycoplasmas such as *M. genitalium* [20].

In several cases, we had evidence of transient growth, but the germ could not be propagated continuously in synthetic medium. However, we have been able to isolate *M. pirum* from 2 AIDS patients, using short-term cultures of PHA-activated PBLs [21] (fig. 7). This finding is interesting per se since *M. pirum* had only been previously shown to be present in cultures of human cancerous cells, but its host was unknown. Our isolation indicates that man (and possibly related primates) is the natural host of this mycoplasma. We have also shown that proteins of *M. pirum* are related to those of *M. pneumoniae* and *M. genitalium.* Its morphology is also similar. We have also isolated a strain of *M. fermentans* from a short-term culture of lymphocytes from an asymptomatic HIV-infected individual [21]. This strain is similar to that observed by Lo. The mycoplasma strains thus isolated have been propagated without difficulty in synthetic SP4 medium and characterized by restriction map analysis of its ribosomal DNA. *M. pirum* organisms can readily infect lymphocytes from healthy donors, in which they rapidly induce a cytopathic effect. Electron microscopy suggests that the mycoplasmas are internalized in cytoplasmic vacuoles before cell death (fig. 8).

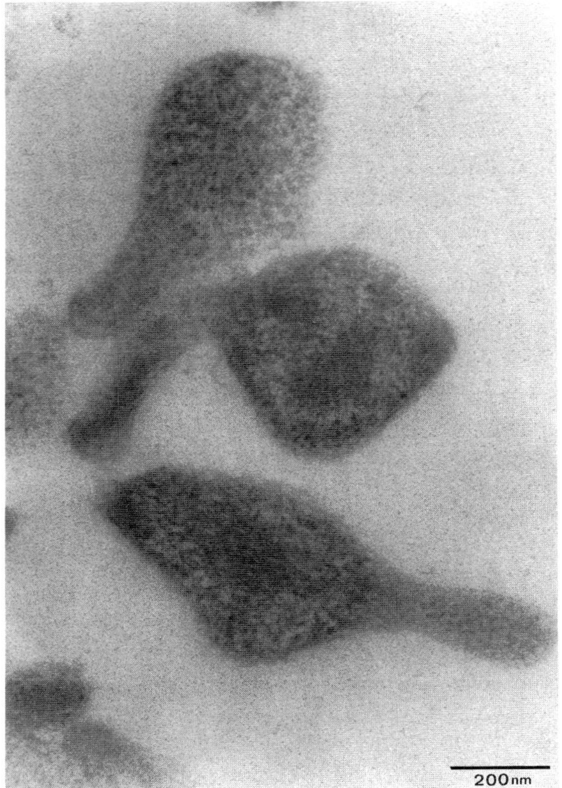

Fig. 7. Electron micrograph of an ultrathin section (uranyl acetate staining) of *M. pirum* isolated from an AIDS patient: note the characteristic pear-shaped form.

Indirect Evidence of Mycoplasma-Like Organisms Associated with Blood Cell Elements

Owing to the difficulties of directly isolating these organisms by culture in synthetic medium, we used indirect techniques for this research: (a) ^{3}H-uracil labelling: Uracil is readily incorporated into prokaryotic organisms and not in eukaryotic cells. We have used this property to specifically label mycoplasmas from patient's lymphocytes and healthy donors and band them to density equilibrium in sucrose gradients: mycoplamas have characteristic densities ranging from 1.18 to 1.24. Using this technique, we frequently detected mycoplasma peaks in short-term cul-

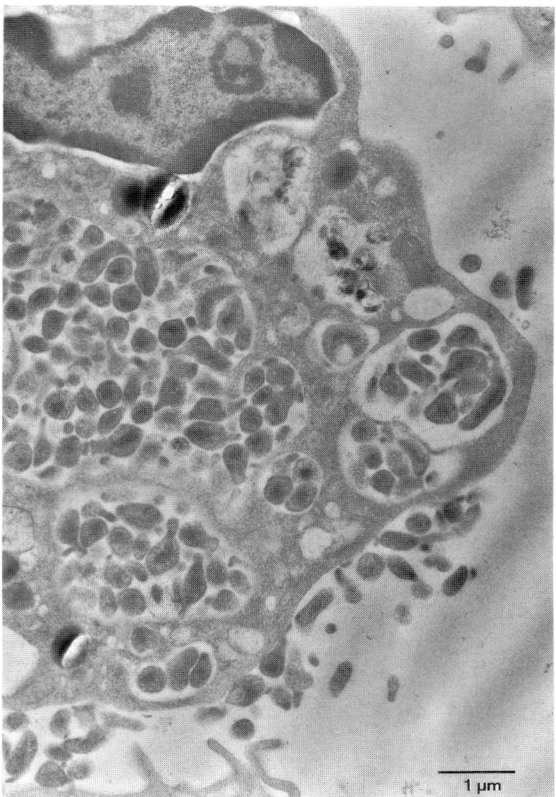

Fig. 8. Electron micrograph of an ultrathin section (uranyl acetate staining) of lymphocytes (PHA-activated) infected in vitro with *M. pirum:* note the numerous intracellular vacuoles filled with the microorganisms.

tures of lymphocytes, sometimes also in cultures of red blood cells. (b) Molecular hybridization with specific probes: The PMC5 probe, cloned from the ribosomal DNA of *M. capricolum,* was used to detect prokaryotic DNA in the DNA extracted from whole lymphocytes, plasma or red cells. Again, we frequently obtained positive hybridization signals from patient material, but also sometimes from healthy HIV-negative donors.

Unfortunately, these two techniques give no information on the species involved. There was not enough hybridized ribosomal DNA to make restriction and Southern blot analysis.

Our current studies are directed towards the detection and characterization of some mycoplasma species, known to be present in AIDS patients *(M. fermentans, M. genitalium, M. pirum)* by using the PCR technique with specific primers. With this approach, the group of Lo [22] has recently reported evidence of *M. fermentans* DNA in the urine of some AIDS patients, and none in a control group.

Search for a Specific Immune Response

Using the whole microorganism proteins, in a Western blot technique, we found the antibody response quite variable from one individual to another and no correlation with HIV infection. Using a peptide supposed to be the binding site of *M. genitalium* adhesine (see below) in an ELISA assay, we found a general antibody response against this peptide, with no significant difference in HIV-positive and HIV-negative individuals. However, there was a tendency for a lowering of this antibody response at late stages of AIDS.

Primate Models

We are studying, in collaboration with Dr. J. Heeney (TNO Primate Center, Delft, Holland) the state of endogenous mycoplasma infection in rhesus macaques and its eventual changes after SIV infection and AIDS. Using the PMC5 ribosomal probe, we found an increased hybridization signal with the leukocyte DNA of 5 out of 6 infected monkeys (7–9 months postinfection).

This result suggests an expansion of prokaryotic DNA in the infected macaques, prior to disease. Unfortunately, the assignment of this signal to a particular mycoplasma species could not be done. Monkeys are infected with mycoplasma species which may differ substantially from those infecting men. For instance, it is known that *M. genitalium* cannot infect rhesus monkeys. However, we tried to inoculate *M. pirum* into monkeys, some of which were infected for several months by SIV. We noticed a good antibody response against the adhesine peptide of *M. genitalium* (the antibody against *M. pirum* adhesine cross-reacts with this peptide) in the control uninfected animals. However, there was no response in the 2 SIV-infected animals.

This indicates that despite the fact there was not yet a general immune depression at this stage of SIV infection, there was already a selective defect of the immune response against this mycoplasma antigen.

Table 2. Influence of *M. fermentans* on TNF secretion by human monocytes

Cells 10^6/ml	Addition 48 h culture	TNF secreted	
		U/ml	pg/ml
Monocytes	None	2.8	140
	M. fermentans	4.4	220
	M. fermentans extract, 10 μg/ml	22.0	1,100

In vitro Studies on the Interaction between Mycoplasma and HIV

We have already reported that, besides the diminution of cytopathic effect by agents reducing the mycoplasma load of some HIV strains, we could find a 'helper effect' of mycoplasma extracts on HIV replication [23]. Assuming that the mycoplasma proteins which interact with the parasitized cell (adhesines) were probably involved in this effect, we prepared a rabbit antibody against the binding site of the adhesine of *M. genitalium*.

This antibody was shown to reduce to a large extent HIV replication in CEM cells or T lymphocytes in a specific manner. It is therefore possible that some HIV strains are helped in their replication by intracellular signals coming from interaction of the adhesine of a *M. genitalium*-related mycoplasma with the cell membrane. It is also possible that this interaction will release cytokines which are known to trigger HIV replication in some cell lines [24]. Indeed, monocytes infected by *M. fermentans* and *M. pirum* or treated by *M. fermentans* extracts show production of IL-6 and TNF (table 2).

However, some other HIV-1 strains were enhanced by the same antibody preparation, suggestive of a negative effect of mycoplasma on HIV replication. We are thus encouraged to study more in depth the interaction between HIV and this class of particular microorganisms at the in vitro and in vivo levels. Figure 9 summarizes our current hypothesis on the subject.

In conclusion, our data suggest that AIDS is caused by a complex interplay between a retrovirus – HIV – and factors leading to abnormal activation of T lymphocytes, mostly T4 lymphocytes, followed by their death by apoptosis. Among these factors, mycoplasmas are possible candidates, perhaps acting in a stage more precocious than that allowing expansion of other opportunistic agents.

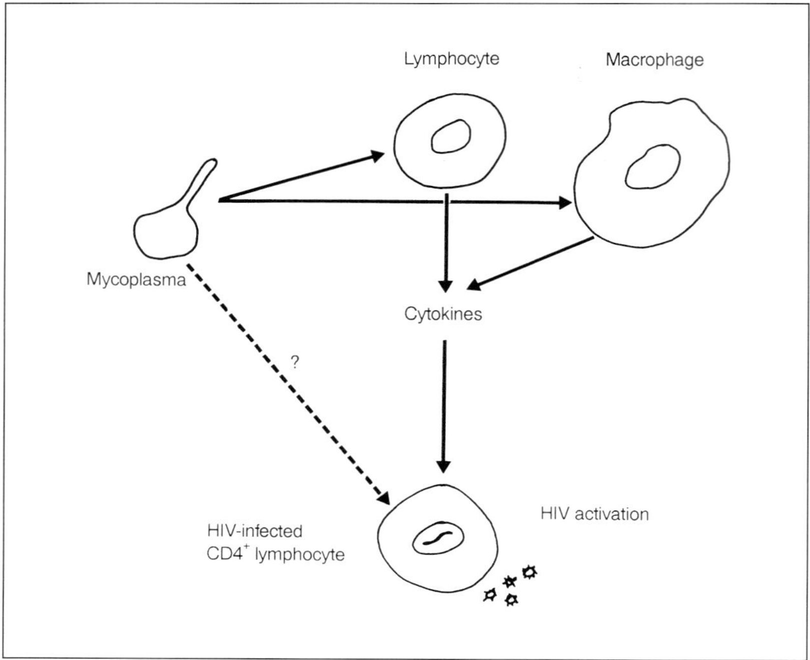

Fig. 9. Hypothetical scheme of the possible mechanisms by which mycoplasms can stimulate HIV in AIDS patients.

References

1 Levacher M, Hulstaert F, Tallet S, Ullery S, Pocidalo JJ, Bach BA: CD8, CD38 cellular changes are independent prognostic indicators of clinical disease progression in HIV-1 infection. VIIth Conf on AIDS, Florence, 1991. Abstract Book, p 284, abstr WB 2408.

2 Montagnier L, Guétard D, Rame V, Olivier R, Adams M: Virological and immunological factors of AIDS pathogenesis; in Fond M Mérieux (ed): 4th Cent-Gardes Coll, 1989, pp 11–17.

3 Smith CA, Williams GT, Kingston R, Jenkinson EJ, Owen JJT: Antibodies to CD3/T cell receptor complex induce death by apoptosis in immature T cells in thymic cultures. Nature 1989;337:181–184.

4 Gougeon ML, Olivier R, Garcia S, Guétard D, Dragic T, Dauguet C, Montagnier L: Mise en évidence d'un processus d'engagement vers la mort cellulaire par apoptose dans les lymphocytes de patients infectés par le VIH. CR Acad Sci Paris 1991;312: 529–537.

5 Groux H, Monte D, Bourrez JM, Capron A, Ameisen JC: L'activation des lymphocytes T CD4+ de sujets asymptomatiques infectés par le VIH entraîne de déclenchement d'un programme de mort lymphocytaire par apoptose. CR Acad Sci Paris 1991;312:599–606.
6 Terai C, Kornbluth RS, Pauza CD, Richman DD, Carson DA: Apoptosis as a mechanism of cell death in cultured T lymphoblasts acutely infected with HIV-1. J Clin Invest 1991;87:1710–1715.
7 Laurent-Crawford AG, Krust B, Muller S, Rivière Y, Rey-Cuillé MA, Béchet JM, Montagnier L, Hovanessian A: The cytopathic effect of HIV is associated with apoptosis. Submitted. See also, VIIth Int Conf on AIDS, Florence, 1991, Abstract Book, p 27.
8 Fultz PN, Stricker RB, McClure HM, Anderson DC, Switzer WM, Horaist C: Humoral response to SIV/SMN infection in macaque and Mangabey monkeys. J AIDS 1990;3:319–329.
9 Newell MK, Haughn LJ, Maroun CR, Julius MH: Death of mature T cells by separate ligation of CD4 and the T cell receptor for antigen. Nature 1990;347:286–289.
10 Ameisen JC, Capron A: Cell dysfunction and depletion in AIDS: The programmed cell death hypothesis. Immunol Today 1991;12:102–105.
11 Cole BC, Atkin CL: The Mycoplasma arthritidis T-cell mitogen, MAM: A model superantigen. Immunol Today 1991;12:271–276.
12 Duke RC, Cohen JJ: IL-2 addiction: Withdrawal of growth factor activates a suicide program in dependent T cells. Lymphokine Res 1986;5:289.
13 Ascher MS, Sheppard HW: AIDS as immune system activation: A model for pathogenesis. Clin Exp Immunol 1988;73:165–167.
14 Walker BD, Chakrabarti S, Moss B, Paradis JT, Flynn T, Durno AG, Blumberg RS, Kaplan JC, Hirsch MS, Schooley RT: HIV-specific cytotoxic T lymphocytes in seropositive individuals. Nature (Lond) 1987;328:345–348.
15 Rivière Y, Tanneau-Salvadori F, Regnault A, Lopez O, Sansonetti P, Guy B, Kieny MP, Fournel JJ, Montagnier L: Human immunodeficiency virus-specific cytotoxic responses of seropositive individuals: Distinct types of effector cells mediate killing of targets expressing gag and env proteins. J Virol 1989;2270–2277.
16 Sadat-Sowti, Autran B, Idziorek T, Manuel N, Oksenhendler E, Katlama C, Debré P: A lectin-binding soluble factor released by CD8+CD57+ lymphocytes from AIDS patients inhibits T cell cytotoxicity. VIIth Int AIDS Conf, Florence, 1991. Abstract Book, p 1171, abstr WA 1168.
17 Lemaître M, Guétard D, Montagnier L, Zérial A: Protective activity of tetracycline analogs against the cytopathic effect of the human immunodeficiency viruses in CEM cells. Res Virol 1990;141:5–16.
18 Lo SC, Tsai S, Benish JR, Shih JWK, Wear DJ, Wong DM: Enhancement of HIV-1 cytocidal effects in CD4+ lymphocytes by the AIDS-associated mycoplasma. Science 1991;251:1074–1076.
19 Saillard C, Carle P, Bové JM, Bebear C, Lo SC, Shih JWK, Wang RYH, Rose DL, Tully JG: Genetic and serologic relatedness between mycoplasma fermentans and a mycoplasma recently identified in tissues of AIDS and non-AIDS patients. Res Virol 1990;141:385–395.

20 Tully JG, Whitcomb RF, Williamson DL: Pathogenic mycoplasmas: cultivation and vertebrate pathogenicity of a new spiroplasma. Science 1977;195:892–894.
21 Blanchard A, Ferris S, Tham TN, di Rienzo AM, Guétard D, Montagnier L: Characterization of two mycoplasmas isolated from the blood of HIV-seropositive patients. VIIth Int Conf on AIDS, Florence, 1991, Abstract Book, vol 2, p 69.
22 Lo SC: ASM Meeting, Dallas, May 1991.
23 Montagnier L, Berneman D, Guétard D, Blanchard A, Chamaret S, Rame V, Van Rietschoten J, Mabrouk K, Bahraoui E: Inhibition de l'infectiosité de souches prototypes du VIH par des anticorps dirigés contre une séquence peptidique de mycoplasme. CR Acad Sci Paris 1990;311:425–430.
24 Fauci AS: The human immunodeficiency virus: Infectivity and mechanisms of pathogenesis. Science (Wash) 1988;239:617–622.

Luc Montagnier, MD, Department of AIDS and Retroviruses, Institut Pasteur, 28, rue du Dr. Roux, F–75724 Paris Cedex 15 (France)

Rossi GB, Beth-Giraldo E, Chieco-Bianchi L, Dianzani F, Giraldo G, Verani P (eds): Science Challenging AIDS. Basel, Karger, 1992, pp 71–106

Molecular Biology of the AIDS Virus: Ten Years of Discovery – Hope for the Future[1]

William A. Haseltine

Division of Human Retrovirology, Dana-Farber Cancer Institute, Boston, Mass., USA

Lessons from the Past

It is 10 years since the first cases of AIDS were reported. The last 10 years has been a decade of discovery. It has also been a decade of misunderstanding, of missed opportunity. As we face the second decade of the AIDS epidemic, it is time to review and learn from the past to prepare for the future.

We are now in the midst of a worldwide pandemic. We come to know the face of our enemy well. The full range of modern analytic tools has been brought to bear on this disease, including the sciences of epidemiology, immunology, virology, molecular biology, structural biology, and pharmacology. Today we know what causes the disease. We know how the disease is transmitted. We can identify the victims of this disease many years before they become ill.

Despite such knowledge we will stand, as individuals and as a species, almost unprotected from the ravages of this epidemic. How can a single microorganism evade the most advanced medical treatments and evade organized efforts to stop its spread.

To understand the current state of affairs, let us remember what we thought we knew, and when we thought we knew, it in the light of what we know today.

[1] Supported in part by NIH grants AI 24845 (NCDDG), AI 24755, AI 29873, AI 28193 and AI 28691.

Ten years ago 5 young American men were diagnosed as having a new severe disease. The first report of AIDS caused barely a ripple in the passive waters of the public health establishment. During the next year we learned that the disease affected primarily promiscuous gay men and that the disease was acquired. Whether the disease was acquired as a result of exposure to a new infectious agent or as a consequence of behavior was uncertain.

What do we know today? We now know that the 5 young men were part of what was already a massive worldwide epidemic. We now know that in 1981, between 150,000 and 200,000 Americans were already infected with a virus that would either kill them or put them at death's door. These 5 young men were the first signal that by 1981 more than one million people worldwide were already infected. Over half of those infected in 1981 have since either died or have become extremely ill. We have learned that the AIDS infection spreads silently. Neither its victims or public health authorities know when it arrives. Our reaction to this epidemic has been, and still is, dramatically out of proportion to its magnitude.

By 1982 we knew that the disease was transmitted by sex among men. We also learned that the disease could be transmitted by blood, from a person who at the time of donation appeared to be healthy but later became ill, and that the disease could be transmitted by mother to child. We also learned that women as well as men could contract the disease.

It was in these early years that many mistaken notions regarding this disease became fixed in the public mind. These fixed notions have greatly slowed public awareness of the seriousness of this disease and continue to hamper government and private efforts to control the epidemic. It was early in the first decade that the mistaken notion that the disease was only acquired by homosexual sex, more specifically by receptive anal intercourse, became widely accepted. It was also during this early period that the mistaken idea that transmission of the virus during sex required blood-to-blood transmission became fixed in the public mind.

Today we know that these early ideas are incorrect. We know that AIDS is a venereal disease – a venereal disease that kills. AIDS transmission resembles that of most other sexually transmitted diseases. Infection can be readily transmitted by vaginal sex, either from men to women, or women to men. Infection can occur upon exposure to seminal or vaginal fluids of intact sexual mucosa; witness specific cases of infection by artificial insemination and animal transmission of AIDS-like viruses across intact, vaginal and rectal mucosa. The virus is transmitted very rapidly among promiscuous sexual populations. The initial concentration of virus in the American homosexual male population is not so much a preference of transmission by rectal intercourse, but rather a consequence of sexual promiscuity in this population. The virus spreads rapidly in any promiscuous population, regardless of whether promiscuous behavior is heterosexual or homosexual. In this respect, AIDS is similar to all other venereal diseases, spreading most rapidly where the population is most promiscuous.

In 1983, the virus which caused the disease was first sighted. In 1984 it was proved that this virus, now called HIV-1, was the cause of the disease.

The discovery that the disease was caused by a virus immediately raised hopes that diagnostic tests, a vaccine, and a cure would soon be forthcoming. The first hope was quickly realized. In 1985, a reliable diagnostic test was developed. The search for a cure and a vaccine continues.

The availability of a diagnostic test for infection began to reveal the true magnitude of the epidemic. We learned that in the United States for every 5 people sick with AIDS, at least 95 more were infected. In other words, as many as 400,000–500,000 people were infected in the United States by 1985.

This observation led to the early hope that only a small fraction of those infected would ultimately become ill. Unfortunately, such hopes proved to be unfounded. Experience has shown that almost all those who test positive for infection will ultimately become ill.

The test has permitted us to estimate the magnitude of infection. A rough estimate places the magnitude of the infection in 1981 at between one and two million people. Today it is estimated that 10–20 million people are infected. By the end of the decade, all of these people will have either died or become extremely ill.

There is now no population which is free of this disease. The true horror of what is happening is now just dawning. We are in the midst of the greatest epidemic of this century, an epidemic, which may become the most serious disease in recorded history.

It is our hope that knowledge will defend us from this epidemic, that knowledge of this organism which causes the disease will lead both to a treatment and to a vaccine. However, despite knowledge of the virus, we do not yet have an effective vaccine, and treatment for the disease is only palliative.

To understand the challenge before us, we must understand the virus itself.

Vaccines: Problems and Prospects

Vaccines work by preparing the immune system to respond quickly to infections. The immune system has the capacity to remember what foreign substances were previously encountered. Memory shortens the time of immune response from 2–3 weeks to 2–3 days. Infections that are established can be quickly eliminated in the vaccinated person by a combination of antibodies and killer cells. Vaccines work to prevent spread of the infecting virus and subsequent disease.

By 1986 we knew that the challenge of an AIDS vaccine was more difficult than that for many microorganisms. We knew that people infected with the virus make a good immune response, both humoral antibody responses and cell-mediated responses to the virus [1–4]. However, despite the good immune response the virus infection is not cleared and disease progression is inexorable. We are now beginning to understand how the AIDS virus effectively evades the immune response.

Silent Infection

The AIDS virus is a retrovirus. The life cycle of retroviruses involves conversion of RNA to DNA and insertion of viral DNA into the host genome. This process is called establishment of infection (fig. 1). The virus then reproduces by copying the viral DNA into RNA and using the RNA to make new viral proteins. New virus buds from the surface of the living cell (fig. 2).

The exceptional feature of the virus life cycle which contributes to difficulties in making vaccines is the insertion of the viral genetic material into the host cell DNA, a process called integration. Once the virus genetic information is integrated into the DNA of the host cell, it remains part of the host cell as long as that cell lives. This is the fundamental reason why infection is not cleared by the immune response. A person infected with HIV-1 remains infected for life.

An infected cell may harbor virus genetic information but may not express it. In other words, infection can be established without resulting

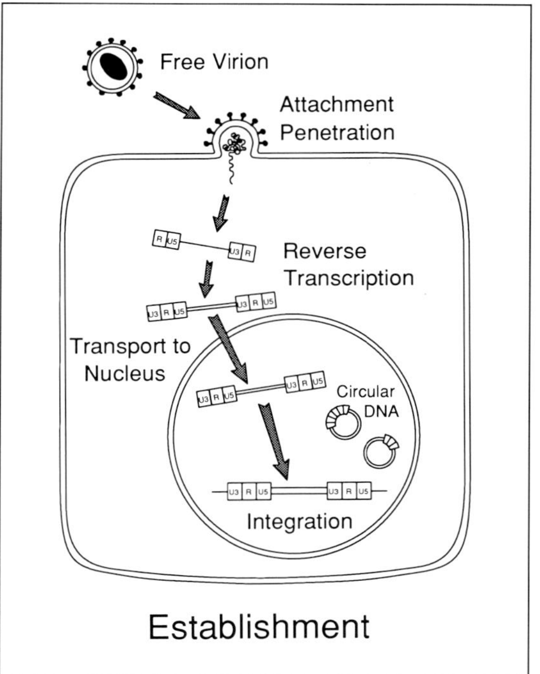

Fig. 1. Establishment of infection in the HIV-1 lifecycle. Steps of establishment are: (1) virus binding to a cell receptor; (2) penetration of a cell membrane; (3) provirus synthesis; (4) migration of the provirus to the nucleus; (5) integration.

expression. Such an infected cell is invisible to the immune system. The immune system cannot recognize the DNA which lies silent within the cell. The immune system recognizes only viral proteins when they are made. The existence of silently infected cells means that in some instances, actual infection may go undetected. It also means that privileged reservoirs of infected cells can exist.

We now know that expression of virus proteins is a highly regulated process. The first step in virus expression is activation of the infected cell. Viruses are not made in the majority of infected cells which are at rest. The infected, immunologically invisible cell, must first be stimulated to become a target for the immune system. If the cell is not stimulated, the DNA remains silent and invisible.

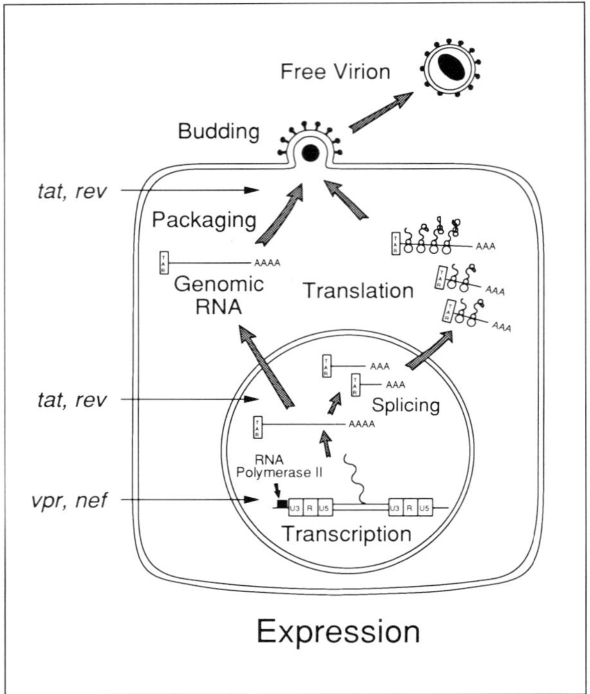

Fig. 2. Virus expression in the HIV-1 lifecycle. Steps of virus expression include: (1) transcription of proviral DNA by cellular RNA polymerase II; (2) messenger RNA processing; (3) messenger RNA transport and genomic RNA transport; (4) synthesis of viral proteins; (5) assembly of virion particle; (6) budding and maturation of the virus particles.

Partial Expression

In our previous studies of retroviruses, we learned that the only requirement for expression was RNA synthesis. Not so for HIV-1. We now know that the expression phase of HIV-1 must be divided into two sub-stages. HIV-1 requires two and possibly three virus proteins for expression. These proteins are called *tat, rev,* and *nef.*

In the absence of the virus protein *tat,* very little full-length RNA is made. The rate of initiation of infection is low and the RNA which is made is short. *tat* permits rapid accumulation of full-length RNA.

Full-length RNA must accumulate to permit synthesis of the virus protein *rev.* In the absence of *rev,* the RNA that is made is rapidly altered

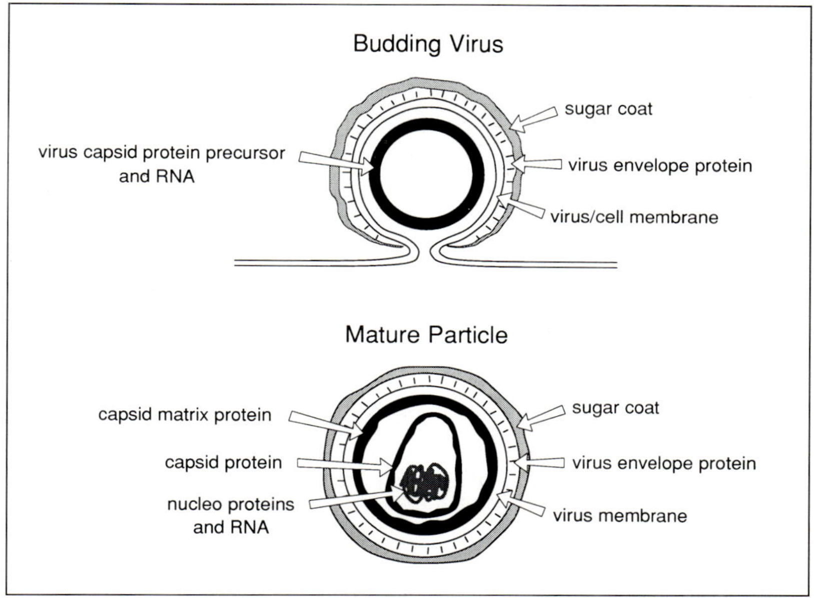

Fig. 3. Diagram of virus in process of budding.

by the cellular splicing apparatus to short forms of the RNA. The short forms of the RNA serve to make *tat, nef,* and the *rev* protein itself. The short RNAs do not make virus structural proteins including the proteins necessary to make the inner capsid of the virus and the surrounding envelope protein.

The early stage of expression is characterized by synthesis of viral RNA, accumulation of short viral RNA species, but little or no synthesis of viral protein. The cells in this intermediate state of expression are largely invisible to the immune system.

The final stage of replication requires high levels of *rev* and possibly *nef* expression. High levels of these proteins permit RNA to accumulate and synthesis of virus particles to be initiated.

Sugar Coat

For the most part the immune system recognizes surface structures. The outer part of the virus and virus-infected cells is covered by a virus specified protein called the envelope protein (fig. 3).

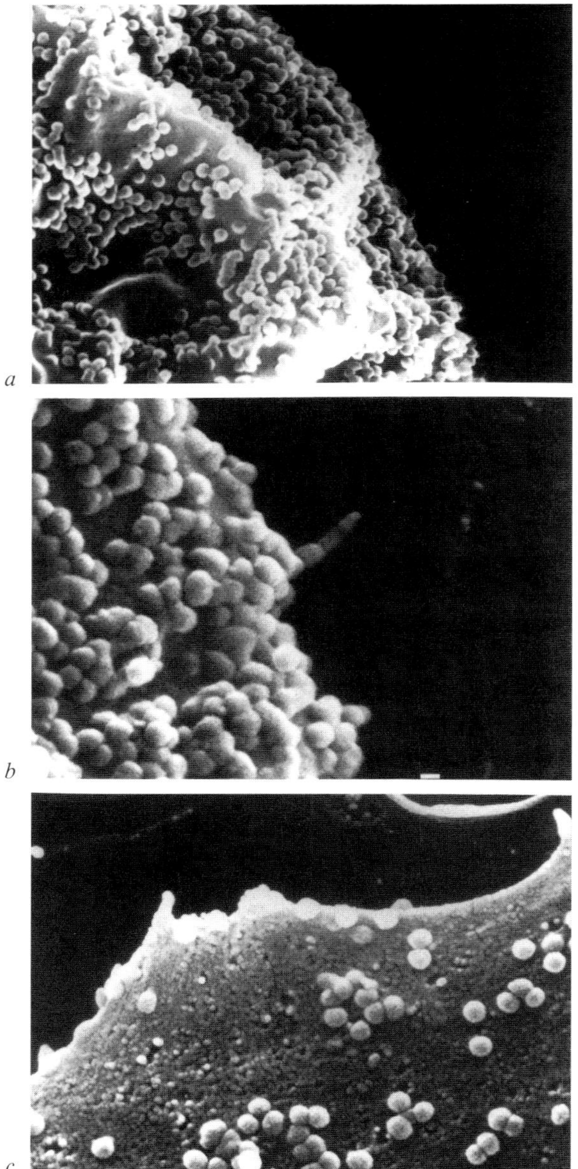

Fig. 4. Scanning electron micrographs of HIV-1 virus particles accumulating at the surface of the COS-7 cells transfected with p6 mutant proviruses.

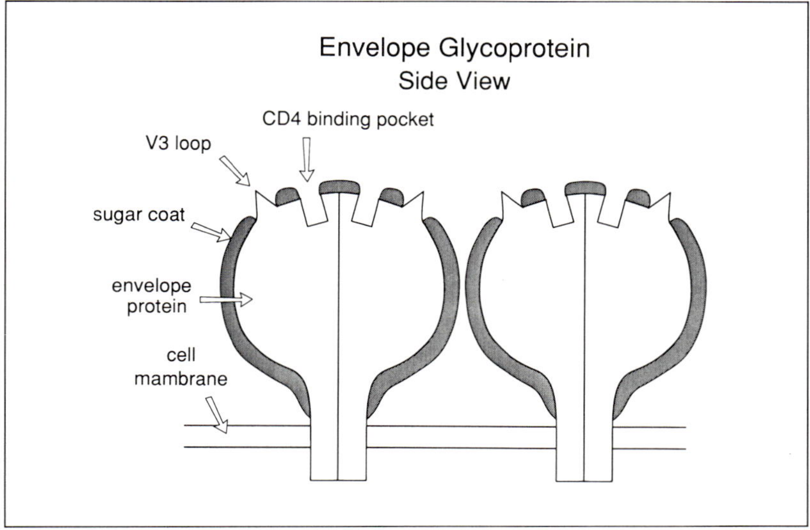

Fig. 5. Schematic view of the envelope glycoprotein.

The envelope protein of HIV-1 is itself covered by sugars (fig. 4a–4c) [5, 6]. These sugars mask much of the protein's surface, effectively screening it from recognition by antibodies or T cells. The sugars also interfere with the process whereby immune response is made. Peptides modified by sugars are not readily presented by MHC Class II to other immune cells, a key step to immune recognition. In other words, the virus is engineered to be invisible, even when it is expressed. The HIV-1 is a stealthy virus.

Variation

Screening of the surface of the virus particle by sugars is not complete. Some regions of envelope proteins are exposed at the surface. Some of these regions are sensitive to neutralizing antibodies. The most pronounced region is the V3 loop of the surface of the virus (fig. 5). The V3 loop of the envelope glycoprotein is the most prominent exposed region at the surface (fig. 6). Antibodies to this region neutralize the virus [7–9].

The virus escapes neutralization by antibodies to V3 by variation. Rapid variation is a consequence of the life cycle of the virus (fig. 7) [10–13]. As RNA is converted to the first single strand of DNA, and as the first single strand of DNA is converted to the second strand of DNA, mistakes

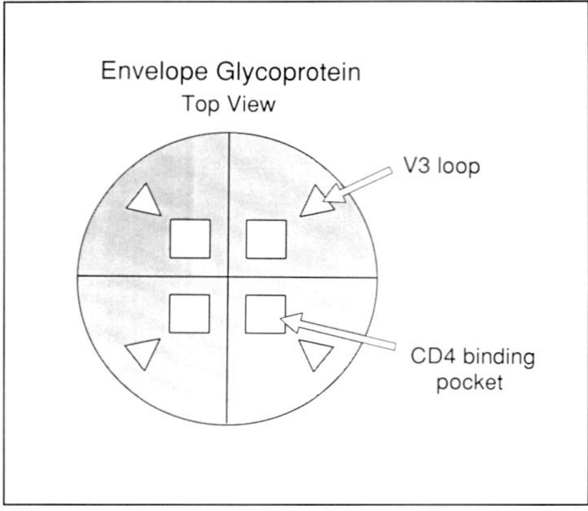

Fig. 6. Schematic of top view of the envelope glycoprotein.

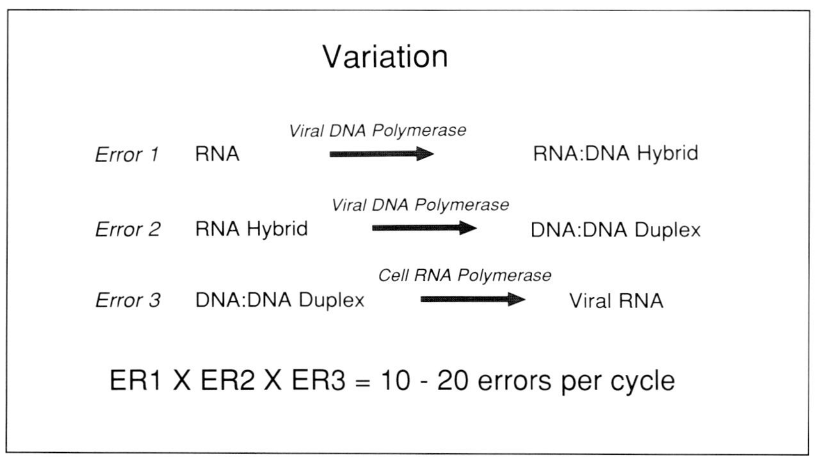

Fig. 7. Virus variation accumulates through the replication cycle.

in copying the genetic code are made. Mistakes are also made when the DNA is transcribed into RNA. The errors in these three steps of replication are multiplicative. The result is that 10–20 mistakes are made in a single complete infectious cycle. The rate of virus variation is compounded by multiple rounds of replication of the virus.

Cycle 1	$Virus_1 \rightarrow Virus_2$
Cycle 2	$Virus_2 \rightarrow Virus_3$
Cycle 3	$Virus_3 \rightarrow Virus_4$

Error:
EC1 x EC2 x EC3 .... = 35% variation

Fig. 8. HIV-1 isolates may vary from one another in primary sequence by up to 35%.

Virus variation keeps the virus one step ahead of the immune system. This is one of the reasons that infection continues despite a vigorous immune response. The intense variation which occurs within an individual has the further consequence that multiple immunological variants of the virus exist within an infected population. RNA sequences may vary maximally from one another by 75%. Variation amongst HIV-1 isolates may vary from one another in primary sequence by up to 35%, half of the possible maximum (fig. 8).

Mucosal Infection

A final and perhaps more daunting difficulty for vaccinations stems from the mode of infection. Sexual transmission of HIV-1, the major route of transmission today, occurs at mucosal surfaces. Over the past year we have gained a much better insight of how HIV-1 may be transmitted across intact mucous membranes (fig. 9). Erik Langhoff in my laboratory, elaborating the early work of Stella Knight in England [14], has shown that the dendritic cell, a mature form of the tissue Langerhans cell [15], is extremely sensitive to infection by HIV-1. Infected dendritic cells can produce prodigious quantities of virus without themselves being killed. The natural role of the Langerhans and dendritic cells is to detect antigens at mucosal surfaces [15–17]. Such antigens are detected and processed by the Langerhans cells. The Langerhans cells are thought to migrate to the circulation as dendiritc cells and thence to the lymph nodes where they nurture T cells and are effective antigen-presenting cells [14, 15].

Evidently, HIV-1 has adapted itself to this natural mode of immune surveillance. Langerhans cells present at the surface of sexual mucosa may

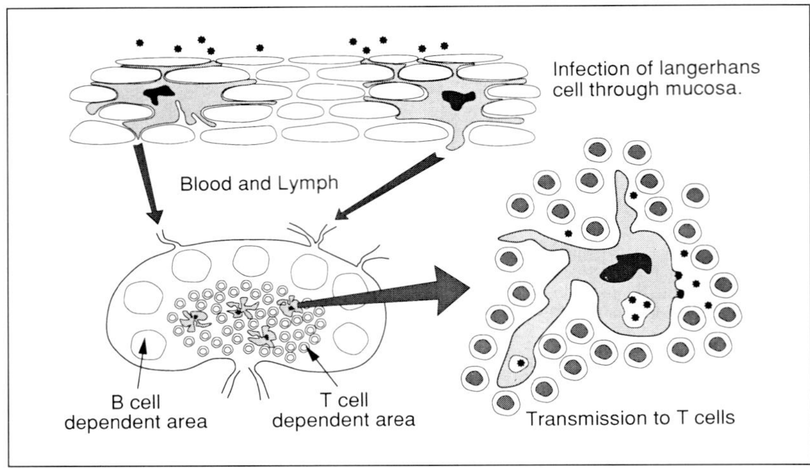

Fig. 9. Dendritic cell hypothesis for sexual mucosal transmission of HIV-1.

be infected by exposure to virus and to virus-infected cells present in semen and vaginal fluids. The dendritic cells transfer the infection to the lymph node where T cell populations are infected.

M Cells

Working with Marian Neutra at Children's Hospital at Harvard, we have also gained insights into possible routes of rectal infection (fig. 10). The surface of intestinal and rectal mucosa contains immune surveillance organelles. These organelles screen intestinal antigens and mount IgA responses. The microarchitecture of such organelles consists of an M cell surrounding a small pocket of lymphoid tissue [18]. The M cells selectively transport antigens and microorganisms to the lymphoid tissue where they are recognized [18–20]. We have demonstrated that the M cells of rats and mice selectively absorb HIV-1 and transport it intact, and unidirectionally, to the underlying lymphoid tissue. This underlying lymphoid tissue contains B cells, T cells, and macrophages which may be sensitive to infection. It is likely that M cells provide a port of entry for virus into the membrane across and intact rectal mucosa [19].

Mucosal infection via either the Langerhans/dendritic cells or the M cell possess a serious problem for vaccination. Once the virus enters the immune system via this route, it can escape elimination by the immune

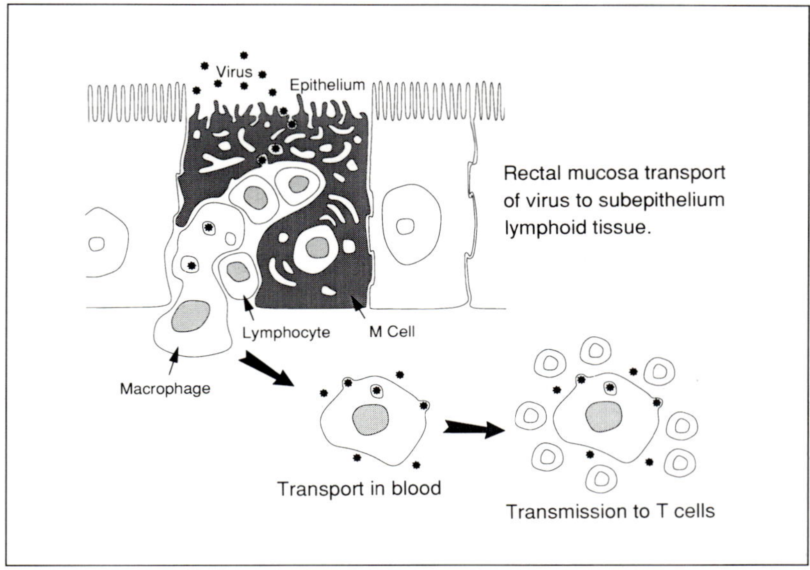

Fig. 10. M cell model for rectal infection by HIV-1.

system and establish persistent and ultimately lethal infection. This observation may explain why monkeys vaccinated for SIV may be immune to challenge via exposure to low amounts of virus by blood but are not immune to vaginal or rectal challenge.

How might these difficulties in vaccine development of an AIDS vaccine be overcome?

The focus on vaccine development must be on raising protective immune responses to conserved and not to variable regions of the viral protein. In particular, the focus must be on raising antibodies to the conserved, exposed regions of the envelope glycoprotein.

We have learned that the HIV-1 viral proteins contain invariant regions of amino acid sequence as well as regions which are highly conserved amongst independent strains. Invariant and highly conserved regions correspond to important functions of the viral proteins. Invariant and highly conserved regions of the virus proteins are found in the capsid proteins in regions required for virus capsid assembly. The replicative enzymes of the virus, the DNA polymerase, the ribonuclease H, the protease, and the integrase also contain invariant and highly conserved

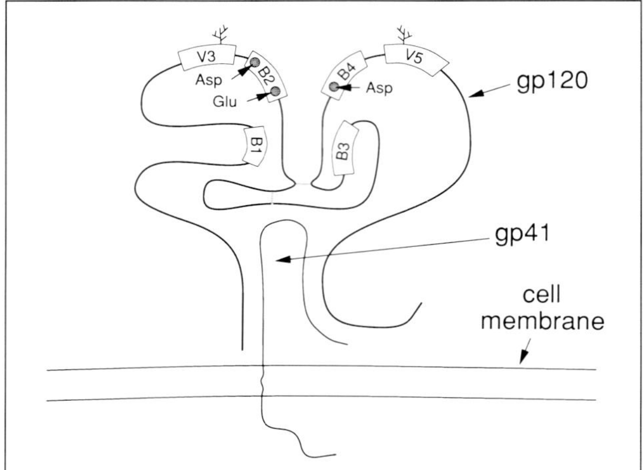

Fig. 11. Diagram of the region of the HIV-1 envelope glycoprotein which binds to CD4.

regions. These are the regions of active sites of these enzymes. Additionally, the active sites of the virus regulatory proteins *tat, rev, vif, vpr, vpu,* and *nef* also contain invariant and highly conserved regions. Each of these regions can serve as a target for an immune response which will inactivate multiple strains of virus.

The CD4 Binding Site

Significantly, the envelope glycoprotein of HIV-1 contains regions which are the most variable amongst viral sequences and also regions which are amongst the most conserved [13]. The conserved regions of the HIV-1 envelope glycoprotein correspond to the CD4 binding region, the regions of gp41, the regions of association between the envelope subunits, gp120 and gp41, and the fusion regions (fig. 11). [21, 22]. Of these invariant and conserved regions, the region of the envelope glycoprotein which binds to the cell surface protein CD4 is most likely to be a good target for neutralization. The CD4 recognition site must be exposed at the surface of the envelope protein as it is this region which binds to the receptor.

We have learned much about the CD4 binding region of the envelope glycoprotein. My colleague, Joseph Sodroski at Dana-Farber Cancer Institute, has shown that four discontinuous regions of the exterior glycoprotein

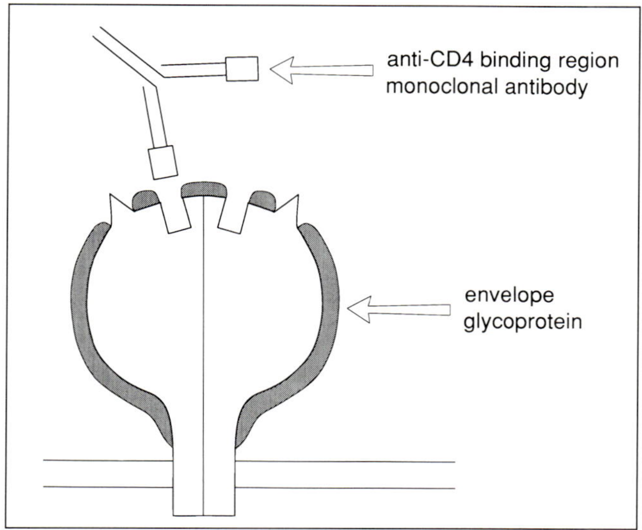

Fig. 12. Diagram of monoclonal antibody interacting with the binding pocket of the envelope glycoprotein.

of HIV-1 fold into a binding pocket for CD4 [23]. The exterior region of this binding pocket is hydrophilic. The interior of this region is hydrophobic. The amino acid residues within the binding pocket are highly conserved amongst independent isolates. Some of the amino acids in this pocket are invariant. The amino acids at the surface of the binding pocket are variable.

Working with Marshal Posner, now at Harvard University, and Joseph Sodroski has also shown that humans can make antibodies which recognize this binding pocket. Marshal Posner and others have isolated monoclonal antibodies from an infected HIV-1 person which compete for CD4 for binding to the virus. Joseph Sodroski has shown that one such monoclonal antibody interacts with the binding pocket of gp120 (fig. 12) [Thali and Sodroski, unpubl. observations]. Precise mapping of the sites of interaction of this human monoclonal antibody with the envelope glycoprotein show that this antibody recognizes the same four discontinuous regions of the envelope protein as are required for CD4 binding. The same amino acids of three of the four conserved regions are required for recognition of this monoclonal antibody as are required for recognition of CD4.

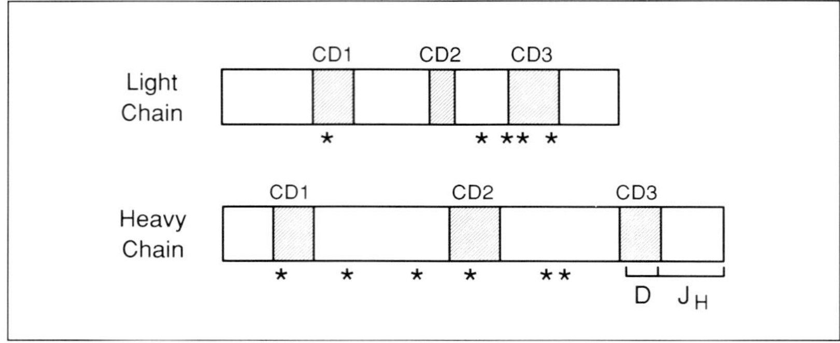

Fig. 13. Diagram of human anti-CD4 binding region monoclonal antibody.

In the fourth region necessary for binding, the monoclonal antibody recognizes amino acids which are near, but not precisely at, the site of those required for CD4 binding.

Wayne Marasco in my laboratory at Dana-Farber Cancer Institute has determined the primary amino acid sequence of this antibody (fig. 13). He finds that the complementarity determining regions in the heavy and light chains of this antibody differ only slightly in their framework and their complementarity determining regions from the germ line V region genes from which these rearranged genes were derived [24, 25].

This work shows that humans do have the immune repertoire necessary to make antibodies which mimic the region of CD4 which binds to the virus. Unfortunately, such antibodies are rarely elicited easily during natural infection.

The challenge for the future is clear. We must learn how to elicit broadly neutralizing antibodies of this type with high frequency. It may be difficult for viruses to evade neutralization by such antibodies. Viruses which no longer bind such antibodies may also be defective for recognition of CD4 and therefore defective for attachment to cells.

An additional challenge for vaccine development is to raise effective mucosal immunity. Secretion of IgA at mucosal surfaces can prevent some microorganisms from reaching the surface of the mucosa and from establishing infection. We are just beginning to learn how to elicit mucosal immunity. Exposure of antigens to M cells, the same cells which may contribute to rectal infection of HIV-1, also elicits IgA responses. It is our hope that by learning how to elicit appropriate IgA responses, we may coat the virus with antibodies that either inactivate them or prevent them from reaching the mucosal surface. The challenge of raising and maintaining such immune responses will require fundamental advances in the science of immunology.

AIDS Treatment: The Past and Future

In the absence of a vaccine, we must rely upon treatment to contain the ravages of this disease. At present, treatment for AIDS is palliative, not curative. Lethal complications of the disease are delayed by treatment but are not avoided.

The discovery that a virus, HIV-1, causes AIDS refocused attempts to treat the disease. Prior to the discovery of the viral origin of the disease, the only treatment available was treatment for the symptoms for the opportunistic infections and opportunistic cancers which arise in profoundly immunosuppressed patients.

In 1984 this focus changed. The virus itself became the preferred target for therapy. At the time antiviral therapy was initiated, the rational basis for such treatments had not been established. In the early days of treatment, it was not known that the continued decline in the $CD4^+$ T cell population, and subsequent decline in immune function could be attributed to continued virus replication.

Today we know that virus replication is a prerequisite to progress of the disease. We understand that to stop progression of the disease, it is necessary to stop virus replication.

The initial drug used, AZT, was rationally selected based on knowledge that infection of retroviruses requires conversion of RNA to DNA. AZT is a member of a broad family of nucleoside analogues, which when phosphorylated and incorporated into a growing DNA chain, terminate DNA synthesis. The chain-terminating nucleoside analogues are especially effective for retroviruses as the viral DNA polymerase cannot remove these analogues from DNA once they have been incorporated. The toxicity of such drugs for normal cells is relatively low as cellular DNA polymerases can remove such nucleoside analogues once incorporated in the DNA.

Studies in AZT patients show that the levels of circulating virus decrease during AZT treatment. Suppression of virus replication by AZT also slows the decline in $CD4^+$ T cell number. The drug does not generally reduce damage done to the immune system. Despite the successes of AZT and other members of this nucleoside analog family, such drugs by themselves are unlikely to cure the disease. HIV-1 infection is life-long. For this reason, treatments must be life-long. AZT and the other nucleoside analogues are too toxic to be used for many years. In fact, some people do not tolerate these drugs at all.

A second problem is the development of virus resistance to nucleoside analogues. Strains of virus resistant to AZT appear in treated patients within 1 or 2 years. The bright side of development of clinical resistance is proof of the effectiveness of AZT. Viruses do not develop resistance during treatment to drugs which are ineffective. The dark side of this picture is the limitation of AZT effectiveness.

At present, AZT is the only drug approved in most countries for treatment of the virus itself. It is expected that other nucleoside analogues will be approved for treatment of AIDS shortly.

What does the future hold for treatment of the disease? An explosion of knowledge regarding the details of the virus life cycle has spawned numerous opportunities for the development of new classes of viral therapies. Each necessary step in the virus life cycle provides an opportunity for therapeutic intervention. The current state of rational antiviral drug development is summarized briefly here.

Binding

Virus infection is initiated by binding of HIV-1 to CD4 on the surface of normal cells [26, 28]. Binding requires recognition of the CD4 protein by gp120. Peptides, polypeptides and peptide mimetic compounds which resemble CD4 block binding [29, 31]. Human CD4-like monoclonal antibodies also block binding. Although these drugs work very well to block replication of HIV in the test tube, as yet they do not slow progressive disease.

Lessons learned from studies with CD4 mimetic inhibitors may be useful for study of other drugs. We have learned that laboratory strains differ from natural isolates in their sensitivity to CD4-like drugs. These studies indicate that antiviral drugs should be tested on primary cells using a variety of natural isolates.

Fusion

The first step in virus entry involves fusion of virus and cell membranes [32]. Fusion is mediated by the envelope glycoproteins [33, 34]. Certain antibodies inhibit the membrane fusion reactions but do not inhibit CD4 binding [9, 35]. Drugs which mimic fusion inhibitors may limit infection. However, no such drugs have yet been described.

Penetration

The virus must penetrate the matrix of cellular proteins which underlie the plasma membrane. The means by which the virus accomplishes this task is not yet known. Studies of virus penetration deserve further exploration.

Conversion of RNA to DNA

Once the virus capsid enters the cytoplasm the viral RNA is converted to DNA by the concerted action of the viral DNA polymerase and virus ribonuclease H proteins [36, 38]. Nucleoside analogues inhibit replication [39–44]. As mentioned before, development of toxicity and development of resistance limit the efficacy of this class of drugs [45].

A new class of drugs, similar to previously described benzodiazapine and benzodiazepine-like drugs, inhibit HIV-1 DNA polymerase [45]. The site of inhibition is different from that at which the nucleoside analogues act. For this reason, these drugs act synergistically with AZT [45]. Currently, at least four different classes of drugs of this type are in clinical trials. These drugs have little known toxicity. However, these compounds do not inhibit the DNA polymerase activity of closely related viruses of HIV-2 and SIV-2 [45]. Moreover, resistance to these compounds can be produced in tissue culture. For these reasons it is likely that resistance will limit the effectiveness of these drugs.

The viral ribonuclease activity is critical to conversion of viral RNAs to DNA. The three-dimensional structure of this enzyme has recently been determined by X-ray crystallographic analysis [46]. No specific inhibitors of this enzyme have been described. Ribonuclease H appears to be an excellent candidate for future antiviral drug development.

Integration

The conversion of viral RNA to DNA occurs in the cytoplasm of the infected cell. The product of this reaction is a complex, called a preintegration complex, which contains the full-length linear form of viral DNA bound tightly to several integrase proteins [47–49]. The viral DNA must integrate into the host DNA for replication to proceed.

The integration reaction is mediated by the integrase protein which has multiple activities including binding to the ends of the viral DNA, precise excision of two nucleotides from each 3′ end of the viral DNA, introduction of a staggered 5′ nick into target DNA, and ligation of viral and host DNA termini [50–53]. Although no inhibitors of this complex enzyme have yet been described, integrase should be an excellent target for antiviral drug development.

Chris Farnet in my laboratory has recently developed in vitro reaction conditions which permit rapid and efficient integration of viral DNA present in the preintegration complex [47, 48]. Such in vitro reactions should facilitate the search for inhibitors of the integration.

Circularization

In the nucleus the viral DNA undergoes conversion to circular forms as well as integration into the host cell DNA. The circularization reactions compete with integration. At least three types of circles are formed, 1-LTR circles formed by homologous recombination of the redundant termini of the linear viral DNA, 2-LTR circles formed by simple end-to-end ligation of the linear viral DNA, and permuted circles containing 2-LTRs which are formed by autointegration of the virus termini into the viral genome itself (fig. 14) [36, 37].

Chris Farnet has also developed in vitro reaction conditions which permit formation of all three types of circles [Farnet and Haseltine, in preparation]. Nucleotides as well as nucleoside analogues accelerate formation of such circles in vitro and decrease the amount of viral DNA integrated into target DNA molecules. Compounds which increase the rate of circularization and decrease the rate of integration should be useful as antiviral drugs.

Transcription Initiation

Cellular factors are required for transcription initiation. The HIV-1 LTR contains sequences which bind transcription factors in resting as well as in activated T cells, macrophages and dendritic cells. In resting T cells these factors repress transcription initiation.

Yichen Lu in my laboratory in the Dana-Farber Cancer Institute has isolated a previously unidentified cellular factor, YC-1, which binds to a sequence required for silencing the transcription of HIV-1. We suspect that this novel cellular proteins, as well as other cellular proteins bind to the LTR and prevent transcription initiation in resting cells (fig. 15).

A series of cellular proteins have also been identified which bind to the enhancer of HIV-1 [54, 55]. The NF-κB is one such enhancer binding protein. This protein is not present in an active form in resting T cells.

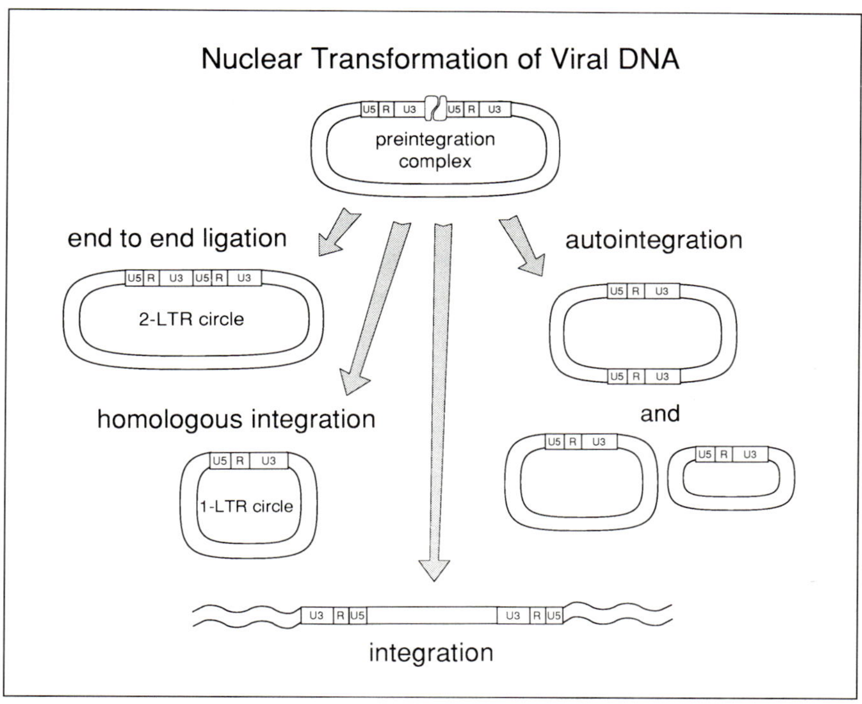

Fig. 14. Structures of the viral DNA forms found in an infected cell. The viral preintegration complex, formed in the cytoplasm of the cell shortly after infection, contains linear viral DNA and the viral integrase protein. The pre-integration complex moves to the nucleus, where several reactions can occur. Some of the linear viral DNA integrates into the host genome to form the provirus. In addition, 1-LTR circles, simple 2-LTR circles, and circular products of autointegration can be found among the unintegrated viral DNA forms in the nucleus.

Stimulation of the T cell activates NF-κB binding activity and increases the rate of transcription initiation.

Modulation of transcription factors provide an opportunity to regulate the rate of HIV-1 replication. Drugs which increase the activity of cellular proteins which repress HIV-1 promoter activity, as well as drugs which decrease the activity of cellular proteins which increase the rate of HIV-1 activity, should have antiviral properties. Such drugs will also modulate the immune response and may be useful for control of autoimmune and other diseases of immunological origin.

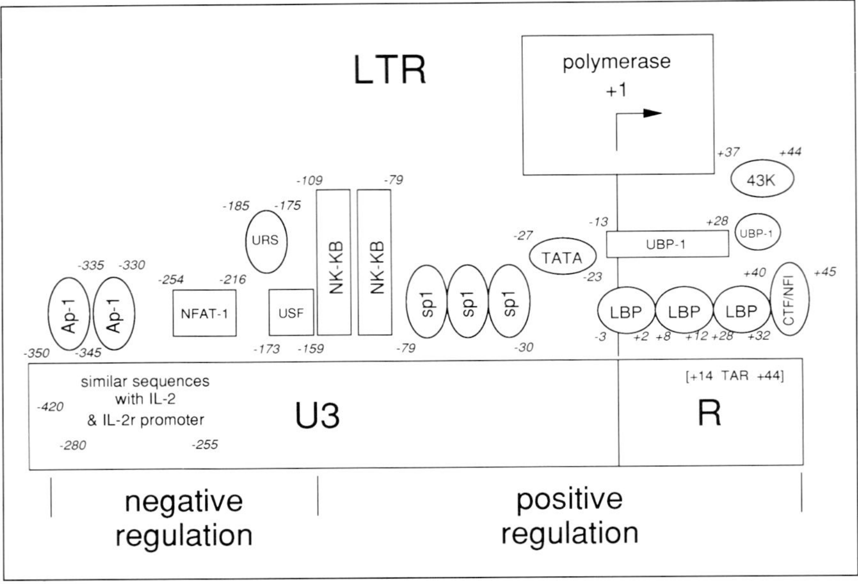

Fig. 15. The relative location and similarity of some of the sequences in the LTR to consensus sequences which are known to bind cellular proteins raise the possibility that these proteins also regulate HIV-1 proviral LTR directed RNA transcription. The location of the negative regulatory gene and the NRE is not shown.

Virus Gene vpr

The product of the virus gene *vpr* [56–58] is packaged within the virus particle itself. The *vpr* protein is made late in virus replication and is present in multiple copies in the virus particle [59, 60].

The *vpr* protein acts in *trans* to increase the rate of transcription of a variety of promoters including that of HIV-1 [57]. It is conceivable thqt *vpr* plays an important role in virus replication [61]. In the absence of the virus transactivator, the HIV-1 LTR is a very poor promoter. Very little full-length RNA accumulates. In the absence of full-length RNA, the *tat* protein cannot be made. Virus replication does not proceed without *tat.*

The *vpr* may act as a transactivator carried into the newly infected cells along with the virus capsid. The *vpr* protein may permit increased transcription of the newly integrated provirus. Its role may be to jump start the HIV-1 promoter, that, is to permit a sufficient level of the primary transcript to accumulate so that *tat* itself can be made.

If this is the case, inhibitors of *vpr* should slow replication of the virus. Anti-*vpr* inhibitors may have a powerful antiviral effect during natural transcription. Although an assay for the transacting affect of *vpr* has been described, no inhibitors have been described to date.

The nef *Protein*

The *nef* protein can be made from RNAs which accumulate both early and late in the life cycle. Early in the life cycle *nef* is made from an extensively spliced mRNA species. Late in infection *nef* may also be made from the *env* RNA. Some experiments suggest that *nef* slows replication [62–64]. Other experiments indicate that *nef* may be required for HIV-1 in primary cells [65, 66]. At present, it is premature to predict the effect on virus replication of anti-*nef* drugs.

The Viral Transactivator tat

The viral transactivator *tat* plays an important role in HIV-1 replication [67]. The *tat* protein binds to the 5′ end of the nascent RNA [68–71] via an RNA stem-loop structure called TAR [72]. Cellular proteins also bind to TAR. This RNA protein complex is essential for the synthesis of abundant amounts of full-length viral RNA (fig. 16).

A family of benzodiazepine analogues has been described which inhibits *tat* activity. These compounds inhibit replication of HIV-1 as well as the related viruses HIV-2 and SIV in culture. These benzodiazepine analogues have low toxicity. These compounds are different from the benzodiazepine analogues which inhibit viral DNA polymerase. The broad spectrum of activity of these compounds for a variety of transactivators indicates that resistance to these drugs may develop slowly.

The rev *Protein*

The *rev* protein is essential for synthesis of virus structural proteins [73, 74]. This protein binds to a sequence in the viral RNA called RRE. The *rev* protein binds to a complex secondary stem-loop structure within this sequence (fig. 17) [75–80].

The *rev* protein should be an excellent target for antiviral drug therapy. There is a quantitative relationship between the concentration of *rev* protein and the amount of virus structural proteins made [81]. The only inhibitor of this activity discussed to date is an antisense oligonucleotide in addition to the so-called trans-dominant negative mutants of the *rev* protein itself [82–84].

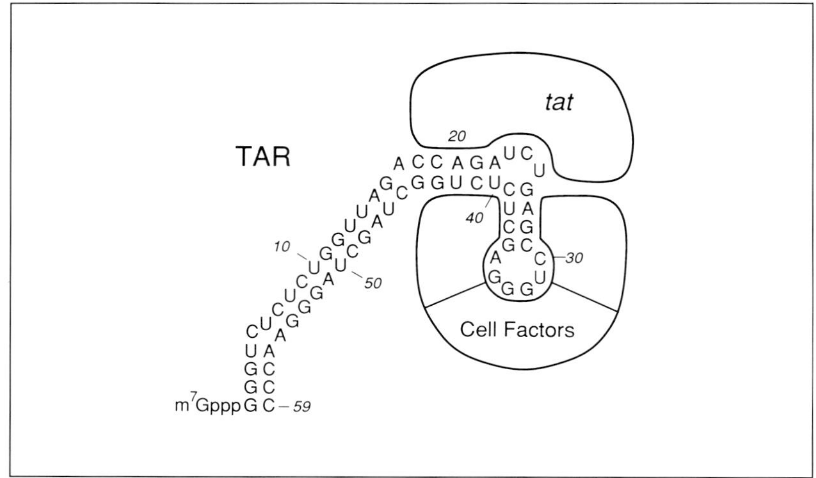

16

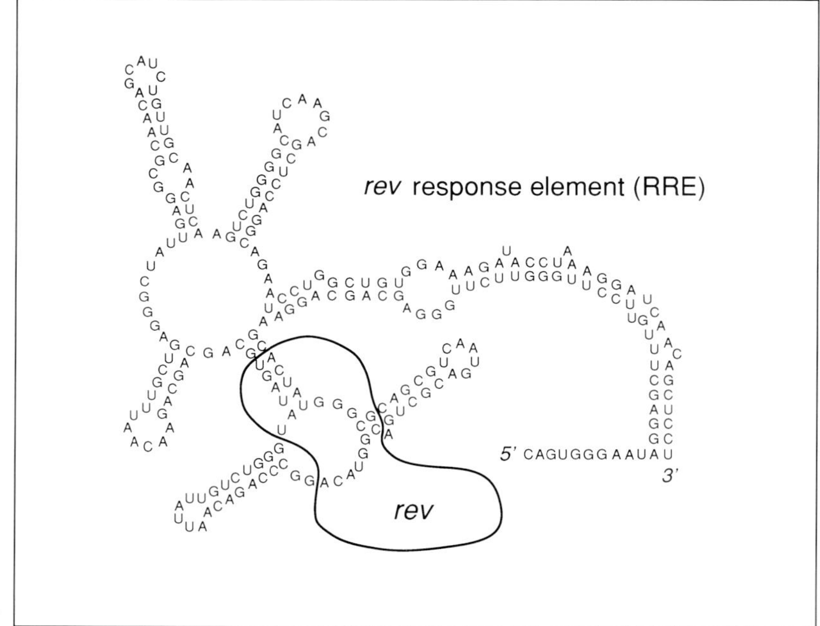

17

Fig. 16. Binding of *tat* and cellular proteins to TAR RNA. The primary and secondary structure of TAR RNA of the HIV-1 HXB2 strain is shown.

Fig. 17. Binding of the HIV-1 *rev* protein to RRE RNA. The PRE RNA of HIV-1 HXB2 strain is shown.

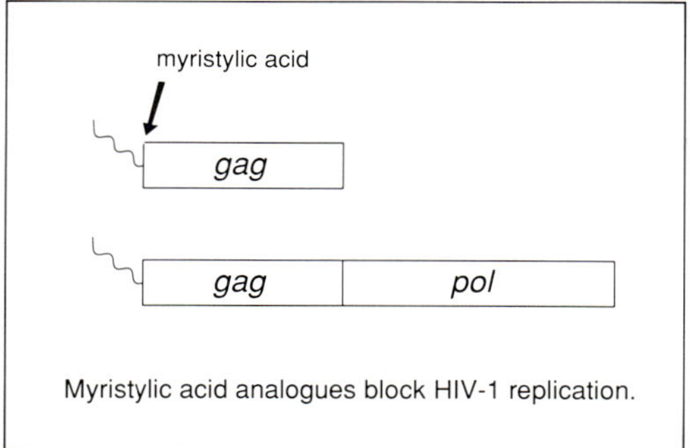

Fig. 18. Myristilic acid analogues which block HIV-1 replication are being analyzed for antiviral activity.

Myristylation

The HIV-1 capsid is assembled form two precursor proteins which are translated in a 20:1 ratio [85]. The abundant, shorter precursor specifies the structural proteins of this capsid [86], whereas the larger, less-abundant precursor contains precursor forms of the viral enzyme activities [87].

The initial stage of assembly requires association of the precursor protein with the inner surface of the plasma membrane. Association with the membrane requires that the amino terminus of the precursor be modified by posttranslational addition of a 14-carbon atom fatty acid, myristic acid [88–90]. In the absence of such modification, the capsid precursor remains in the cytoplasm and viruses are not produced. Myristic acid analogues which block myristylation of the virus capsid precursors are being evaluated as antiviral drugs [91, 92] (fig. 18).

RNA Encapsidation

The viral RNA is encapsidated into the virus particle as it assembles. Capture of the viral RNA requires specific recognition and binding of viral RNA by the capsid precursor [93, 94]. The virus RNA contains a sequence near the 5′ terminus required for encapsidation [94–96]. This sequence is called the packaging site, or Psi site. Deletion of all or part of the Psi sequence markedly reduces RNA encapsidation [94–96].

A region near the carboxy terminus of the capsid precursor is believed to interact with the Psi sequence. This sequence contains two zinc finger-like structures. Heinrich Gottlinger in my laboratory at Dana-Farber Cancer Institute has shown that mutants which disrupt the zinc finger severely impair the ability of the virus to encapsidate viral RNA [in preparation].

A specific interaction between the Psi site and the nucleocapsid protein p15 which is derived from the carboxy terminus of the capsid precursor has been demonstrated in vitro [97]. Drugs which inhibit this interaction should inhibit virus replication.

vif *and* vpu

Two viral proteins made late in infection, *vif,* and *vpu,* affect late stages in the virus assembly. The *vpu* protein increases the rate of assembly and release of HIV-1 capsid and envelope proteins [98–100]. The *vif* protein increases the ability of the released particles to infect some cell types (ET-54,55) [101].

Both of these proteins are reported to affect *env* protein processing. *vpu* is reported to dissociate an *env* CD4 complex which may form in the Golgi [Strebel, personal commun.]. *vif* is reported to trim the carboxy terminus of the transmembrane protein gp41 [102]. Further knowledge of the means by which these proteins act is required for rational development of antiviral drugs.

Virus particle maturation depends upon the activity of a virus specified protease [88, 103, 104]. The protease is made and incorporated into the assembling virus particle as part of the precursor of the replicative enzymes. The protease is believed to be activated upon completion of particle assembly. The viral protease cleaves both the capsid precursor and the replicative enzyme precursor [105, 106]. In the absence of the protease activity, immature particles are released from the infected cells [88, 104]. The immature particles are not infectious (fig. 19).

A family of protease inhibitors has been derived. Many of these protease inhibitors have been designed based on the known three-dimensional structure of the protease, determined by X-ray crystallographic analysis [107–109]. Several second-generation protease inhibitors have also been designed as a result of an understanding of the interactions of protease inhibitors which is based on the known three-dimensional structure of inhibitors. The current generation of protease inhibitors are peptide analogues of the substrate [110–115]. These inhibitors contain tetrahedral molecules, usually carbon atoms, at the active site. At present most of these inhibitors do not have favorable pharmacological properties. Most of these are insoluble and difficult to deliver orally.

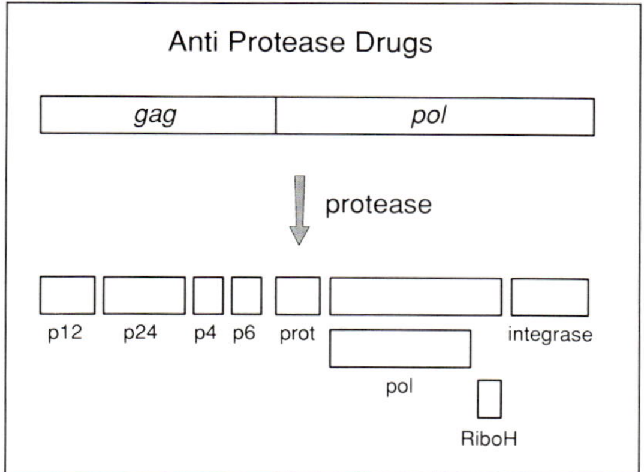

Fig. 19. Virus assembly requires protease activity. Antiprotease drugs are being analyzed to inhibit the formation of infectious particles.

Release

The final stage in viral particle maturation is release of the particle from the exterior of the cell membrane. Release of the particles may be a rate-limiting step for infeciton of many cell types. After budding, many particles remain attached to the surface of the infected cell via a tether. Mutations in the carboxy terminus of the capsid precursor protein affect release of the particles from the surface of the cell [115]. Deletion of these sequences severely impairs particle release. It is likely that such mutations subtly alter the configuration of the assembled capsid and preclude final closure of the particle, a prerequisite for release. These observations indicate that compounds which interfere with particle closure and release may also inhibit replication.

Gene Therapy

Gene therapy offers an exciting new prospect for therapy of AIDS. The principle of gene therapy for AIDS is to render the population of cells which are normally susceptible to HIV-1 resistant to infection or to virus replication, by insertion of antiviral genes.

One approach which Joseph Sodroski, Mark Poznansky, and I have developed is to use HIV-1 itself as a tool for gene therapy. The rationale for this approach is that the modified HIV-1 will seek out appropriate target cells in the infected person.

Packaging Vectors
capsid
gag
LTR
pol
Δ env
env
LTR
Δ psi
env
LTR
env
LTR
empty virus particle

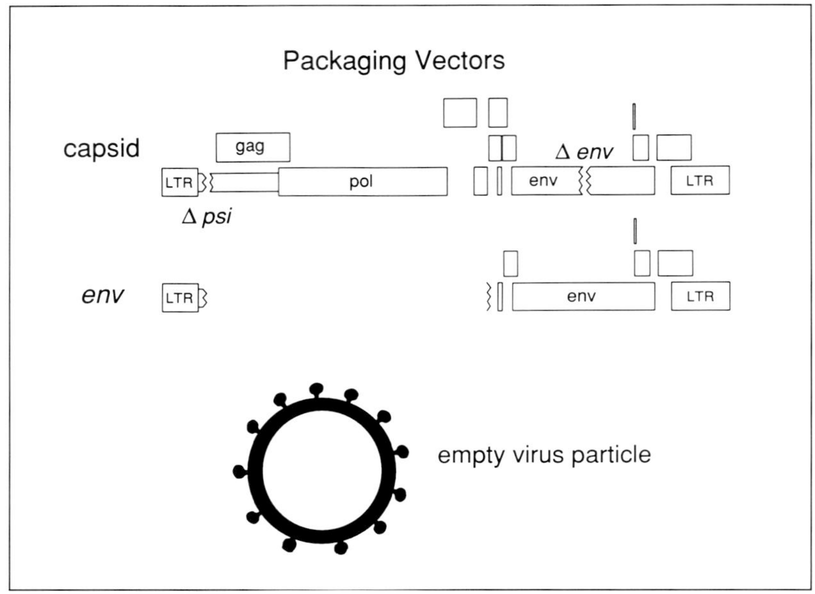

20

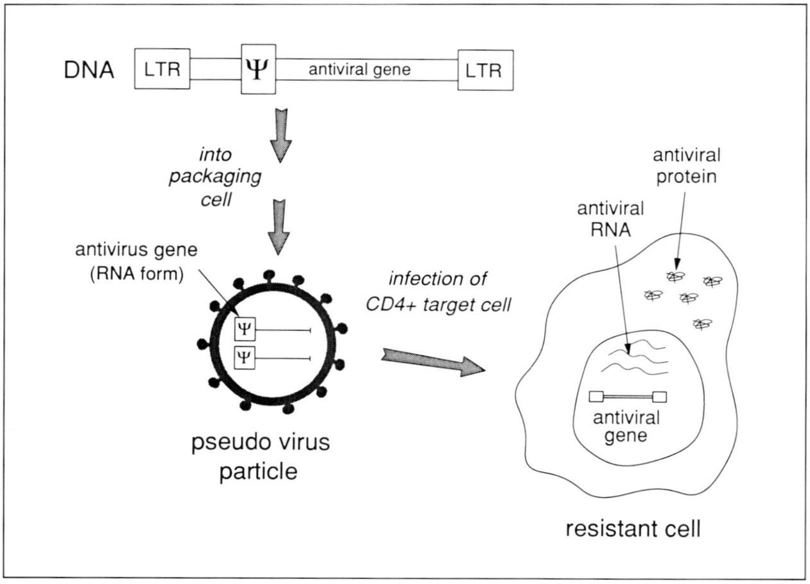

DNA
LTR
Ψ
antiviral gene
LTR
into
packaging
cell
antivirus gene
(RNA form)
Ψ
Ψ
infection of
CD4+ target cell
antiviral
protein
antiviral
RNA
antiviral
gene
pseudo virus
particle
resistant cell

21

Fig. 20. Diagram of HIV-1 vector designed for gene transfer.
Fig. 21. Mechanism for using the HIV-1 vector for antiviral gene therapy.

The strategy we follow is relatively straightforward. An empty carrier particle is made from two expressor plasmids. One of the expressor plasmids specifies the capsid and replicative enzymes as well as regulatory genes of HIV-1. The second plasmid expresses the envelope glycoprotein and *rev* genes. The RNA which specifies the capsid and envelope genes is devoid of Psi sequences. The virus particles made from these plasmids alone are devoid of viral RNA (fig. 20) [95].

The antiviral gene of choice is used to replace all HIV-1 coding sequences on a third plasmid. This gene also contains an intact Psi site. Co-expression of these three plasmids results in production of HIV-1 capsids into which the antiviral gene alone is incorporated in an RNA form. Upon infection, this gene is converted to DNA and inserted into the host genome (fig. 21).

Mark Poznansky at the Dana-Farber Cancer Institute has successfully demonstrated that a minimal HIV-1 vector can be used to transfer genes into HIV-1 target cells [117]. He has also demonstrated that an antisense HIV-1 gene inserted into susceptible target cells using HIV-1 as a vector greatly reduces the sensitivity of this population to infection by HIV-1 itself. A variety of antiviral genes can be inserted into this vector including transdominant structural and regulatory proteins, antisense genes, and antiviral ribozymes.

Gene therapy, although in its infancy, holds promise as an exciting new approach for HIV-1 treatment.

References

1 Allen JS, Colligan JE, Darin F, McLane MF, Sodroski JG, Rosen CA, Haseltine WA, Lee TH, Essex M: Major glycoprotein antigens that induce antibodies in AIDS patients are encoded by HTLV-III. Science 1985;228:1091–1094.

2 Robey WG, Safai B, Oroszian S, Arthur LO, Gonda MA, Gallo RC, Fishinger PJ: Characterization of envelope and core structural gene products of HTLV-III with sera from AIDS patients. Science 1985;225:593–595.

3 Lyerly H, Matthews T, Langlois A., Bolognesi D, Weinhold K: Human T-cell lymphotropic virus III-B glycoprotein (gp120) bound to CD4 determinants on normal lymphocytes and expressed by infected cells serves as target for immune attack. Proc Natl Acad Sci USA 1987;84:4601–4605.

4 Lyerly H, Reed D, Matthews T, Langlois A, Ahearne P, Petteway S, Weinhold K: Anti-gp120 antibodies from HIV of seropositive individuals mediate broadly reactive anti-HIV ADCC. AIDS Res Hum Retrovirus 1987;3:409–422.

5 Matthews TJ, Weinhold KJ, Lyerly HK, Langlois AJ, Wigzell H, Bolognesi D: Inter-

action between the human T-cell lymphotropic virus type IIIB envelope glycoprotein gp120 and the surface antigen CD4: Role of carbohydrate in binding and cell fusion. Proc Natl Acad Sci USA 1987;84:5424–5428.

6 Kozarsky K, Penman M, Basiripour L, Haseltine W, Sodroski J, Krieger M: Glycosylation and processing of the human immunodeficiency virus type 1 envelope protein. Acquir Immune Defic Syndr 1989;2:163–169.

7 Skinner M, Langlois AJ, McDanal C, McDougal J, Bolognesi D, Matthews J: Neutralizing antibodies to an immunodominant envelope sequence do not prevent gp120 binding to CD4. J Virol 1988;62:4195–4200.

8 Linsley P, Ledbetter J, Kinney-Thomas E, Hu S: Effects of anti-gp120 monoclonal antibodies on CD4 receptor binding by the *env* protein of human immunodeficiency virus type 1. J Virol 1988;62:3695–3702.

9 Freed E, Myers D, Risser R: Identification of the principal neutralizing determinant of human immunodeficiency virus type 1 as a fusion domain. J Virol 1991;65: 190–194.

10 Ratner L, Haseltine W, Patarca R, Livak KJ, Starcich B, Josephs SF, Doran ER, Rafalski A, Whitehorn, EA, Baumeister K, Ivanoff L, Petteway SR Jr, Pearson ML, Lautenberger JA, Papas TS, Ghrayeb J, Chang NT, Gallo RC, Wong-Staal F: Complete nucleotide sequence of the AIDS virus, HTLV-III. Nature 1985;313:277–284.

11 Muesing MA, Smith DH, Cabradilla CD, Benton CV, Lasky LA, Capon DJ: Nucleic acid structure and expression of the human AIDS/lymphadenopathy retrovirus. Nature 1985;313:452–458.

12 Wain-Hobson S, Sonigo P, Danos O, Cole S, Alizon M: Nucleotide sequence of the AIDS virus, LAV. Cell 1985;40:9–17.

13 Starcich BR, Hahn B, Shaw G, McNeely P, Modrow S, Wolf A, Parks E, Parks W, Josephs S, Gallo R, Wong-Staal F: Identification and characterization of conserved and variable regions in the envelope gene of HTLV III/LAV, the retrovirus of AIDS. Cell 1986;45:637–648.

14 Patterson S, Knight SC: Susceptibility of human peripheral blood dendritic cells to infection by human immunodeficiency virus. J Gen Virol 1987;68:1177–1181.

15 Schuler G, Steinman RM: Murine epidermal Langerhans cells mature into potent immunostimulatory dendritic cells in vitro. J Exp Med 1985;161:526–546.

16 Austyn JM: Lymphoid dendritic cells. Immunology 1987;62:161–170.

17 Austyn JM, Kupleic-Weglinski JW, Hankins DF, Morris PJ: Migration patterns of dendritic cells in the mouse. Homing to T cell-dependent areas of spleen, and binding with marginal zone. J Exp Med 1988;167:646–651.

18 Weltzin R, Lucia-Jandris P, Michetti P, Fields BN, et al: Binding and transepithelial transport of immunoglobulins by intestinal M cells: Demonstration using monoclonal IgA antibodies against viral proteins. J Cell Biol 1989;108:1673–1685.

19 Wolf JL, Rubin DH, Finberg R, Kauffman RS, et al: Intestinal M cells: A pathway for entry of reovirus into the host. Science 1981;212:471–472.

20 Sharpe AH, Fields BH: Pathways of viral infections. Basic concepts derived from the reovirus model. N Engl J Med 1985;312:486–497.

21 Kowalski M, Potz J, Basiripour L, Dorfman T, Goh WC, Terwilliger E, Dayton A, Rosen C, Haseltine W, Sodroski J: Functional regions of the envelope glycoprotein of human immunodeficiency virus type 1. Science 1987;237:1351–1355.

22 Lasky LA, Nakamura G, Smith DH, Fennie C, Shim-saki C, Patzer E, Berman P, Gregory T, Capon DJ: Delineation of a region of the human immunodeficiency virus type 1 gp120 glycoprotein critical for interaction with the CD4 receptor. Cell 1987;50:975–985.

23 Olshevsky UE, Helseth E, Furman C, Li J, Haseltine WA, Sodroski J: Identification of individual HIV-1 gp120 amino acids important for CD4 receptor binding. J Virol 1990;64:5701–5707.

24 Lee, KH, Matsuda F, Kodaira M, Honjo T: A novel family of variable region genes of the human immunoglobulin heavy chain. J Mol Biol 1987;195:761–768.

25 Radaux V, Chin PP, Loige JA, Carson DA: A conserved human germline Vk gene directly encodes rheumatoid factor light chains. J Eng Med 1986;164:2119–2124.

26 Dalgleish AG, Beverley PC, Clapham PR, Crawford DH, Greaves MF, Weiss RA: The CD4 (T4) antigen is an essential component of the receptor for the AIDS retrovirus. Nature 1984;312:763–767.

27 Klatzmann D, Champagne E, Chamaret S, Gruest J, Gutetard D, Hercend T, Gluckman JD, Montagnier L: T-lymphocytes T4 molecule behaves as the receptor for human retrovirus LAV. Nature 1984;312:767–768.

28 McDougal JS, Mawle A, Cort SP, Nicholson KA, Cross GD, Scheppler-Campbell JA, Hicks D, Sligh J: Cellular tropism of the human retrovirus HTLV III/LAV. I. Role of T cell activation and expression of the T4 antigen. J Immunol 1985;135: 3151–3162.

29 Nara PL, Hwang KM, Rausch DM, Lifson JD, Eiden LE: CD4 antigen-based antireceptor peptides inhibit infectivity of human immunodeficiency virus in vitro at multiple stages of virus life cycle. Proc Natl Acad Sci USA 1989;86:7139–7143.

30 Capon DJ, Chamow SM, Mordenti J, Marsters SA, Gregory T, Mitsuya H, Byrn RA, Lucas C, Wurm FM, Groopman JE, Broder S, Smith DH: Designing CD4 immunoadhesins for AIDS therapy. Nature 1989;337:525–531.

31 Lifson J, Hwang KM, Nara P, Fraser B, Padgett M, Dunlop, Eiden L; Synthetic CD4 peptide derivations that inhibit HIV infection and cytopathicity. Science 1988;241: 712–714.

32 Stein B, Gonda S, Lifson J, Penhallow R, Bensch K, Engleman E: pH-independent HIV entry into CD4-positive T cells via virus envelope fusion to the plasma membrane. Cell 1987;49:659–668.

33 Lifson JD, Feinberg MB, Reyes GR, Rabin L, Banapour B, Chakrabarti S, Moss B, Wong-Staal F, Steimer KS, Engleman EG: Nature 1986;323:725–728.

34 Sodroski JG, Goh WC, Rosen C, Campbell K, Haseltine WA: Role of the HTLV-III/LAV envelope in syncytium formation and cytopathicity. Nature 1986;322:470–474.

35 Takeuchi Y, Akutsu M, Murayama K, Shimizu N, Hoshino H: Host range mutant of human immunodeficiency virus type 1: Modification of cell tropism by a single point mutation at the neutralization epitope in the env gene. J Viol 1991;65:1710–1718.

36 Varmus H. Retroviruses. Science 1988;240:1427–1433.

37 Varmus HE, Swanstrom R: Replication of retroviruses; in Weiss R, Teich N, Carmus H, Coffin J (eds): RNA Tumor Viruses. Cold Spring Harbor, Cold Spring Harbor Laboratory, 1982;pp 369–512.

38 Varmus HE, Swanstrom R: Replication of retroviruses; in Weiss R, Teich N, Var-

mus H, Coffin J (eds): RNA Tumor Viruses, ed 2. Cold Spring Harbor, Cold Spring Harbor Laboratory, 1985; pp74–134.

39 Furman PA, Barry DW: Spectrum of antiviral activity and mechanism of action of zidovudine. Am J Med 1988;85(suppl 2A):176–181.

40 Mitsuya H, Weinhold KJ, Furman PA, et al: 3′-Azido-3′deoxythimidine (BW A509U): An antiviral agent that inhibits the infectivity and cytopathic effect of human T-lymphotropic virus type III lymphadenopathy-associated virus in vitro. Proc. Natl Acad Sci USA 1985;82:7096–7100.

41 Mitsuya H, Broder S: Inhibition of in vitro infectivity and cytopathic effect of human T-lymphotropic virus type III/lymphadenopathy associated virus (HTLV III/LAV) by 2′-3′-dideoxynucleosides. Proc Natl Acad Sci USA 1986;83:1911–1915.

42 Ahluwalin G, Conney DA, Mitsuya H, et al: Initial studies in the cellular pharmacology of 2′,3′-dideoxyinosine, an inhibitor of HIV infectivity. Biochem Pharmacol 1987;36:3797–3800.

43 Balzarini J, Kang GJ, Dalal M, Herdewlin P, DeClerq E, Broder S. The anti-HTLV III (anti-HIV) and cytotoxic activity of 2′,3′-didehydro-2′,3′-dideoxyribonucleosides. Mol Pharmacol 1987;32:162–167.

44 Lin TS, Schinazi RF, Prusoff WM: Potent and selective in vitro activity of 3′-deoxythymidine-2′-ene (3′-deoxy-2′-3′-didehydrothymidine) against human immunodeficiency virus. Biochem Pharmacol 1987;36:2713–2718.

45 Larder BA, Darby G, Richman DD: HIV with reduced sensitivity to zidovudine (AZT) isolated during prolonged therapy. Science 1989;234:1731–1734.

46 Davies JF, Hostomska Z, Hostomsky Z, Jordan SR, Matthews DA: Crystal structure of the ribonuclease H domain of HIV-1 reverse transcriptase. Science 1991;252: 88–95.

47 Farnet CM, Haseltine WA: Integration of human immunodenficiency virus type 1 DNA in vitro. Proc Natl Acad Sci USA 1990;87:4164–4168.

48 Farnet CM, Haseltine WA: Determination of the viral proteins present in the HIV-1 preintegration complex. J Virol 1991;65:1910–1915.

49 Bowerman B, Brown PO, Bishop JM, Varmus HE: A nucleoprotein complex mediates the integration of retroviral DNA. Genes Dev 1989;3:469–479.

50 Bushman FD, Fujiwara T, Craigie R: Retroviral DNA integration directed by HIV integration protein in vitro. Science 1990;249:1555–1558.

51 Craigie R, Fujiwara T, Bushman F: The IN protein of Moloney murine leukemia virus processes the viral DNA ends and accomplishes their integration in vitro. Cell 1990;62:829–837.

52 Katz RA, Merkel G, Kulkosy J, Leis J, Skalka AM: The avian retroviral IN protein is both necessary and sufficient for integrative recombination in vitro. Cell 1990;63: 87–95.

53 Bushman FD, Craigie R: Activities of human immunodeficiency virus (HIV) integration protein in vitro: Specific cleavage and integration of HIV DNA. Proc Natl Acad Sci USA 1991;88:1339–1343.

54 Nabel G, Baltimore D. An inducible transcription factor activates expression of human immunodeficiency virus in T cells, Nature 1987;326:711–713.

55 Bohnlein B, Siekevitz M, Ballard DW, Lowenthal JW, Rimsky L, Bogerd H, Hoffman J, Wano Y, Franza BR, Greene WC: Stimulation of the human immunodeficiency virus type 1 enhancer by the human T-cell leukemia virus type I *tax* gene

product involves the action of inducible cellular proteins. J Virol 1989;63:1578–1586.

56 Wong-Staal F, Chanda PK, Ghrayeb J: Human immunodeficiency virus: The eighth gene. AIDS Res Hum Retroviruses 1987;3:33–39.

57 Cohen EA, Terwilliger EF, Jalinoos Y, Provix J, Sodroski JG, Haseltine WA: Identification of HIV-1 vpr product and function. J Acquir Immune Defic Syndr 1990;3:11–18.

58 Ogawa K, Shibata R, Kiyomasum T, et al: Mutational analysis of the human immunodeficiency virus vpr open reading frame. J Virol 1989;63:4110–4114.

59 Cohen EA, Fehni G, Sodriski JG, Haseltine WA: J Virol 1990;64:3097–3099.

60 Yu XF, Matsuda M, Essex M, Lee TH: Open reading frame vpr of simian immunodeficiency virus encodes a virion-associated protein. J Cirol 1990;64:5688–5693.

61 Hattori N, Michaels F, Fargnoli K et al: The human immunodeficiency virus type 2 vpr gene is essential for productive infection of human macrophages. Proc Natl Acad Sci USA 1990;87:8080–8084.

62 Terwilliger E, Sodroski JG, Rosen CA, Haseltine WA: Effects of mutations within the 3′ orf region of HTLV-III/LAV on replication and cytopathogenicity. J Virol 1986;60:754–760.

63 Luciw PA, Cheng-Mayer C, Levy JA: Mutational analysis of the human immunodeficiency virus: The orf-B region down-regulates virus replication. Proc Natl Acad Sci USA 1987;84:1434–1438.

64 Cheng-Mayer C, Lannello P, Shaw K, Luciw PA, Levy JA: Differential effects of nef on HIV replication: Implications for viral pathogenesis in the host. Science 1989;246:1629–1632.

65 Terwilliger E, Langhoff E, Gabuzda D, Haseltine WA: Variable effects of nef upon HIV-1 replication in vitro. In press.

66 Kestler H, Ringler D, Mori K, Panscall D, Sehgal P, Daniel M, Desrosier R: Cell 1991;65:651–662.

67 Sodroski J, Patarca R, Rosen C, Wong-Staal F, Haseltine W: Location of the *trans*-acting region of the genome of the human T-cell lymphotropic virus type III. Science 1985;229:74–77.

68 Müller WEB, Okamoto T, Reuter P, Ugardovie D, Schroder HC: Functional characterization of tat protein from human immunodeficiency virus. J Biol Chem 1990;265:3803–3808.

69 DIngwall C, Ernberg I, Gait MJ, Green SM, Heaphy S, Karn J, Lowe AD, Singh M, Skinner MA. HIV-1 *tat* protein stimulates transcription by binding to a U-rich bulge in the stem of the TAR RNA structure. EMBO J 1990;9:4145–4153.

70 Weeks KM, Ampe C, Schultz SC, Steitz TA, Crothers DA: Fragments of the HIV-1 tat protein specifically bind TAR RNA. Science 1990;249:1281.

71 Roy S, Delling U, Chen C-H, Rosen CA, Sonenberg N: A bulge structure in HIV-1 TAR RNA is required for that binding and tat-mediated trans-activation. Genes Dev 1990;4:1365–1373.

72 Rosen CR, Sodroski JG, Haseltine WA: The location of *cis*-acting regulatory sequences in human T lymphotropic virus type III (HTLV-III/LAV) long terminal repeat. Cell 1985;41:813–823.

73 Terwilliger E, Burghoff R, Sia R, Sodroski JG, Haseltine W, Rosen C: The *art* gene

product of the human immunodeficiency virus is required for replication. J Virol 1988;62:655–658.

74 Sadaie MR, Bentner T, Wong-Staal F: Site-directed mutagenesis of two transregulatory genes (*tat*-III,*trs*) of HIV-1. Science 1988;293:210–213.

75 Zapp ML, Green MR: Sequence-specific RNA binding by the HIV-1 *rev* protein. Nature 1989;342:714–716.

76 Daly TJ, Rusche JR, Malone TE, Frankel AD: Circular dichroism studies of the HIV-1 *rev* protein and its specific RNA binding site. Biochemistry 1990;129:9791–9795.

77 Olsen HE, Cochrane AW, Dillon PJ, Nalin CM, Rosen CA: Interaction of the human immunodeficiency virus type 1 *rev* protein with structured region in *env* mRNA is dependent upon multimer formation mediated through a basic stretch of amino acids. Genes Dev 1990;1357–1365.

78 Dayton ET, Powell DM, Dayton AI: Functional analysis of CAR, the target sequence for *rev* protein of HIV-1. Science 1989;246:1625–1629.

79 Malim MH, Tiley LS, McCarn DF, Rusche JR, Hauber J, Cullen BR: HIV-1 structural gene expression requires binding of the *rev trans*-activator to its RNA target sequence. Cell 1990;60:675–683.

80 Heaphy S, Dingwall C, Ernberg I, Gait MJ, Green SM, Karn J, Lowea D, Singh M, Skinner MA: HIV-1 regulator of viron expression (*rev*) protein binds to an RNA stem-loop structure located within the *rev*-responsive element region. Cell 1990;60: 685–693.

81 Pomerantz RJ, Trono D, Feinberg MB, Baltimore D: Cells nonproductively infected with HIV-1 exhibit an aberrant pattern of viral RNA expression: A molecular model for latency. Cell 1990;61:1271–1276.

82 Venkatesh LK, Chinnadurai G: Mutants in a conserved region near the carboxy-terminus of HIV-1 *rev* identify functionally important residues and exhibit a dominant negative phenotype. Virology 1990;178:327–330.

83 Mermer B, Felber BK, Campbell M, Pavlakis GN: Identification of *trans*-dominant HIV-1 *rev* protein mutants by direct transfer of bacterially produced proteins into human cells. Nucl Acids Res 1990;18:2037–2044.

84 Malim MH, Bohnlein S, Hauber J, Cullen BR: Functional dissection of the HIV-1 *rev trans*-activator derivation of the *trans*-dominant repressor of *rev* function. Cell 1989;58:205–214.

85 Wilson W, Braddock M, Adams SE, Rathjen PD, Kingsman SM; Kingsman AJ: HIV expression strategies: Ribosomal frameshifting is directed by a short sequence in both mammalian and yeast systems. Cell 1988;55:1159–1169.

86 Mervis RJ, Ahmad N, Lillehoj EP, Raum MO, Salazar FHR, Chan HW, Venkatesan S: The *gag* gene products of human immunodeficiency virus type 1: Alignment within the *gag* open reading frame, identification of posttranslational modifications, and evidence for alternative *gag* precursors. J Virol 1988;62:3993–4102.

87 Peng C, Chang NT, Chang TW: Identidication and characterization of human immunodeficiency virus type 1 *gag-pol* fusion protein in transfected mammalian cells. J Virol 1991;65:2751–2756.

88 Gottlinger HG, Sodroski JG, Haseltine WA: Role of capsid precursor processing and myristoylation in morphogenesis and infectivity of human immunodeficiency virus type 1. Proc Natl Acad Sci. USA 1989;86:3781–3785.

89 Bryant M, Ratner L: Myristoylation-dependent replication and assembly of human immunodeficiency virus 1. Proc Natl Acad Sci USA 1990;87:523–527.

90 Pal R, Reitz MS Jr, Tschachler E, Gallo RC, Sarngadharan MG, Veronese FDM: Myristoylation of *gag* proteins of HIV-1 plays an important role in virus assembly. AIDS Res Hum Retroviruses 1990;6:721.

91 Bryant ML, Heuckeroth RO, Kimata JT, Ratner L, Gordon JI: Replication of human immunodeficiency virus 1 and Moloney murine leukemia virus is inhibited by different heteroatom-containing analogs of myristic acid. Proc Natl Acad Sci USA 1989;86:8655–8659.

92 Bryant ML, Ratner L, Duronio RJ, Kishore NS, Devadas B, Adams SP, Gordon JI: Incorporation of 12-methoxydodecanoate into the human immunodeficiency virus 1 *gag* polyprotein precursor inhibits its proteolytic processing ad virus production in a chronically infected human lymphoid cell line. Proc Natl Acad Sci USA 1991;88: 2055–2059.

93 Gorelick RJ, Nigida SM Jr, Bess JW Jr, Arthur LO, Henderson LE, Rein A: Noninfectious human immunodeficiency virus type 1 mutants deficient in genomic RNA. J Virol 1990;64:3207–3211.

94 Aldovini A, Young RA: Mutations of RNA and protein sequences involved in human immunodeficiency virus type 1 packaging result in production of noninfectious virus. J Virol 1990;64:1920–1926.

95 Lever A, Gottlinger H, Haseltine W, Sodroski J: Identification of a sequence required for efficient packaging of human immunodeficiency virus type 1 RNA into virions. J Virol 1989;63:4085–4087.

96 Clavel F, Orenstein JM: A mutant of human immunodeficiency virus with reduced RNA packaging and abnormal particle morphology. J Virol 1990;64:5230–5234.

97 Darlix J-L, Gabus C, Nugeyre M-T, Clavel F, Barre-Sinoussi F: *Cis* elements and *trans*-acting factors involved in the RNA dimerization of the human immunodeficiency virus HIV-1. J Mol Biol 1990;216:689–699.

98 Strebel K, Klimkait T, Martin MA: A novel gene of HIV-1, *vpu*, and its 16-kilodalton product. Science 1988;241:1211–1223.

99 Terwilliger, EF, Cohen EA, Lu Y, Sodroski JG, Haseltine WA: Functional role of the human immunodeficiency virus type 1 *vpu*. Proc Natl Acad Sci USA 1989;86:5163–5167.

100 Klimkait T, Strebel K, Hoggen MD, Martin MA, Orenstein JM: Human immunodeficiency virus type 1-specific protein *vpu* is required for efficient virus maturation and release. J Virol 1990;84:621–629.

101 Strebel K, Daugherty D, Clouse K, Cohen D, Folks T, Martin MA: The HIV ‘A’ (sor) gene product is essential for virus infectivity. Nature 1987;328:728–730.

102 Guy B, Geist M, Dott K, Spehner D, Kieny M-P, Lecocq J-P: A specific inhibitor of cysteine proteases impaires a vis-dependent modification of human immunodeficiency virus type 1 env protein. J Virol 1991;65:1325–1331.

103 Kohl NE, Emini EA, Schleif WA, Davis LJ, Heimbach JC, Dixon RAF, Scolnick EM, Sigal IS: Active human immunodeficiency virus protease is required for viral infectivity. Proc Natl Acad Sci USA 1988;85:4686–4690.

104 Peng C, Ho BK, Chang TW, Chang NT: Role of human immunodeficiency virus type 1-specific protease in core protein maturation and viral infectivity. J Virol 1989;63: 2550–2556.

105 Debouck C, Gorniak JG, Strickler JE, Meek TD, Metcalf BW, Rosenberg M: Human immunodefidiency virus protease expressed in *Escherichia coli* exhibits autoprocessing and specific maturation of the gag precursor. Proc Natl Acad Sci USA 1987;84: 8903–8906.

106 Mous J, Heimer EP, Le Grice SFJ: Processing protease and reverse transciptase from human immunodeficiency virus type 1 polyprotein in *Escherichia coli.* J Virol 1988; 62:1433–1436.

107 Navia MA, Fitzgerald PMD, McKeever MB, Leu C-T, Heimbach JC, Herber WK, Sigal IS, Darke PL, Springer JP: Three-dimensional structure of aspartyl protease from human immunodeficiency virus HIV-1. Nature 1989;337:615–620.

108 Wlodawer A, Miller M, Jaskolski M, Sathyanarayana BK, Baldwin E, Weber IT, Selk LM, Clawson L, Sohneider J, Kent SBH: Conserved folding in retroviral proteases: Crystal structure of a synthetic HIV-1 protease. Science 1989;245:616–621.

109 Lapatto R, Blundell T, Hemmings A, Overington J, Widerspin A, Wood S, Merson JR, Whittle PJ, Danley DE, Geoghegant KF, Hawrylik SJ, Lee SE, Scheld KG, Hobart PM: X-ray analysis of HIV-1 proteinase at 2.7A resolution confirms structural homology among retroviral enzymes. Nature 1989;342:299–302.

110 Miller M, Schjneider J, Sathyanarayana BK, Toth MV, Marshall GR, Clawson L, Selk L, Kent SBH, Wlodawer A: Structure of complex of synthetic HIV-1 protease with a substrate-based inhibitor at 2.3 Å resolution. Science 1989;246:1149–1152.

111 Erickson J, Neidhart DJ, Van Drie J, Kempf DJ, Wang XC, Norbeck DW, Plantner JJ, Rittenhouse JW, Turon M, Wideburg N, Kohlbrenner WE, Simmer R, Helfrich R, Paul DA, Knigge M: Design, activity and 2.8Å crystal structure of a C_2 symmetric inhibitor complexes to HIV-1 protease. Science 1990;249:527–533.

112 Dreyer GB, Metcalf BW, Tomaszek TA, Carr TJ, Chandler AC, Hyland L, Fakhoury SA, Magaard VW, Moore ML, Strickler JE, Debouck C, Meek TD: Inhibiton of human immunodeficiency virus 1 protease in vitro: rational design of substrate analogue inhibitors. Proc Natl Acad Sci USA 1989;86:9752–9756.

113 Robert NA, Martin JA, Kinchington D, Broadhurst AV, Craig JC, Duncan IB, Galpin SA, Handa BK, Kay J, Krohn A, Lambert RW, Merret JH, Mills JS, Parkes KEB, Redshaw S, Ritchie AJ, Tayler DL, Thomas GJ, Machin PJ: Rational design of peptide-based HIV proteinase inhibitors. Science 1990;248:358–361.

114 McQuade TJ, Tarpley WG, Liu L, Karacostas V, Moss B, Sawyer TK, Heinrikson RL; Tarpley WG: A synthetic HIV-1 protease inhibitor with antiviral activity arrests HIV-like particle maturation. Science 1990;247:454–456.

115 Ashorn P, McQuade TJ, Thaisrivongs S, Tomasselli AG, Tarpley WG, Moss B: An inhibitor of the protease blocks maturation of human and simian immunodeficiency viruses and spread of infection. Proc Natl Acad Sci USA 1990;87:7472–7476.

116 Gottlinger HG, Dorfman T, Sodroski JG, Haseltine WA: Effect of mutations affecting the p6 *gag* protein on human immunodeficiency virus particle release. Proc Natl Acad Sci USA 1991;88:3195–3199.

117 Poznansky M, Lever A, Bergeron L, Haseltine W, Sodroski J: Gene transfer into human lymphocytes by a defective HIV-1 vector. J Virol 1991;65:532–536.

Dr. William A. Haseltine, Dana-Farber Cancer Institute, 44 Binney Street, Boston, MA 02115 (USA)

Rossi GB, Beth-Giraldo E, Chieco-Bianchi L, Dianzani F, Giraldo G, Verani P (eds): Science Challenging AIDS. Basel, Karger, 1992, pp 107–115

Cellular Pathogenesis of Human Immunodeficiency Viruses Types 1 and 2

R.A. Weiss, P.R. Clapham

Chester Beatty Laboratories, Institute of Cancer Research, London, UK

The human immunodeficiency viruses (HIV-1, HIV-2) and related simian immunodeficiency viruses (SIV) cause cellular immune deficiency, wasting and neurological disease. The cell types infected or indirectly affected by the virus helps to explain much about the clinical features of disease, especially the infection of CD4-positive T lymphocytes, and monocytes and macrophages including microglial cells in the brain. Properties of the envelope proteins (gp120, gp41) of HIV determine early events in infection and virus entry that distinguish the particular cell tropism of different HIV strains and groups. While all HIV and SIV strains utilise the CD4 molecule as a high affinity cell surface receptor, HIV-1, HIV-2 and SIV_{mac} exhibit different requirements for cell entry after binding to CD4. Furthermore, HIV-2 strains can infect certain CD4-negative human cells at up to 10% the efficiency of sensitive $CD4^+$ cells. Most HIV-2 strains, however, only infect $CD4^-$ cells after treatment with soluble CD4 molecules.

HIV Pathogenesis at the Cellular Level

HIV can damage cells in several ways leading eventually to systemic illness. First, there are direct cytotoxic effects due to viral infection. Some T lymphocytes infected with HIV may be killed by the virus; other cells can carry the virus in latent form, but the virus becomes activated upon T cell immune activation. With abundant expression of late viral genes and generation of virus particles, the cells die through the cytotoxic effects of the virus, which may trigger apoptosis.

Cells producing envelope antigens may fuse with adjacent $CD4^+$ cells, and the resulting multinucleated syncytia soon die. Syncytium induction is readily seen in culture and to a lesser extent in vivo; it is not clear whether cell fusion is important in HIV pathogenesis. Syncytium induction indicates, however, that uninfected bystander cells can be killed by HIV without becoming directly infected by the virus.

The binding of shed gp120 envelope antigen may also make $CD4^+$ cells a target for antibody-dependent lysis by complement, and antibody-dependent cellular toxicity. Natural killer cells and major histocompatibility complex restricted cytotoxic $CD8^+$ lymphocytes also recognize and kill HIV-infected cells. While these mechanisms of immunological recognition are essential for the control of viral infections, in progression to AIDS in HIV-infected individuals, cytotoxic T cells may eventually destroy the very helper cells they depend on because these cells express HIV antigens.

Apart from cytotoxicity, HIV may cause pathophysiological changes. The attachment of gp120 to CD4, like the binding of monoclonal antibodies (mAb) such as Leu3a, blocks activation of mixed lymphocyte reactions and thus can lead to the loss of immune functions. HIV infection of macrophages is probably less cytotoxic than infection of T lymphocytes, but may adversely affect natural macrophage activity such as phagocytosis and cytokine production. Cytokines may account for the weight loss and general wasting seen in AIDS, possibly mediated through tumour necrosis factor alpha. Atrophy in the brain could also result from cytokine changes in HIV-infected microglial cells, though little is known as yet of normal interactions between brain cell types.

Both the cytotoxic and the pathophysiological effects of HIV will alter the dynamics of lymphocyte and monocyte populations. It is also likely that HIV infects immature T cells in the thymus when they express both CD4 and CD8 antigens, so that fewer progenitor cells mature into T helper lymphocytes. Which of the various pathogenic mechanisms contributes most to the loss of T helper lymphocytes in vivo is not known. Although the number of HIV-infected cells in the peripheral blood at any one time is small, direct cytopathic effects of HIV could still lead to the decline in $CD4^+$ lymphocytes. A minor change in the balance between cell loss and replacement will, over the years of HIV incubation, lead to the crucial depletion of $CD4^+$ cells that precipitates AIDS.

Cellular Tropism of HIV

HIV-1 principally infects $CD4^+$ T lymphocytes, monocyte macrophages and cells of the dendritic-Langerhans lineage. Brain microglial cells are of monocytic origin derived from the bone marrow. Each of these cell types probably becomes infected with HIV-1 via CD4 receptors [14]. Many laboratory-adapted strains of HIV-1 also infect established leukemic T cell lines in culture. Different HIV-1 strains, however, exhibit tropism for T lymphocytes or for macrophages, though most freshly isolated HIV-1 stocks will infect non-adherent peripheral blood lymphocytes (PBL), irrespective of whether they are classified as monocyte or macrophage tropic. The laboratory strains of HIV-1 that adapt to grow well in T cell lines are generally not macrophage tropic.

Recent studies in several laboratories have indicated that the V3 loop of gp120, which is the principal neutralization epitope, plays an important role in determining cellular tropism of HIV-1. A segment of gp120 including V3 distinguishes macrophage and T cell tropic strains [15]. A variant of the Gun-1 isolate of HIV-1 was selected to grow on $CD4^+$ brain cells and lost ability to propagate in some T cell lines [17]. This variant had a single codon change, P to S, in the beta-turn of the V3 loop, which determined its tropism. Thus envelope properties of HIV-1 influence the ability to infect different types of $CD4^+$ cells.

Some human cell lines showing little or no CD4 expression were shown to be susceptible to HIV-1 infection albeit at low efficiency compared to $CD4^+$ cells [4]. These now include astroglia, neuroblastoma, hepatoma, colorectal carcinoma, and B lymphoblastoid cells [2]. As described below, some HIV-2 strains infect certain $CD4^-$ cells quite efficiently. An alternative, currently unidentified receptor may be involved in the binding to and penetration of $CD4^-$ cells [5].

HIV-1 Attachment Is Not Sufficient for Entry

CD4 is the principal high affinity receptor for HIV-1, HIV-2 and SIV [14]. Treatment of cells with antibodies to CD4 competitively blocks infection; treatment of virus with soluble recombinant forms of CD4 (sCD4) strongly neutralizes all established, laboratory strains of HIV-1 tested, and, to a lesser extent, HIV-2, SIV and fresh HIV cultures [4, 7], possibly by detaching gp120 from the virion surface [12]. Transfection and expression

Table 1. Cell and virus specificity of fusion

	Cells	HIV-1	HIV-2	SIV_{mac}
$CD4^+$	Molt4	+++	+++	+++
	Daudi	+++	+++	–
	RD	+++	+++	–
	U87	–	+++	+++
$CD4^-$	HeLa	–	–	–
	Daudi	–	+++	–
	RD	–	+++	–
	U87	–	–	–

of the CD4 gene to human cell lines such as HeLa renders them sensitive to HIV-1 infection and fusion, but expression of human CD4 on the surface of mouse cells did not confer HIV-1 sensitivity although virions bound to the cell surface [10].

We have extended the analysis of HIV sensitivity to a wider variety of cell lines expressing human CD4 [3]. One cell line, the astroglioma U87, out of five human lines tested was resistant to HIV-1. Ten further cell lines derived from monkeys, carnivores, rabbits and rodents, each transfected with human CD4, were similarly resistant to infection although they efficiently bound gp120 (table 1). It appears that CD4 is sufficient for HIV-1 attachment but that other human cell surface molecules are required to trigger events leading to membrane fusion and entry of HIV-1 into cells. An external fusigen, such as polyethylene glycol, permits viral penetration and subsequent replication if added after viral adsorption [3].

In contrast to the restriction of HIV-1, HIV-2 and SIV_{mac} will infect a much broader range of cells expressing human CD4, including human U87 cells, and rabbit and cat cells. Ten laboratory isolates of HIV-1 derived from African and Western people with AIDS each shows an identical plating pattern on the various CD4-transfected cell lines, and six independent strains of HIV-2 also show a pattern characteristic of HIV-2 [3]. These results indicate that the post-CD4 host range of infection in vitro is a consistent property of the envelopes of diverse HIV-1, HIV-2 and SIV strains.

The binding of gp120 to CD4 or to soluble CD4 induces secondary conformational changes which themselves may increase interactions with secondary receptors [1, 13].

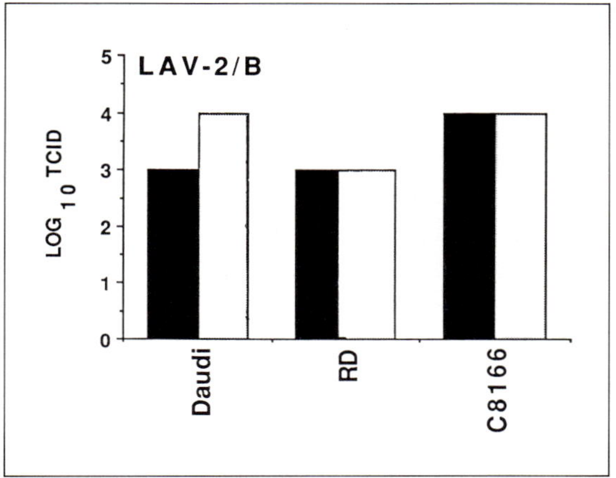

Fig. 1. Titration of LAV-2/B on CD4⁻ Daudi and RD cells and on CD4⁺ C8166 cells.

CD4-Independent Infection of HIV-2

Certain human cell lines can be infected by HIV-1 in a CD4-independent manner [2, 4, 19]. Neither anti-CD4 nor sCD4 block infection but the plating efficiency is at least 1,000-fold lower than that on $CD4^+$ cells. In screening 13 $CD4^-$ human cell lines for HIV-2 infection, we noted three (RD, Daudi and Raji) that were highly sensitive to fusion by LAV-2/B, a substrain of LAV-2 ROD [5]. Table 1 shows that pattern of syncytium formation of HIV-1, HIV-2 and SIV_{mac} on selected $CD4^+$ and $CD4^-$ cell lines. Titration studies revealed that LAV-2/B infects $CD4^-$ RD and Daudi cells almost as efficiently as $CD4^+$ T cells (fig. 1).

Five other strains of HIV-2 were plated on $CD4^-$ RD cells and showed little or no infection and cell fusion. However, when the virus was pretreated with sCD4, massive syncytia ensued (fig. 2). Fusion was blocked by neutralizing the virus with human HIV-2 antiserum.

Table 2 summarizes the observations on syncytial induction by HIV-2 with sCD4. Curiously, treatment with a sCD4-Ig chimeric molecule did not induce fusion and blocked the spontaneous fusion by LAV-2/B. The effect of sCD4-Ig may result from its bivalent structure, as monovalent sCD4 complexed with mAb to the third domain (which does not interfere with gp120 recognition) also blocked cell fusion [5].

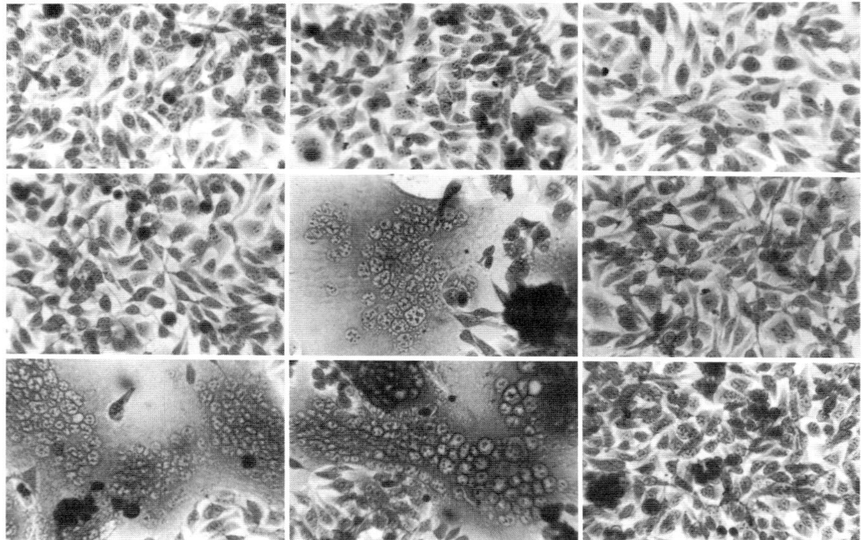

Fig. 2. Syncytium formation in RD cells. Top rows: Cells exposed to HIV-1 RF strain; middle row: cells exposed to HIV-2 CBL-20 strain; bottom row: cells exposed to HIV-2 LAV-2/B strain. Left-hand column: Infection with virus alone; middle column: virus and sCD4; right-hand column; virus, sCD4 and human anti-HIV-2 serum [5].

Conclusions

Our studies of HIV infection of cell lines indicate that accessory molecules are required for membrane fusion following the binding of virions to CD4. These secondary receptors distinguish between HIV-1, HIV-2 and SIV strains. We have postulated that they might be cell membrane proteinases that cleave the V3 loop of gp120 [6], but there is insufficient evidence so far.

In addition to CD4 as the principal binding receptor, other cell surface molecules are involved in virus-cell interactions and in HIV-induced cell fusion [18]. Antibodies to LFA-1 prevent syncytium induction in lymphocytes [9] though not in $CD4^+$ HeLa cells. Fc-receptors induce the uptake and infection of HIV particles lightly opsonized by non-neutralizing antibody, even in $CD4^-$ cells [11]. *Galactocerebrosides* may be involved in infection of neural cells [8], but we do not know whether any of these

Table 2. CD4-Independent RD cell fusion: induction by soluble CD4

Virus	Strain	No sCD4	sCD4	sCD4-Ig
HIV-1	10 strains	–	–	–
HIV-2	LAV-2/B	+++	+++	–
	SBL6669	+	+++	–
	CBL-20	–	+++	–
	CBL-21	–	+++	–
	CBL-22	+	+++	–
	CBL-23	–	+++	–

cellular molecules are involved in the differential tropism of HIV and SIV strains.

We have also shown that the CD4-negative RD, Daudi and Raji cells can be infected by LAV-2/B almost as efficiently as $CD4^+$ cells. Whether the non-CD4 receptor on these cells is the same as the accessory molecules on $CD4^+$ cells sensitive to HIV-2 is not yet known. For example, U87 cells are wholly resistant to HIV-2 in the $CD4^-$ state, are highly sensitive to HIV-2 in the $CD4^+$ state, yet remain resistant to HIV-1. We have not yet determined whether the non-CD4 and post-CD4 receptors for HIV-2 are the same.

The remarkable induction of $CD4^-$ cell fusion by sCD4 treatment of other HIV-2 strains indicates that the recognition site for the receptor or fusigen of these viruses remains cryptic until sCD4 induces conformational changes in the viral envelope. It is not yet known whether sCD4 causes gp120 stripping of HIV-2 in the same way as HIV-1 [12]. The enhancement of infection and the broadening of cell tropism of HIV-2 by sCD4 should indicate caution in applying therapeutic agents based on CD4 molecules for clinical trial in HIV-2 infected subjects. It is interesting to note, however, that bivalent sCD4-Ig blocks HIV-2 infection of $CD4^-$ cells.

Our studies of the differences in cell tropism and CD4 dependency of HIV-1, HIV-2 and SIV are based on experimental systems employing selected laboratory strains of virus and the use of immortalized, established cell lines to test infection. Both the viruses and the cells are, of course, distant from natural infection in vivo. S. Wain-Hobson has well reminded us that 'To culture is to disturb'. Yet it is also the means to probe and dissect the molecular mechanism of viral attachment and entry.

Acknowledgement

The work of our laboratory discussed here has contributions from A. McKnight, D. Blanc, T.F. Schulz, J. Moore and J. McKeating. Smith Kline Beecham and Genentech kindly provided sCD4 constructs. Our research is supported by the Medical Research Council and the Cancer Research Campaign.

References

1 Allan JS: Receptor-mediated activation of immunodeficiency viruses in viral fusion. Science 1991;252:1322.
2 Clapham PR: Human immunodeficiency virus infection of non-haematopoietic cells. The role of CD4-independent entry. Rev Med Virol 1991;1:51–58.
3 Clapham PR, Blanc D, Weiss RA: Specific cell surface requirements for the infection of CD4-positive cells by human immunodeficiency virus types 1 and 2 and by simian immunodeficiency virus. Virology 1991;181:703–715.
4 Clapham PR, Weber JN, Whitby D, Deen K, Sweet RW, Weiss RA: Soluble CD4 blocks the infectivity of diverse strains of HIV and SIV for T cells and monocytes but not for brain or muscle cells. Nature 1989;337:368–370.
5 Clapham PR, McKnight A, Weiss RA: CD4-independent infection by HIV-2: Enhancement by soluble CD4. 1991; in press.
6 Clements GT, Price-Jones MJ, Stephens PE, Sutton C, Schulz TF, Clapham PR, McKeating JA, McClure MO, Thomson S, Marsh M, Kay J, Weiss RA, Moore JP: The V3 loop of the HIV-1 and HIV-2 surface glycoproteins contain proteolytic cleavage sites: A possible function in viral fusion? AIDS Res Human Retroviruses 1991;7:3–10.
7 Daar ES, Li XL, Moudgil T, Ho DD: High concentrations of recombinant soluble CD4 are required to neutralize primary human immunodeficiency virus type 1 isolates. Proc Natl Acad Sci USA 1990;87:6574–6578.
8 Harouse JM, Bhat S, Spitalnik SL, Laughlin M, Stefano K, Silberberg DH, Gonzalez-Scarano F: Inhibition of entry of HIV-1 in neural cell lines by antibodies against galactosyl ceramide. Science 1991;253:320–323.
9 Hildreth JEK, Orentas RJ: Involvement of a leucocyte adhesion receptor (LFA-1) in HIV-induced syncytium formation. Science 1989;224:1075–1078.
10 Maddon PJ, Dalgleish AG, McDougal JS, Clapham PR, Weiss RA, Axel R: The T4 gene encodes the AIDS virus receptor and is expressed in the immune system and the brain. Cell 1986;47:333–348.
11 McKeating JA, Griffiths PD, Weiss RA: HIV susceptibility conferred to human fibroblasts by cytomegalovirus-induced Fc receptor. Nature 1990;343:659–661.
12 Moore JP, McKeating JA, Weiss RA, Sattentau QJ: Soluble CD4 induces shedding of gp120 from virions. Science 1990;250:1139–1142.
13 Moore JP, McKeating JA, Weiss RA, Clapham PR, Sattentau QJ: Receptor-mediated activation of immunodeficiency viruses in viral fusion. Science 1991;252:1322–1333.
14 Sattentau QJ, Weiss RA: CD4: HIV receptor and physiological ligand. Cell 1988;52:631–633.

15 Shioda T, Levy JA, Cheng-Mayer C: Macrophage and T cell-line tropisms of HIV-1 are determined by specific regions of the envelope gp120 gene. Nature 1991;349: 167–169.

16 Takeda A, Sweet RW, Ennis FA: Two receptors are required for antibody-dependent enhancement of human immunodeficiency virus type 1 infection: CD4 and $Fc_{\gamma}R$. J Virol 1990;64:5605–5610.

17 Takeuchi Y, Akutsu M, Murayama K, Shimizu N, Hoshino H: Host range mutant of human immunodeficiency virus type 1: Modification of cell tropism by a single point mutation at the neutralization epitope in the *env* gene. J Virol 1991;65.

18 Ugen KE, Perussia B, Kamoun M, Merva M, Williams MV, Nara P, Kieber-Emmons T, Weiner DB: Inhibition of HIV-1 cellular infection by immunologic reagents. Vaccines 1991;91:115–121.

19 Werner A, Winskowsky G, Cichutek K, Norley SG, Kurth R: Productive infection of both $CD4^+$ and $CD4^-$ human cell lines with HIV-1, HIV-2 and SIVagm. AIDS 1990; 4:537–544.

R.A. Weiss, PhD, Chester Beatty Laboratories, Institute of Cancer Research, Fulham Road, GB–London SW3 6JB (UK)

Rossi GB, Beth-Giraldo E, Chieco-Bianchi L, Dianzani F, Giraldo G, Verani P (eds): Science Challenging AIDS. Basel, Karger, 1992, pp 116–125

AIDS and Related Malignancies

Robert C. Gallo

Laboratory of Tumor Cell Biology, National Cancer Institute,
National Institutes of Health, Bethesda, Md., USA

Though the subject of this session has to do with neoplasias that are associated with AIDS and HIV infections, before I discuss the two major tumors – B cell lymphoma and Kaposi's sarcoma (KS) – I would first like to direct my attention to a few other areas of research that are now ongoing in our group. In particular, I wish to select those areas of research which I think hold promise for our future help to HIV-infected people.

The Molecular Biology of HIV

We have continuing efforts on the replication cycle of the virus, on its molecular makeup, and on molecular mechanisms involved in the disease induced by HIV. The essential feature of the life cycle of a retrovirus is its ability to integrate its DNA form which in the case of HIV is one major factor responsible for the so-called period of latency that we see in HIV infection at the cellular level. But a peculiar result came out at the onset of this field. In early 1984, George Shaw, a postdoctoral in our laboratory, observed that primary tissue from infected people had substantial amounts of unintegrated DNA. That was a novel finding in our experience, and we had no explanation for it.

Shortly after, Jean-Claude Chermann and his coworkers, then at the Pasteur Institute, discovered that resting T cells could be infected. We had almost no precedent in retrovirology that a cell not making DNA can be infected. He and also Daniel Zagury and my coworkers and myself then showed that when that cell was stimulated by immune activation, the virus life cycle would complete itself. The mechanism for infection of the 'resting' T cell remained elusive.

Shortly after that, Irving Chen and his coworkers at UCLA found DNA in the cytoplasm for considerable periods of time in resting T cells. They are still involved in the analysis of that DNA. Only yesterday we heard of results from Mario Stephenson's laboratory which agree with those previous studies but also made the interesting finding that DNA can stay in a resting cell for a few weeks. In addition, they showed for the first time some in vivo correlations. Asymptomatic people had more of this unintegrated DNA.

Franco Lori in our laboratory will soon describe in one of these sessions that HIV virions carry heterogeneous short segments of DNA, tightly complexed to reverse transcriptase (at least some of the virions). What's the significance of all of this? The virions carrying DNA suggest the possibility of episomal replication – perhaps related to the phenomenon of long-lasting cytoplasmic DNA found in resting T cells. Also, viral DNA, carried within virions, may explain how macrophages, though cells not making DNA, can nonetheless be infected. At a practical level, this means that the pools of nucleotides may be critical and rate-limiting to the early events of HIV infection. Since resting T cells have very low levels of enzymes involved in the synthesis of nucleotides, this suggests to me that one new future target of anti-HIV therapy might be to reevaluate those enzymes that are rate-limiting (such as ribonucleotide reductase) in building the nucleotide precursors for making DNA and RNA.

A second aspect of HIV molecular biology ongoing in our laboratory is its exploitation for gene therapy and focuses on the second stage of the replication of HIV. It is well known since the work of Haseltine and his coworkers in 1984 and Wong-Staal in our group at that same period that an essential first step in the expression of the viral RNA from the integrated provirus is mediated by the *tat* gene product, the TAT protein.

TAT interacts with a sequence named TAR in the virus LTR. So TAT interacts with these specific sequences to turn on and/or complete the transcription of the rest of the viral genome. Though the mechanistic details are complex and not yet completely understood (they will be described by others in these proceedings), this is an essential step in HIV replication.

Julianna Lisziewicz in our group has taken advantage of that observation, and she has made multimers of TAR – 50 mers – and transfected this construct (with upstream HIV LTR regulatory sequences) into uninfected cells in the laboratory, uninfected cells that are the potential target cells for HIV. The integrated TAR sequences are not expressed – unless that cell is infected by HIV. When that cell is infected by HIV, the HIV provirus integrates, TAT is made. TAT will activate the 50 mers of TAR, and will be sequestered by the expression of the 50 mers of TAR. Thus, little or no HIV replication will occur because little or no TAT is available for activation of the HIV provirus containing only one TAR sequence.

Such fanciful and theoretical experiments have now been put to practical test and are beginning to work in laboratory settings. We would like to combine this with other approaches of gene therapy with the ultimate goal of applying it to HIV-infected people's bone marrow stem cells and reinfusing those stem cells in the hope that they'll be permanently protected and repopulate the bone marrow. I think this is at least theoretically doable, and we will pursue this work as fast as we can.

Cofactors in the Progression of AIDS

I would now like to turn to the third area of ongoing research in our laboratory, and that is the question of cofactors, which Robin Weiss discussed.

Since early 1984 we were convinced that there is one single cause of AIDS, HIV. Cofactors can be defined as essential. I do not believe there is any evidence now (and I do not think there will be) of any *essential* cofactor for development of AIDS other than HIV. But that doesn't mean that there aren't things that could make the process go faster, and in my view we should maintain an awareness for such factors or else we miss possibilities of helping HIV-infected people.

But most human disease have factors which may contribute to the rate by which the disease develops, not only the primary causative agent. Are there hints for such *nonessential* cofactors in AIDS? Weiss referred to reports from Miami suggesting that HTLV-I may be such a cofactor. He critically commented that this is because they deal with HTLV-I in Miami, i.e. HTLV-I is endemic there, and it was noted that AIDS occurs elsewhere without HTLV-I; for example, in London where the presence of HTLV-I is not very significant. But the right comparison is not Miami to London. Because the rates of progression to AIDS are roughly comparable in various areas in the world does not mean these other agents are not *nonessential* cofactors. The right comparison is between those infected with HTLV-

I in Miami and those not (both, of course, being infected with HIV) and determining the rate of developing AIDS in them. Several epidemiological studies in different parts of the world indicate that HTLV-I is correlated with faster progression. Moreover, mechanisms have already been elucidated to explain how that may occur. I would tentatively conclude, therefore, that HTLV-I is a nonessential cofactor in AIDS. But HTLV-I cannot be terribly important to the AIDS epidemic because less than 10% of HIV-1-infected people in the world have HTLV-I infection.

I'd like to turn away from HTLV-I to the virus Weiss facetiously stated was of interest in Bethesda, namely human herpes virus number 6 or HHV-6. It is, of course, true that we are interested in HHV-6 because it was found in our laboratory in 1986. But it seems to me to be a mistake not to pay attention to the first known *ubiquitous* human *T lymphotropic* virus of any class of virus, particularly when it can kill T cells. But we cannot get the answer to its possible role in AIDS progression epidemiologically by antibody testing because almost everyone has HHV-6 infection. Virtually every patient we've looked at with AIDS has HHV-6 infection. Therefore, it's a very difficult epidemiologic problem. (Antigen assay development could solve this problem.)

But let me argue another way. Would any of us deny that cytokines are important in HIV disease progression? We would quickly agree that it's wise to study TNF, IL-6, and other cytokines in AIDS progression. Since the HHV-6 is so common, isn't it prudent to consider it almost endogenous? If so, shouldn't we consider a replicating T lymphotropic endogenous agent with the capacity to alter cells of the immune system as possibly as important as a cytokine in AIDS progression? In fact, there are several in vitro experimental studies which suggest to me that HHV-6 may be involved.

(1) Actively replicating HHV-6 kills CD4-positive T cells, as well as HIV kills these cells. But normally, HHV-6 replication is contained.

(2) HHV-6 can infect the same T cell as HIV, and – at least some strains of HHV-6 can activate latent HIV. (There are at least two groups of HHV-6, A and B. The prototype A strain is the GS strain we isolated in 1986. Most of these in vitro studies were with the GS prototype.)

(3) Paolo Lusso in our laboratory found that HHV-6 can turn on the gene for CD4 in T8 cells. To my knowledge, this is the first biological factor we know that can turn on CD4, and those T8 cells are now targets for HIV. They are CD4-positive, CD8-positive, and can become infected and killed by HIV.

(4) HHV-6 down-regulates CD3, also shown by Lusso in collaboration with Malnati of NIAID.

(5) A recent result from Dharam Ablashi working in our group indicates that HHV-6 activates the expression of the cytokines TNF and IL-1, which in turn promote more HIV expression, and these same cytokines can also enhance proliferation of KS spindle cells.

Finally, studies reported at these meetings from Lawrence Corey's group in Seattle show that HHV-6-infected CD4-positive T cells are lost late in AIDS progression, consistent with the view that the presence of HHV-6 fosters T4 cell decline though obviously not proving it.

I think one of three kinds of experimental results is needed to 'nail down' an HHV-6 linkage to AIDS progression: a chimp study in which it is shown that AIDS develops after *HIV-1 plus HHV-6;* a drug which selectively inhibits HHV-6 and in clinical trials reduces the rate of AIDS progression; and/or an expansion of the Seattle type studies mentioned above based on antigen (or virus isolation) and not antibody determinations.

A fourth area of research in our laboratory relates to HIV variation, the immune response, and experimental attempts to develop an HIV vaccine. These will not be reviewed here. Instead, I would like to finish this brief overview with a discussion of our studies and concepts on the neoplasias associated with HIV infection.

Associated Malignancies

In our view, the pathogenesis of AIDS-associated B cell lymphoma is very likely as in Burkitt lymphoma (BL): the concomitant immune stimulation with immune deficiency. Immune stimulation leads to signals for B cell activation and proliferation, which when combined with impaired function of CTLs results in an imbalance frequently associated with B cell lymphomas. B cells perhaps may have a built-in mechanism more prone to become neoplastic, to transform, because the DNA of a B cell is 'dynamic' (the immunoglobulin gene shuffling that occurs throughout life).

In BL, holoendemic malaria is clearly associated with the increase in this tumor in parts of Africa. This provides a chronic antigenic stimulation for B cell activation and a mitogenic signal which should increase the probability of an accident like the *c-myc* translocation that occurs in BL.

At the same time, malaria can inhibit CTL immune function. In so doing, the population of Epstein-Barr virus (EBV)-immortalized B cells can expand, increasing further the probability of B cell lymphoma. If we take malaria out of the equation and substitute HIV, we have the same conditions but in even greater force. HIV gives persistent protein stimulation of the immune system for B cell proliferation. HIV-infected cells also release cytokines – some of which are B-cell growth factors – both events can lead to stimulation of mitosis. The immune suppression and CTL dysfunction that can occur after HIV infection may lead to an expansion in the EBV immortalized B cell population, giving us the tremendous increase in B cell lymphoma in HIV-infected people.

Earlier I said I wanted the theme of my talk to have to do with areas which might help HIV-infected people. But what does this B cell lymphoma discussion have to do with therapy? How do we do treatment research today for these cancers? We base it on older protocols in chemotherapy for other B cell lymphomas. But what we need most is an animal model, and recently Peter Biberfeld and his colleagues in Stockholm developed just such an animal model. The animal model developed in the last few years was already summarized at this meeting. Remarkably, more than one-third of cynomolgus monkeys infected with a certain strain of SIV developed marked immune deficiency and B cell lymphoma with a life period of 5–15 months. This lymphoma seems to be very similar in pathology and clinical course to the human B cell lymphoma of HIV infection. Presumably, SIV infection, leading eventually to CTL dysfunction and an expansion of immortalized B cells (with another EBV-like virus), is not directly causing the tumor. (Sequences of SIV are not found in the majority of the tumor cells.) Its mechanism is indirect and not as a transforming virus (such as HTLV-I in T cell leukemia). Recent studies in our laboratory by Sandra Colombini showed that these tumors have an EBV-like virus integrated in a clonal fashion. Similar and more extensive results are also reported at this meeting from a collaborative study of Biberfeld with the Austrian group. I won't, therefore, dwell on this any further other than to note that Dharam Ablashi in our group has isolated the virus, and, as we expected, it has properties like human EBV. Therefore, we have every reason to think the pathogenesis of this tumor is going to be similar or the same as in man and that this animal model system merits attention for drug development for treating people with B cell lymphoma. Parenthetically, to my knowledge this is the first proof that a virus can cause malignancy by an indirect mechanism, i.e. not by infecting progenitors of the tumor cells but by infecting other cell types, a concept I have been long attracted by and one which I favor in KS.

I next want to turn to KS which will be addressed in more detail by my coworker Barbara Ensoli. Until recently, the study of the pathogenesis of KS was limited. Most studies focused on epidemiology and clinical-pathological aspects.

Epidemic KS is, of course, associated with HIV but there are reasons to think we may have missed another sexually transmitted virus. This is principally from epidemiologic studies highlighted by Harold Jaffe and his colleagues at the Center for Disease Control. One impressive but not very large study indicates that women who are HIV-infected from bisexuals have a greater incidence of developing KS than a woman HIV-infected from sex with a hemophiliac or a drug addict. Though this is certainly strongly suggestive, it is not a completely convincing argument because the route of infection can often determine the outcome and nature of disease, and the root of infection of a woman from bisexuals may not be exactly the same (in a population study) as it is from hemophiliacs or of i.v. drug addicts. In any case, to date there is no strong evidence for linkage of any novel or well-known microbe to KS other than HIV in the epidemic form. If it is, it remains to be shown.

Let's direct our attention to the mechanisms of pathogenesis involved in KS no matter the cause. If we can understand the pathogenetic mechanisms, can we help people who have KS? What was needed for progress in this direction was an experimental laboratory system. We think it may now be available. A few years ago, my coworkers and I reported the development of an in vitro culture system for the spindle cells of KS. As most of you know, the spindle cells are believed to be the tumor cell of KS. Pathologists think they are probably derived from the cells forming the walls of the vasculature – endothelia or smooth muscle, or perhaps both. The rest of the KS tumor consists of a mixture of fibroblasts, infiltrating white cells, and edema.

We did some empirical work over a period of a year or two ('we' is chiefly Shuji Nakamura when he was a visiting scientist in my laboratory from Kummamoto University, S.Z. Salahuddin, and our collaborator Peter Biberfeld when he was a visiting scientist with us). And, indeed, we were able to grow these spindle cells and show that they resemble endothelial cells in some characteristics, but in other ways they resemble more smooth muscle cells. Our tentative conclusion is that they are a primitive mesenchymal cell capable of going in either direction. I also would like you to think of them as an 'emergency cell' because we now know that these spindle cells have receptors for a wide range of cytokines, which can induce them to undergo rapid proliferation. We have not seen other cell types with so many receptors for so many different growth factors. But when an abrasion from a wound is examined histologically, it appears like early KS. What is needed to rapidly heal the wound? We can speculate a cell type which can indicate rapid development of new blood vessels and itself rapidly proliferate in response to urgent signals. The KS spindle cell may be this kind of cell.

The key feature of KS is, of course, angiogenesis. And the question was: Do those spindle cells grown in our lab make factors which make blood vessels grow? In the first assay we did (the chicken chorioallantoic membrane assay), we found a remarkable increase in formation of new blood vessels, but much more interesting was the next experiment with nude mice inoculated with spindle cells. Within 5–10 days, a tumor or a lesion developed at the site of inoculation which had the features of early KS. Proliferating spindle cells, leukocytes, fibroblasts and edema were evident. We expected that this tumor would be of human origin (like in many nude mouse models, when a human tumor is inoculated). We were surprised to find that the tumors were of mouse origin. In other words, the human spindle cells cultured in our laboratory produce some things that rapidly induce a KS-like lesion in the mouse of mouse origin.

We measured the biological factors that are produced by these cultured cells in a variety of biological activity assays. I've already noted that they produce angiogenic activities and induce KS-like lesions in the nude mouse. Adriana Albini, when at NIH, discovered that the cells that we sent to her also released chemoinvasive, as well as chemotactic activities. Other assays performed in our laboratory showed that the cultured spindle cells also produced growth factor for themselves, as well as for normal fibroblasts and normal endothelial cells, IL-1 activity and GM-CSF activity.

I have indicated that another key feature of KS is edema. Shinsaku Sakurada, a visiting scientist in my laboratory from Tokyo, began working on this problem about one and one-half years ago. Meanwhile, our collaborator, Parkash Gill at the University of Southern California in Los Angeles, showed in clinical studies that the edema of KS extends far beyond the visible tumor. He and his surgical colleagues also noted that when the KS lesion is resected only at the visible tumor borders, the tumor returns. But if they resect the lesion out to where the edema extends (determined by dye studies) then the local tumor doesn't return. Thus, we think the edema from increased vascular permeability can be important not only for producing local obstruction and suffering, but also perhaps provides cytokines to adjacent cells that are not yet involved in the local KS lesion. Sakurada's studies exploited our nude mouse assay. He inoculated the spindle cells in certain places, and he gave the animal Evans blue dye, and the simple assay of just looking by eye or spectrophotometrically measuring the concentration of dye revealed heightened vascular permeability which occurred in two phases, an early phase which also occurs with control cells due to histamine release and a late phase occurring only with the KS spindle cells which is not due to histamine.

The late phase edema appears to be due to cytokine release. IL-1, GM-CSF and tVPF (tumor cell-derived vascular permeability factor) are all expressed in these cells (the latter first shown by Herbert Weich) and each is known to directly or indirectly increase vascular permeability. We also know that these spindle cells produce high amounts of basic fibroblast growth factor (bFGF) and/or related molecules, high levels of IL-1, IL-6, and GM-CSF and lesser amounts of TGF-beta and acidic fibroblast growth factor (aFGF). Of these, bFGF, IL-1, and IL-6 are powerful growth factors for the spindle cells themselves.

What is the role of HIV in all of this? Why is there such a massive increase in incidence of KS after HIV infection? Chronic activation of immune cells will occur from HIV proteins which will in turn lead to the production of many cytokines. HIV infections of macrophage and T cells also induce the release of many cytokines. Included in these cytokines are very powerful growth factors like TNF for the spindle cells, IL-1 which I noted are also released by the growing spindle cells, GM-CSF and IL-6, again also released by the spindle cells. We now know from Lunardi-Iskander in our laboratory that the spindle cells also have receptors for IL-2, which I think is a novel finding for mesenchymal cells. IL-2 is, of course, also released by activated T cells. Since the KS spindle cells contain receptors for IL-2, this cytokine may also be involved in promoting their growth. One of the most active cytokines promoting KS spindle cell growth we have found to date is a 30,000 MW protein also released by activated CD4-positive T cells. We have noted this factor in our earliest KS studies, but in recent months it has been successfully purified, sequenced, and identified as oncostatin M by Sarngadharan, DeVico, Copeland and Oroszlan in collaboration with our laboratory. Oncostatin M is a little studied cytokine chiefly known for its capacity to inhibit growth of some tumor cells in vitro.

Finally, Barbara Ensoli and others in our group discovered that the TAT protein of HIV is released by acutely infected T4 cells and is still another growth factor for the KS spindle cells, and it is active at extremely low concentrations. Normal cells do not respond to TAT, but pretreatment of normal endothelial cells or smooth muscle cells with cytokines from chronically activated T cells make these normal cells TAT responsive (see Ensoli paper). To summarize, we think of KS as a phenomenon of spindle cell formation and chronic activation → proliferation. Avenues which lead to these events favor KS. By providing TAT and by inducing cytokines which also produce growth promotion for the spindle cells and which make normal cells TAT responsive, HIV is a key factor in KS development. Once formed, activated, and proliferating, these spindle cells in turn release cytokines which create blood vessel proliferation, edema, and other inflammatory signs.

We agree that other exogenous factors or factor are likely to be involved but remain elusive. The lack of HIV in endemic African KS, in classical KS and in iatrogenic KS have long ago demonstrated that there are other etiological mechanisms (perhaps analogous to the different leukemias).

Male prevalence might be explained by differences in hormonal influences on spindle cells. Preference for gay men might be due to greater immune cell activation in this group compared to other large groups with HIV infection studied to date. These remain open issues.

Without these answers we can attempt to exploit the experimental system to look for inhibitors of KS development. In collaboration with Drs. Osada and Tanaka and others at the Daiichi Pharmaceutical Company, in Japan, we have found that a polysulfated proteoglycan (MW about 30,000) called SP-PG significantly inhibits the KS lesion development in nude mice. The effect is due to a cytostatic effect on the KS spindle cells. We hope that this will be brought to the clinic in the near future.

Robert C. Gallo, MD, Laboratory of Tumor Cell Biology,
National Cancer Institute, National Institutes of Health,
Building 37, Room 6A-09, Bethesda, MD 20892 (USA)

Rossi GB, Beth-Giraldo E, Chieco-Bianchi L, Dianzani F, Giraldo G, Verani P (eds): Science Challenging AIDS. Basel, Karger, 1992, pp 126–140

Present Outlook for Clinical Trials

Ian V.D. Weller

Genito-Urinary Medicine, University College and Middlesex School of Medicine, London, UK

Introduction

This overview of the impact that clinical trials have made on clinical practice in the management of HIV-infected patients in the developed world is made from the clinic rather than the phase I trial laboratory. Furthermore, the advances that have been made in paediatric infection, in many instances as a result of adult clinical trial experience and the management of HIV-related tumours, with a paucity of large randomised studies, will not be considered.

In the developed world in the last 4 years the improvement in survival of patients with early and late HIV disease has been due to many factors. Two further placebo-controlled trials of zidovudine in early symptomatic disease have demonstrated efficacy. Physicians and patients appear to be comfortable in using this agent at various stages of the symptomatic phase of HIV infection.

Over and above antiretroviral therapy earlier diagnosis as a result of earlier patient self-referral, physician training and improved diagnostic techniques have also contributed to the improvement in survival. Primary prophylaxis of *Pneumocystis carinii* pneumonia (PCP), herpes simplex virus infections and both intentional and perhaps inadvertent prophylaxis for fungal infection have also contributed. There have been advances in treatment and new treatments for PCP such as nebulised pentamidine, early high-dose steroids and alternative drug regimens. We are now more confident that clindamycin/pyrimethamine is an efficacious second-line therapy for toxoplasmosis.

As far as viral infections are concerned two agents, ganciclovir and phosphonoformate are being widely used for CMV infection. Maintenance therapy buys time and an oral preparation of ganciclovir looks promising. As far as peripheral CMV retinopathy is concerned when to start treatment is a question that has not been answered. As with HIV, drug-resistant herpes viruses appear to be a growing problem. The triazoles are making an important contribution to the management of fungal infections. Advances have also been made in more difficult areas with studies of combinations of anti-mycobacterial drugs showing symptomatic relief and a short-term decrease in bacteraemia in atypical mycobacterial infections. New drugs such as the new macrolide antibiotics look promising in a number of protozoal and the atypical mycobacterial infections.

Inroads have been made into the mechanisms of zidovudine resistance and the prevalence in symptomatic and asymptomatic individuals. We still do not know in what way the emergence of resistance is associated with loss of clinical effect and the use of certain dose regimens. More convenient assays and therapeutic strategies to overcome it are required.

There is still uncertainty as to when to intervene in asymptomatic HIV infection. In the long term is early or deferred therapy the best approach? The optimum dose of zidovudine is still not yet established.

Combination chemotherapy with other drugs, mainly dideoxynucleosides, is being evaluated with the aims of improving tolerance by reducing drug dose, retaining or enhancing efficacy and hopefully preventing or delaying the emergence of viral resistance. In the future more attractive combinations will be used with agents acting at other sites of the replicative cycle. However, these have yet to complete full phase I/II evaluation.

There is increasing pressure to use surrogate laboratory markers and more proximal endpoints because of the long natural history of HIV infection, the time it takes to conduct efficacy studies using clinical endpoints and the large number of candidate drugs that will be available in the near future. However, it is clear that the markers that we have may not be robust enough yet to fulfil this aim.

Zidovudine in Symptomatic Disease

The first placebo-controlled trial of zidovudine in symptomatic disease in 282 patients with AIDS (first attack PCP) and AIDS-related complex (ARC) demonstrated that over 4–6 months zidovudine significantly decreased mortality, the incidence and severity of life-threatening opportunistic infection and constitutional symptoms [1, 2]. There were associated changes in surrogate markers. There was a small transient increase in CD4 counts and an approximate 90% reduction in serum P24 antigen levels in those on full dose. With this information and licensing of the drug, physicians rapidly extended its use to those with other signs and symptoms of advanced disease and indeed moved the use of the drug to those with milder symptomatic disease.

This last year has seen the results of two further placebo-controlled trials of zidovudine in earlier symptomatic disease. AIDS Clinical Trials Group protocol 016 (ACTG 016) examined the effect of zidovudine in HIV-infected patients with one or two HIV-related features, namely, oral candida, oral hairy leucoplakia, herpes zoster, weight loss, rash or intermittent diarrhoea [3]. Enrolled patients had CD4 counts between 200 and 800 cells per mm^3. The results applied to the 513 patients with CD4 counts less than 500 as very few endpoints in those with counts over 500 were reached. The major endpoints were severe ARC and AIDS. The median follow-up period was 11 months (range 0.2–23). In that time 34 of the 253 placebo recipients had first clinical events compared to 12 of 260 zidovudine recipients. The overall event rate was 11.2/100 person years for the placebo group and 4.5 for the zidovudine group.

The results of a second trial conducted by the Veterans Administration Group (VA cooperative study 298) is not yet published but a summary of the results was communicated at a public meeting in the United States in March this year [Hamilton, Simberkoff, Hartigan and VA Cooperative Group, 1991]. This is a placebo-controlled study in symptomatic disease short of AIDS. Initially, patients with CD4 counts between 100 and 500 were recruited. However, early on in the study, because of the Centres for Disease Control (CDC) recommendations concerning PCP prophylaxis, recruitment was limited to patients with CD4 counts between 200 and 800. A total of 338 patients were recruited. The switch point for open zidovudine therapy was when the CD4 count fell below 200 on 2 occasions. The major endpoints were AIDS and death. The nature of the enrolled patients and the design of the study are different from ACTG protocol 016. In the late zidovudine arm n = 168 (i.e. placebo until the switch point) there were 44 AIDS progression and 66 deaths. In the early zidovudine arm (n = 170) there were 25 AIDS progression and 44 deaths. Although both of these studies demonstrate a delay in progression from mild to moderate to more severe symptomatic disease neither of them were able to show a significant effect on mortality. The net effect of these studies on clinicians and patients is that they are now more comfortable prescribing or taking antiretroviral therapy when illness first appears.

Survival in Patients with Symptomatic Disease

A number of studies have demonstrated an improved survival of patients with symptomatic disease since the introduction of zidovudine using either historical or non-treated comparison groups [4–7]. This improvement is probably due to many factors in addition to antiretroviral therapy.

Increased patient and physician education leading to earlier self-reporting of disease and improved standards of care has almost certainly played a part in the earlier diagnosis of life-threatening complications. In addition, improved diagnostic techniques, particularly in the area of non-invasive tests, such as induced sputum for the diagnosis of pneumocystis and MRI scanning for the diagnosis of cerebral toxoplasmosis have contributed. The CDC recommendations of 1989 for primary and secondary prophylaxis for PCP were that it should be given to patients who have had one episode of PCP and for those HIV-infected patients who have not, if the CD4-positive lymphocyte cell count (CD4) is below 200 mm^3 or less than 20% of total lymphocytes on at least two counts. In addition, patients with CD4 counts less than 100 mm^3, or CD4 cells less than 10% and patients with oral thrush or persistent fever were identified as being at particularly high risk [8].

These recommendations were made because there was a strong association between baseline CD4 counts and the incidence of PCP demonstrated in cohort studies such as the multicentre AIDS cohort (MACS) study [9]. Survival analysis showed that patients whose counts were less than 200 mm^3 had attack rates of 6, 12 and 36 months of 13, 24 and 39%, respectively. Oral thrush and persistent fever were additional independent predictors of the development of PCP.

Secondly, a retrospective study of 49 episodes of PCP showed that the majority of patient's CD4 counts were below 200 (most below 100) and below 20% (most below 10%) within 60 days before the episode of pneumocystis [10]. Thirdly, it was clear from a comparative dose study of zidovudine in patients with recent pneumocystis that antiretroviral therapy with zidovudine alone was not sufficient to prevent a fairly high attack rate for second episodes of PCP [11]. As a result of these recommendations, physicians have used either oral co-trimoxazole or nebulised pentamidine widely in patients with CD4 lymphocyte counts persistently below 200.

In addition, suppressive acyclovir therapy is used for recurrent herpes simplex virus infections. Similarly, the systemic azoles anti-fungal drugs, e.g. fluconazole and ketoconazole, are used more widely for treatment and maintenance therapy in patients presenting with oral candida. As a result, such broad-spectrum prophylaxis will reduce not only the incidence of pneumocystis, herpes and candida infections but inadvertent cross-prophylaxis may be occurring. For example, the systemic azole compounds may be having an effect on preventing cryptococcal infections and acyclovir suppressive therapy may be affecting the emergence of other herpes virus infections.

As a result of therapeutic trials we have seen the arrival of new therapies for protozoal infections.

For PCP, a review of five trials of adjunctive high-dose corticosteroid therapy has led to a consensus that such therapy reduces severe hypoxaemia respiratory failure and mortality if used early in severe PCP [12]. Other alternative treatments particularly for mild-to-moderate PCP have also been shown to have efficacy, namely a combination of dapsone and trimethoprim and clindamycin and primaquine. More information on the latter combination will be presented at this conference. How well these combinations compare to conventional therapy is not yet known.

In cerebral toxoplasmosis two studies have been unable to detect a difference in efficacy between clindamycin/pyrimethamine and the gold standard of therapy sulphadiazine/pyrimethamine [13, 14].

In fungal infections the systemic azole drugs are being used widely for oral candidiasis and this may reduce the incidence of oesophageal disease. Fluconazole appears to be an effective alternative therapy and maintenance treatment for cryptococcal meningitis compared to amphotericin B with or without 5-flucytosine although different dose regimens of both treaments need to be assessed [15, 16].

A number of newer drugs, namely the new macrolides, azithromycin and clarithromycin and BW586 C80, a hydroxy napthoquinone, are being evaluated in a variety of infections.

As we become more successful in suppressing some of the commoner infections of the immune deficiency induced by HIV and prolong survival, we may witness a change in the spectrum of disease, perhaps a higher incidence of lymphoma, neurological problems and CMV infections.

Drugs such as ganciclovir and phosphonoformate are used widely to treat CMV retinopathy. In retinopathy the median time to relapse without treatment is 21 days but with maintenance therapy approximately 70 days.

In other words, with maintenance therapy we are buying time. One of the problems with maintenance therapy is that a long-term intravenous line has to be established and this brings with it the problem of sepsis, particularly if neutropenia occurs. Promising results are being obtained with an oral formulation of ganciclovir and studies comparing this to intravenous maintenance therapy are about to start [17]. As far as CMV retinopathy is concerned it is still not clear whether one should treat early or defer therapy until more central areas of the retina are involved. A trial which attempts to answer this question will be presented at the conference. Although it seems clear that CMV is a pathogen in the eye, its role is less certain in gastrointestinal disease and disease of the lung. A placebo-controlled trial, the preliminary results of which were communicated at last year's conference, only showed a marginal benefit in CMV colitis and the place of maintenance therapy in this situation is unclear [18]. With the widespread use of acyclovir particularly for suppressive therapy in anogenital herpes simplex virus infections and the wider use of ganciclovir in CMV, herpes virus resistance may become a prominent clinical problem. This will make it more important to find other drugs which are effective.

Zidovudine Resistance

Viruses isolated in vitro with reduced sensitivity to zidovudine were first described in 1989 [19]. These were described in symptomatic patients treated with zidovudine. The mechanism of this resistance is still not entirely clear but it is associated with conserved point mutations of the reverse transcriptase gene producing amino acid substitutions in positions 67, 70, 215 and 219. A fifth point mutation leading to a substitution at position 41 is being characterised [Larder and Kellam, personal commun.; 20]. Information is accumulating on the prevalence of this phenomenon in patients treated at different disease stages. In symptomatic patients these viruses seem to become dominant fairly quickly, emerging around 6 months with the majority of patients having some degree of decreased sensitivity by 1 year. In asymptomatic individuals these viruses are slower to appear and the level of resistance lower such that at 12 months of therapy an estimated 30% of asymptomatic individuals may have resistant viruses but the level of resistance of these isolates is much lower than those seen in symptomatic individuals [21, 22].

There are some key questions about antiretroviral resistance which remain unanswered. The first is does it matter? No study has yet demonstrated a clear association between the emergence of isolates with reduced sensitivity and a worsened prognosis for the patient. In other words, does virus resistance explain the perceived window of zidovudine's clinical efficacy. If resistance does matter then clinical virologists will need more convenient assays to detect it. In the future these assays might be used to tailor antiviral strategy for individual patients. We still do not know whether these mutants can be transmitted from one individual to another and what implications this might have. We need more information on the relationship between the emergence of this phenomenon and the dose and frequency of administration of antiretrovirals. There is an increasing fashion for lower doses of zidovudine and in the long term this may encourage the more rapid emergence of strains with reduced sensitivity, particularly in viraemic patients.

Although there is no published account of virus strains with reduced sensitivity to the dideoxynucleosides, namely, dideoxycytidine (ddC) and dideoxyinosine (ddI or didanosine) (see below), this cannot be far away. In particular, what mutations of the reverse transcriptase gene account for it? There seems to be no cross-resistance between zidovudine and other nucleoside analogues other than those with a 3′-azido moiety. Does the same apply to these drugs or will cross-resistance be demonstrated. Finally, although scientific information is lacking, most of us would feel that resistance does matter and therefore how do we prevent it occurring?

Zidovudine in Asymptomatic Infection

The licence for zidovudine has now been extended in many countries in the developed world for use in patients who are well but infected with HIV. The basis for this move was the results of ACTG protocol 019, a placebo-controlled study of zidovudine in two doses, 500 and 1,500 mg a day [23]. The study enrolled 3,200 patients between 1987 and 1989. The endpoints of the study were severe ARC and AIDS and patients were stratified according to CD4 count at entry. The median follow-up was about 14 months (range 5–27). As with ACTG protocol 016 the results apply to the 1,338 patients with CD4 lymphocytes counts less than 500 as very few endpoints were reached in those with counts greater than 500. In this time a total of 58 progressions to AIDS occurred. 33 were seen in the placebo group compared to 11 and 14 in the low- and high-dose groups, respectively. A similar pattern was seen when considering AIDS with ARC together as first clinical events. In other words, the progression rate to AIDS and advanced ARC in patients with baseline CD4 counts below 500 was approximately halved from 7.6/100 patient years in the placebo group to 3.6 and 4.3 in the low- and high-dose groups, respectively. Although the differences between the placebo and treatment arms were significant, no significant difference was seen between the two doses. However, there was a clear difference in associated haematological toxicity, both anaemia and neutropenia, and for this reason an independent Data and Safety Monitoring Committee recommended the use of the low dose in patients with CD4 lymphocyte counts less than 500.

The results of this trial are of course important but have to be put into perspective. A simple way of looking at the results of this study is to imagine that a given physician has 400 asymptomatic patients with CD4 counts less than 500. If the physician offers treatment to 200 and all accept at the end of just over 1 year (the median follow-up of the trial) 8 will still progress to AIDS or ARC and 192 will remain well. If the physician defers treatment until illness appears then 15 will progress in just over 1 year and 185 will remain well. Treamtent then over this period of time prevents 7 progressions in just over 200 patient years. That is the benefit of therapy but the cost concerns the patients that remain well on treatment after a year. Is this efficacy maintained and for how long? If a decline in efficacy is seen, is this accounted for by the emergence of zidovudine resistance such that when an individual becomes symptomatic, treatment is not as effective. Indeed, could the emergence of zidovudine resistance in asymptomatic individuals compromise potential benefit from combination chemotherapy? Combination therapy may be no different from monotherapy in these individuals. What is the cost of long-term toxicity, haematological, other side effects such as myopathy and perhaps as yet unexpected adverse events.

There is much uncertainty surrounding early intervention and it is hoped that European trials such as the MRC/INSERM Concorde study, a placebo-controlled trial of 1 g of zidovudine in asymptomatic HIV infection, may provide further information. As of April 1991 the intake to this study was 1,718 patients with a follow-up of 2,285 patient years. 126 individuals have progressed to AIDS and there have been 50 deaths. 244 individuals have been put on open zidovudine because of a persistently low CD4 count following a change in the protocol in October 1989 in the light of the ACTG 019 data. There will be a data freeze at the end of June 1991 when 3,000 patient years of follow-up should have been reached (the time for the single planned interim analysis) and data from this should be available for the Data and Safety Monitoring Board in the Autumn of this year.

It is likely therefore that a number of factors will influence physicians who will be pragmatic and pro-term practice the art of medicine rather than the science. Such factors which might influence the decision around any individual might include the age of that person, the length of HIV infection and the appearance of very early symptoms and signs. In addition trends not individual values in a cocktail of laboratory markers, such as P24 antigenaemia, serum β_2-microglobulin and/or neopterin as well as the CD4 lymphocyte count will be used to assess immune status. Finally, and most importantly, the decision will be influenced by the comfort level that the physician and patient have with a balanced interpretation of the scientific information that we have to date.

Zidovudine Dose

The 1-gram dose being used in trials in asymptomatic infection in Europe and Australia was chosen to compliment the studies in the USA which have to some extent concentrated on 500, 600 and 1,500 mg. The optimum dose of zidovudine is not yet established. In ACTG 019 no difference was demonstrated in terms of clinical efficacy between the low (500 mg/day) or high (1,500 mg/day) zidovudine doses in asymptomatic individuals with CD4 counts less than 500. But the lower dose was less toxic.

ACTG protocol 02 compared two doses of zidovudine, 1,500 mg daily, or 1,200 mg daily for 1 month followed by 600 mg daily, in patients with one episode of PCP [11]. In a median follow-up period of 26 months, the number of new opportunistic infections, malignancies, change in P24 antigen levels and CD4 counts were similar in both groups. Estimated survival rate at 18 and 24 months was superior in the low-dose arm. Survival curves of the two groups separated and then converged. Because of dose reductions and interruptions in therapy because of toxicity, the high dose arm received a median cumulative dose of zidovudine of only 204.5 vs. 169.1 g in the low-dose arm. However, the reason for the apparent difference in survival is unclear. A significantly greater proportion of subjects in the low-dose group received pneumocystis prophylaxis earlier in the study although the authors felt that this did not account for the difference in survival.

More recently in an open pilot study with six arms to compare zidovudine 300, 600 or 1,500 mg/day orally alone or in combination with acyclovir 4.8 g daily in patients with ARC and CD4 counts between 200 and 500, no additive or synergistic anti-HIV effect of this combination was observed [24]. However, the trial did provide some interesting information with respect to a low dose of zidovudine and antiviral effect. The 300-mg dose of zidovudine appeared to have similar antiviral effects, as measured by effect on P24 antigenaemia and plasma virus titres. to doses of 600 and 1,500 mg. The high number of withdrawals in the 12-week assessment period meant that the number of individuals actually studied, particularly in the high-dose arm was rather small. Nevertheless, this interesting observation needs to be confirmed in a larger number of patients. In addition, further low-dose comparison studies, to examine clinical efficacy, are needed before there is any move towards routinely prescribing such low doses. It is not yet clear what the relationship might be between low dose and the development of resistance and furthermore the effects of such low doses, if any, on the development of HIV-related disease of the nervous system.

Combination Chemotherapy

The aims of combination chemotherapy are to retain or enhance efficacy with lower doses of drugs, with little or no toxicity overlap, thereby improving tolerance. If there is no cross-resistance and other mechanisms of resistance are involved in the combination additons to zidovudine then such therapy may prevent or delay the emergence of resistant viruses. We will see the early results of phase I studies of newer agents in the coming year, namely, other nucleoside and non-nucleoside reverse transcriptase inhibitors, protease inhibitors, glucosidase inhibitors, etc. However, we must remember that most phase I studies bury drugs. The real options currently for combination chemotherapy are other dideoxynucleosides, namely ddC and ddI. Trials are also in progress to assess the value of α-interferon and acyclovir.

Many thousands of patients have received ddI on open therapy and many hundreds have received both ddI and ddC in trials which are still in progress. However, we should stand back at this point in time and ask ourselves what we really know about these two drugs. Both appear to cause promising changes in surrogate markers (CD4, P24 and weight) in patients on open therapy. Both cause a dose-dependent peripheral neuropathy, which in most cases is reversible and ddI is associated with a pancreatitis in between 1 and 2% of the patients treated with advanced disease which is sometimes fatal. Do they work? We have no good scientific information about their clinical efficacy, in particular whether they are better or worse than the gold standard, namely zidovudine. We await with interest the results of important head-on comparison studies currently in progress.

Surrogate Markers of Therapeutic Outcome

HIV infection like a number of other diseases is a chronic process with a long asymptomatic phase before clinically measurable 'hard' endpoints occur, namely symptomatic disease and death. A number of these endpoints are being suppressed by antimicrobial prophylaxis and an earlier use of zidovudine is leading to a prolonged survival. With an increasing number of candidate drugs to be assessed in phase III clinical efficacy studies there is a pressure on clinical investigators to define more proximal endpoints, namely surrogate biological markers which will predict outcome in a therapeutic trial [25, 26]. The question is: Can we get a quicker correct answer? Candidate markers which are being used in trials are markers which measure either immune status, immune activation or the amount of virus. Markers of immune status consist of CD4 number but CD4 function using an anti-CD3 proliferative assay is also being assessed. Serum β_2-microglobulin and neopterin are non-specific markers of immune activation. A decrease in serum P24 antigen has been used as an indirect measure of antiviral activity but it is not very sensitive. A minority of asymptomatic individuals and only 50–60% of patients with symptomatic disease are positive. New methods of increasing sensitivity of P24 assays by disassociating immune complexes of P24 and anti-P24 are being assessed. More direct measures of the level of virus in plasma or cells using culture or PCR have also been developed and changes have been demonstrated with antiviral therapy. Many of these markers particularly the old ones rather than the new have been shown in natural history studies to be associated with a poor or better outcome depending on their level. With therapy the question we are asking is whether a given change in one of these markers predicts outcome. A natural history marker is usually but not always a good marker of therapeutic outcome [25, 26]. A lowered serum cholesterol and levels of blood pressure are taken as surrogate markers for drugs which will prevent cardiovascular disease and mortality, rather than evaluating these drugs in trials which have to wait for long-term clinical endpoints. Similarly, bone density and total body calcium are taken as surrogate markers for drugs active against osteoporosis rather than waiting for fractures. Change in tumour size is taken as a surrogate marker for drugs in cancer therapy and intra-ocular pressure is a surrogate marker for long-term visual function in drugs active in glaucoma. All these surrogate markers have served as a basis for regulatory agencies' approval. However, there are examples where an observed change in an accepted natural history marker with therapy does not fit with the expected outcome. Cardiac arrhythmias around the time of a myocardial infarction predict early death. Two drugs, namely flecanide and encainamide, although having an effect in suppressing arrhythmias were associated with an increase in mortality in a recent study [26]. Furthermore, hypoalimentation regimens that lead to weight gain or at least a decrease in weight loss in patients with cancer were unexpectedly associated with a decreased survival.

In addition, there is a danger that we read too much into our current markers and forget that we might not be measuring the right thing. With current antiviral therapy in patients with asymptomatic or symptomatic infection the change that we see in CD4 lymphocyte count is a modest but transient increase with a return to baseline. Serum P24 decreases and after a variable time it increases back towards baseline. There may be a biological parameter that we are not measuring, an increase or decrease which may be associated with a significant effect on outcome over and above that predicted by CD4 lymphocyte and P24 antigen changes. For example, this may be CD4 function or some other immunological status measure or cell-free viraemia. As a result, we may run with drugs that turn out to be inferior and discard those that are superior to standard treatment.

The ideal surrogate marker must predict response (morbidity/mortality) ideally at all stages of disease. It must be detectable in the majority of individuals at all stages (a major drawback with P24 antigenaemia). It must be related to our concepts of HIV pathogenesis, i.e. it must be biologically plausible or acceptable. It should change towards normal quickly since we are seeking proximal endpoints. If it takes years rather than months or weeks to change nothing would be gained from using such a marker. It should respond in the same way to different drugs and if possible to different drug classes although this is a tall order. Finally, it must be user friendly, such that we can use it in multicentre studies and then in clinical practice and for this reason high standards of standardisation and quality control are required.

Conclusions

In 1991 at this conference, there is still uncertainty around early intervention with zidovudine, still the gold standard antiretroviral therapy. If we stand back from those drugs soon perhaps to be available at the bedside, it is surprising how little we know of them. There will be pressure from surrogate marketeers for clinical scientists to use surrogate marker endpoints in clinical trials, although which markers to use and how to use them is not yet clear. Certainly, the newer candidate markers need to be written in to scientific trials now if we are to evaluate them in terms of how they predict clinical outcome. If we look back on the therapeutic advances in HIV infection there is a temptation to close chapters long before they are completed. The star wars of antiviral therapy is not necessarily just around the corner. If we are to make progress in 1991–1992, we must continue to question the science, cross-examine the hype and then we might move forward faster.

References

1 Fischl MA, Richman DD, Grieco MH, et al: The efficacy of azidothymidine (AZT) in the treatment of patients with AIDS and AIDS–related complex. A double-blind, placebo-controlled trial. N Engl J Med 1987;317:185–191.
2 Richman DD, Fischl MA, Grieco MH, et al: The toxicity of azidothymidine (AZT) in the treatment of patients with AID and AIDS-related complex. A double-blind placebo controlled trial. N Engl J Med 1987;317:192–197.
3 Fischl MA, Richman DD, Hansen N, et al: The safety and efficacy of zidovudine (AZT) in the treatment of subjects with mildly symptomatic human immunodeficiency virus type 1 (HIV) infection. A double-blind, placebo-controlled trial. Ann Intern Med 1990;112:727–737.
4 Rottenberg R, Woelfel M, Stoneburner R, et al: Survival with the acquired immunodeficiency syndrome. N Engl J Med 1987;317:1297–1302.
5 Creagh-Kirk T, Doi P, Andrews E, Nusinoff-Lehrmann S, Tilson H, Hoth D, Barry

BW: Survival experience among patients with AIDS receiving zidovudine. Follow-up of patients in a compassionate plea program. JAMA 1988;260:3009–3015.
6 Williams I, Gabriel G, Cohen H, Williams P, Tedder RS, Machin S, Weller LVD: Zidovudine – the first year of experience. J Infect 1989;18(suppl 1):23–31.
7 Moore RD, Hidalgo J, Sugland BW, Chaisson RE: Zidovudine and the natural history of the acquired immunodeficiency syndrome. N Engl J Med 1991;324:1412–1416.
8 Centers for Disease Control: Guidelines for prophylaxis against pneumocystis carinii pneumonia for persons infected with human immunodeficiency virus. MMWR 1989;38:1–9.
9 Polk BF, Fox R, Brookmeyer R, et al: Predictors of acquired immunodeficiency syndrome developing in a cohort of seropositive homosexual men. N Engl J Med 1987;316:61–66.
10 Masur H, Ognibene FP, Yarchoan RY, et al: CD4 counts as predictors of opportunistic pneumonias in human immunodeficiency virus infected individuals. Ann Intern Med 1989;111:223–231.
11 Fischl MA, Parker CB, Pettinelli C, et al: A randomised controlled trial of a reduced daily dose of zidovudine in patients with the acquired immunodeficiency syndrome. N Engl J Med 1990;323:1009–1014.
12 The National Institutes of Health University of California expert panel for corticosteroids as adjunctive therapy for pneumocystis pneumonia: Consensus statement on the use of corticosteroids as adjunctive therapy for pneumocystis pneumonia in the acquired immunodeficiency syndrome. N Engl J Med 1990;323:1500–1504.
13 Dannemann BR, Israelski DM, Remmington JS: Treatment of toxoplasmic encephalitis in patients with AIDS: preliminary report of the CCTG randomized trial of pyrimethamine plus sulfonamides versus pyrimethamine plus clindamyzin. 28th Interscience Conf Antimicrobial Agents and Chemotherapy, Los Angeles, 1988, abstr 562.
14 Katlama C, Wit S de, Guichard A, et al: 7th Int Conf AIDS, Florence, June 16–21, 1991, abstr WB30.
15 Dismukes W, Cloud G, Thompson S, et al: Fluconazole versus amphotericin B therapy of acute cryptococcal meningitis. 29th Interscience Conf Antimicrobial Agents and Chemotherapy, Houston, September 1989, abstr 1065.
16 Powderly W, Saag M, Cloud G, et al: Fluconazole versus amphotericin B as maintenance therapy for prevention of relapse of AIDS-associated cryptococcal meningitis. 30th Interscience Conf Antimicrobial Agents and Chemotherapy, Atlanta, October 1990.
17 Jacobson MA: Ganciclovir therapy for opportunistic cytomegalovirus disease in AIDS; in Volberding P, Jacobson MA (eds): AIDS Clinical Review 1990. New York, Marcel Dekker, 1990, pp 149–163.
18 Dieterick D, Kotler D, Busch D, et al: Randomised placebo-controlled study of ganciclovir treatment of cytomegalovirus (CMV) colitis in AIDS patients. 6th Int Conf AIDS, abstr FB94.
19 Larder BA, Darby G, Richman DD: HIV with reduced sensitivity to zidovudine (AZT) isolated during prolonged therapy. Science 1989;243:1731–1734.
20 Larder BA, Kemp SD: Multiple mutations in HIV-1 reverse transcriptase confer high-level resistance to zidovudine (AZT). Science 1989;246:1155–1158.

21 Boucher CAB, Tersmette M, Lange JMA, et al: Zidovudine sensitivity of human immunodeficiency viruses from high-risk, symptom-free individuals during therapy. Lancet 1990;ii:336,585–590.
22 Richman D, Grimes JM, Lagakos SW: Effect of stage of disease and drug dose on zidovudine susceptibilities of isolates of human immunodeficiency virus. J Acquir Immune Defic Syndr 1990;3:743–746.
23 Volberding PA, Lagakos SW, Koch MA, et al: Zidovudine in asymptomatic human immunodeficiency virus infection. A controlled trial in persons with fewer than 500 CD4-positive cells per cubic millimetre. N Engl J Med 1990;322:941–949.
24 Collier AC, Bozzette S, Coombs RW, et al: A pilot study of low-dose zidovudine in human immunodeficiency virus infecton. N Engl J Med 1990;323:1015–1021.
25 Valentine FT, Jacobson MA: Immunological and virological surrogate markers in the evaluation of therapies for HIV infection. AIDS 1990;4(Suppl 1):S201–S207.
26 Ellenberg S: Surrogate end points in clinical trials (Leader). Br Med J 1991;302: 63–64.

Prof. Ian V.D. Weller, Academic Department of Genito-Urinary Medicine, UCMSM, 73–75 Charlotte Street, GB–London W1N 8AA (UK)

Rossi GB, Beth-Giraldo E, Chieco-Bianchi L, Dianzani F, Giraldo G, Verani P (eds): Science Challenging AIDS. Basel, Karger, 1992, pp 141–158

AIDS Vaccines

Dani P. Bolognesi

Duke Center for AIDS Research, Duke University Medical Center, Durham, N.C., USA

Introduction

Despite overwhelming obstacles, the quest for a vaccine against HIV is now a full-fledged target goal of the biomedical research establishment. Much like AZT has demonstrated that antiviral agents can be effective against HIV and has opened the field of anti-AIDS drugs, successful trials in animal models with candidate immunogens have established the all important first step of feasibility and paved the way for the development of candidate vaccines, several of which are already being evaluated in man. In this discussion, progress with both simian and human immunodeficiency virus (SIV and HIV, respectively) vaccine research will be reviewed and balanced against the many challenges which still lie ahead. It will conclude that a pathway toward a vaccine against HIV is beginning to form.

SIV Vaccine Research

The AIDS-like disease which develops in certain small subhuman primates following infection with SIV, a virus which is quite similar to HIV, establishes this as a most attractive model for development of an AIDS vaccine. Susceptible animals remain generally plentiful, the time course for infection and disease is relatively brief and a number of diverse SIV isolates exist which can be evaluated as both immunogens and challenge stocks. This system is ideally suited for determining the principles of vaccination against primate lentiretroviruses. However, to what extent the experiences with SIV can be directly translated to HIV remains to be determined.

The experience to date with SIV vaccine trials can be summarized as follows:

(1) Protection can be achieved almost uniformly with inactivated whole virus or cell vaccines provided that the immunization protocol is balanced with the challenge dose.

(2) In contrast, protection against infection with subunits, recombinant vectors or combinations thereof has been observed in only one case.

(3) Correlates of protection are not known but, in the case of killed virus vaccines, appear to be unrelated to the presence of neutralizing antibodies.

(4) Cross-protection between diverse SIV strains has been achieved.

(5) Vaccines that fail to protect against infection can nonetheless have a significant impact on disease onset.

To reconcile these results at this time is problematic for a number of reasons. For instance, using the approach of priming with a vaccinia *env* recombinant and boosting with a recombinant subunit, some experiments have failed to protect [1] while one was successful [2]. However, there were differences in the vectors, the recombinant subunits and the SIV strains which were used both to prepare the immunogens and as the challenge virus. Another perplexity relates to the lack of correlation between protection induced by inactivated vaccines with humoral immune responses which is further accentuated by recent studies showing that passive administration of plasma from both infected and vaccinated animals can indeed confer protection of macaques against SIV challenge [3].

It can be safely assumed that many of these discrepancies as well as the poor performance when subcomponents of the virus are used as immunogens are due to the relatively early phase of experimentation with these elements. It can be anticipated that, as with HIV (see below), when component immunogens, immunization protocols and challenge viruses are improved and standardized, they will likely prove to be successful candidate vaccines. At the same time it would be of great benefit if one could pinpoint the elements contained within inactivated virus or cell preparations which give rise to a protective immune response. Such an approach was initiated by Murphey-Corb et al. [4] demonstrating that the protective activity partitioned with the SIV envelope fraction. Additional fractionation of this material should further illuminate this question.

One of the major stumbling blocks for vaccines against this class of viruses is their extensive variation. It now appears that using inactivated viral and cell vaccines cross-protection between divergent SIV strains can be achieved [5]. However, attempts to extend this to the related but more divergent HIV-2 strains were unsuccessful. As with the protection against virus challenge, both the active elements within the immunogens and the mechanism responsible for protection remain to be defined.

Several other important principles have been derived from the SIV studies over and above the demonstration of vaccine feasibility. One is the establishment of challenge models whereby virus can be introduced across mucosal surfaces. Although there are notable drawbacks in these systems, such as level of virus needed for infection (100–1,000 times more than for systemic infection) the emerging results are nonetheless of interest. In brief, vaccines which have protected against systemic challenge have failed to block intravaginal challenge on the one hand [6], while possibly showing efficacy in prevention against intrarectal infection (where the minimal infectious dose appears to be one order of magnitude lower) [7].

Studies are also underway to evaluate another important milestone in the quest for a successful vaccine; namely, the development of a protocol for evaluating efficacy against challenge with infected cells or mixtures of virus with infected cells. The barrier to be overcome is how to establish a challenge dose where infectivity can be titrated and standardized.

Finally, a potentially very important lesson has emerged from the SIV model. In several of the vaccine studies it was noted that even though protection against infection was not uniformly observed, some of the preparations induced an immune response that significantly delayed the onset of disease. The immunogens that gave rise to such results included killed virus [8], live attenuated virus [9] and possibly peptides [10]. In addition, a distinct but close relative of SIV, HIV-2, which is nonpathogenic in monkeys, also induced a protective response against disease [11], but not infection. In general, the observations in such cases indicate that the vaccinated animals appear to be able to control virus replication, and possibly virus spread much more efficiently than nonvaccinated animals. On the other hand, attempts at vaccination after the animals are already infected has had no demonstrable effect on either viremia or disease [12, 13], indicating that once the virus has established residence it may be very difficult to impact on its pathogenic course, an issue which is central to development of therapeutic vaccines (see below).

At the present time, the mechanism whereby vaccination aids the host in suppressing the virus is not understood. One could hypothesize that its impact is in reducing the infectious dose to a level which is more manageable by the immune system. More important to determine, however, is if vaccine-induced immunity is superior at controlling the virus than the immune response generated naturally to the infection.

HIV Vaccine Studies in Chimpanzees

Outside of man, the chimpanzee remains the only animal species which can be infected with HIV. It is important to differentiate the chimpanzee HIV-1 model from HIV in man and SIV in macaques in a number of respects. First, HIV-1 infection in chimps does not lead to disease. Second, the number of chimpanzees available for research is dwindling which dictates that experiments be limited to a few animals. Third, safety restrictions approach those applied to man making it difficult to incorporate certain experimental vaccinology tools. These limitations have restricted progress in HIV vaccine research per se.

Between 1985 and 1989 a number of vaccine-related studies were conducted in chimpanzees. The immunogens used derived principally from the HIV-1 envelope and included killed virus, subunits, peptides and vaccinia recombinants [14]. None of these approaches protected against infection; but, importantly, only low or no neutralizing antibodies and relatively weak cellular immunity were measurable at the time of challenge.

The poor immunogenicity of these preparations dictated that new tacts be taken. Girard et al. [15] at Pasteur Vaccins employed the prime/boost approach and tried several combinations of whole killed virus, vaccinia recombinant vectors bearing the HIV-1 envelope, envelope and gag viral subunits and even nonstructural viral proteins. In the main, none of these combinations produced protective immunity. However, when animals were boosted with peptides representing the principal neutralizing determinant (PND) of the virus [16], substantial neutralizing antibodies appeared. The animals were therefore challenged with virus and protection was achieved.

Berman et al. [17] at Genentech focused instead on the properties of the immunogen and the immunization scheme. Using a highly pure preparation of gp120, a different immunization schedule and a lower challenge dose, they succeeded in protecting chimps against infection, reversing their own prior experiment that had failed [18]. In the successful study, the titer of neutralizing antibodies, particularly antibodies directed to the PND at the time of challenge, appeared to be the best correlate of protection. Of particular significance is that this study employed immunogens formulated with aluminum hydroxide, an adjuvant which is suitable for use in man.

The apparent correlation between protection and the presence of neutralizing antibodies to the PND in two separate and different studies stands out and is consistent with those which identify this domain as the primary target of neutralizing antibodies in vitro [16]. In an effort to demonstrate that in vitro neutralization is significant in vivo, Emini et al. [19] mixed virus with neutralizing antibodies to the PND and gave the mixture to chimpanzees with the result that no infection resulted. These studies have recently been extended to a bona fide passive immunization experiment with monoclonal antibodies to the PND with complete protection against infection at the 6-month time point [20]. It would thus appear that neutralizing antibodies to the PND are sufficient for protection against experimental infection with the homologous virus, strengthening the notion that the correlation of protection with anti-PND antibodies in the vaccine studies mentioned above was probably meaningful.

Characteristics of the HIV-1 PND

The existence of a PND in HIV-1 was established by several independent but complementary approaches. One investigative team [20] followed the reductionist approach beginning with the entire envelope and used progressively smaller fragments, eventually narrowing down to a 24 amino acid fragment capable of generating neutralizing antibody activity equivalent to that induced by the whole envelope. Moreover its presence in the neutralization reaction resulted in the abrogation of neutralizing activity of anti-envelope antibodies, hence its designation as the PND [21]. The very same region was uncovered by approaches which sought to identify the neutralizing antibody inducing capability of predicted B cell epitopes consisting of hydrophilic domains possessing strong β turns [22]. Finally, overlapping peptides covering the entire envelope gene generated only one fragment capable of inducing neutralizing antibodies which was localized to the same region [23].

The PND is now known to be situated within third variable domain of the exterior gp120 glycoprotein. It is contained within a disulfide linked loop (cysteine *303* to cysteine *337*) [24] and represents a uniquely hypervariable region. This variability is apparently responsible for the predominantly isolate-specific neutralizing antibodies which result upon immunization or in response to natural infection [25]. The variation in this region is extensive among naturally occurring HIV-1 isolates [26] and significant variability can be detected within the population of viruses in a single individual [27, 28]. This extensive variation coupled to the apparent immunodominant nature of the variable domains presents an overwhelming obstacle for vaccine development insofar as this epitope is concerned.

However, not all portions of the PND fall into the variable category, particularly regions representing the crown of the loop (fig. 1). In some instances immunization with peptides representing the PND resulted in neutralizing antibodies which recognized this semiconserved region (e.g. the GPGRAF motif) and such sera were indeed able to cross-neutralize divergent HIV-1 isolates which possessed this sequence of amino acids within their PND [29]. To what extent this domain of the PND can be used to raise meaningful cross-protective immunity against the plethora of viruses in the population remains to be determined.

The relative efficiency of anti-PND antibodies to neutralize HIV-1 in comparison to other sites suggests that this region may represent an important functional domain of the virus. Recent studies demonstrate that single amino acid substitutions at certain sites within the crown of the loop result in noninfectious viruses [30] while others govern target cell preference [31]. Yet other amino acid changes appear to selectively affect the ability of HIV-1-infected cells to fuse with uninfected counterparts [32]. When this latter observation is considered in the light that anti-PND monoclonal antibodies selectively inhibit cell/cell fusion as well as virus infection at a step after binding of the envelope to the CD4 receptor [33, 34], it would appear that this domain is probably involved in the multistep process virus entry, most likely associated with the fusion event. One hypothesis is that cleavage of the PND by proteases on the surface of the target cell represents the 'trigger' for fusion. The critical cleavage sites have been proposed to reside within the semiconserved loop crown [35, 36]. However, while cleavage of the PND has been observed in preparations of gp120 isolated from virions, infected cells or mammalian expression systems (CHO cells), evidence that such cleavage occurs during the processes of infection or cell/cell fusion is lacking. Finally, it has also been suggested that the PND may contribute to the target-cell specificity of HIV-1 [37, 38a, b], perhaps through amino acid substitutions which affect its susceptibility to cleavage by different cell surface proteases.

Whatever represents the precise function(s) of the PND in virus entry it is most likely that its conserved domains play a critical role in the pro-

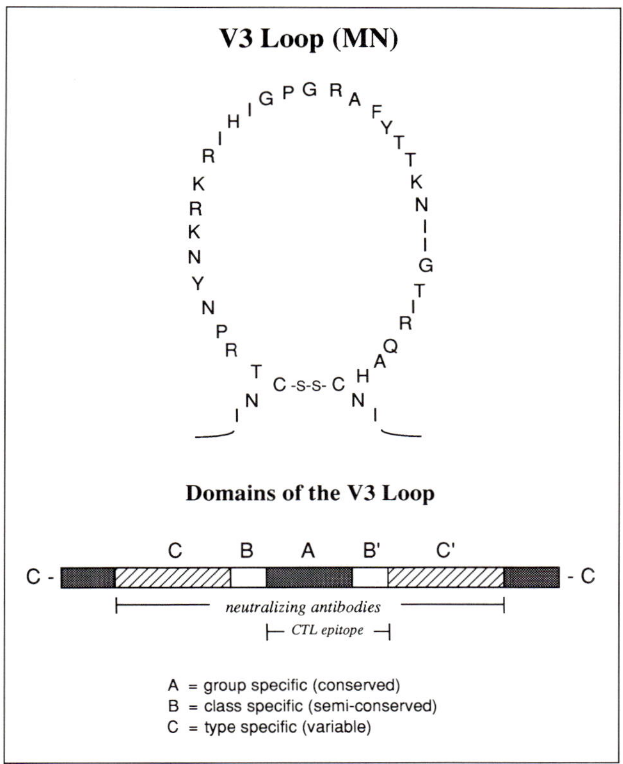

Fig. 1. Depicted is the PND amino acid sequence of the HIV_{MN} isolate. A high percentage of sera from HIV-1-infected individuals from North America recognize this sequence. Variation within this domain occurs primarily at the sides of the loop while the crown and the regions near the cysteines are much less divergent from isolate. Neutralizing antibodies are principally targeted to the sides (isolate restricted) or crown (cross-neutralizing). In addition, a CTL epitope has been mapped within this region [56] and evidence has been obtained that antibodies to the PND can also mediate ADCC [57].

cess. At the same time, variation at other sites may well contribute to conformational features of the loop which are important for both function and susceptibility to immune attack, particularly by neutralizing antibodies [39]. It is widely believed that this variability is one of the means by which the virus can escape from host immune defenses. This feature extends to escape from cellular immune defenses given that a dominant epitope for recognition by cytotoxic lymphocytes also exists within V3 (fig. 1). It would thus appear that an intricate balance exists between variation, structure/function and immune recognition in this unique structure.

Cross-Reactive Immune Responses among Diverse HIV-1 Isolates
As noted above, experimental immunizations with HIV envelope subunits, fragments or peptides primarily results in generating strain-specific neutralizing antibodies to the PND. This extends to experiments with vaccinia *env* recombinant vectors [40] and, most importantly, to infections in chimps [39] and man [41] with HIV-1. The latter observations with live HIV-1 pertain to the initial neutralizing responses to the virus which remain strain specific for at least 1 year and perhaps more (fig. 2). Thereafter, the neutralizing response begins to progressively broaden to other isolates. This shift in responsiveness is accompanied by two important correlates: (1) recognition of other PNDs, and (2) emergence of antibodies which block binding of gp120 and/or virus to CD4. The strength of the latter reactivity shows some correlation with the neutralizing capability of sera from infected individuals. This is consistent with recent findings of human monoclonal antibodies which block binding of gp120 to CD4 and exhibit some degree of cross-neutralization with diverse HIV-1 isolates [42–44]. Furthermore, evidence has been obtained that the epitopes in question are conformational in nature involving several noncontiguous sites on gp120. It remains a mystery why such epitopes appear to be occult in experimental immunizations and during the initial phases of infection. As was done in chimpanzee studies with antibodies to V3 it is now important to determine their role in the control of virus replication and their potential role in protection against infection in vivo.

Vaccine Trials in Man with HIV-1

A number of candidate immunogens are currently undergoing the initial phases of testing in man [45, 46]. To date, such trials have been carried out both in normal volunteers and in individuals already infected with HIV-1. The purpose of the former trials is to investigate the immunogenicity and toxicity of the preparations in a small number of subjects as an obligatory first step toward larger trials in high-risk individuals where vaccine efficacy might be evaluated. On the other hand, the goals of the post-exposure trials are to determine if vaccination is of benefit to the already

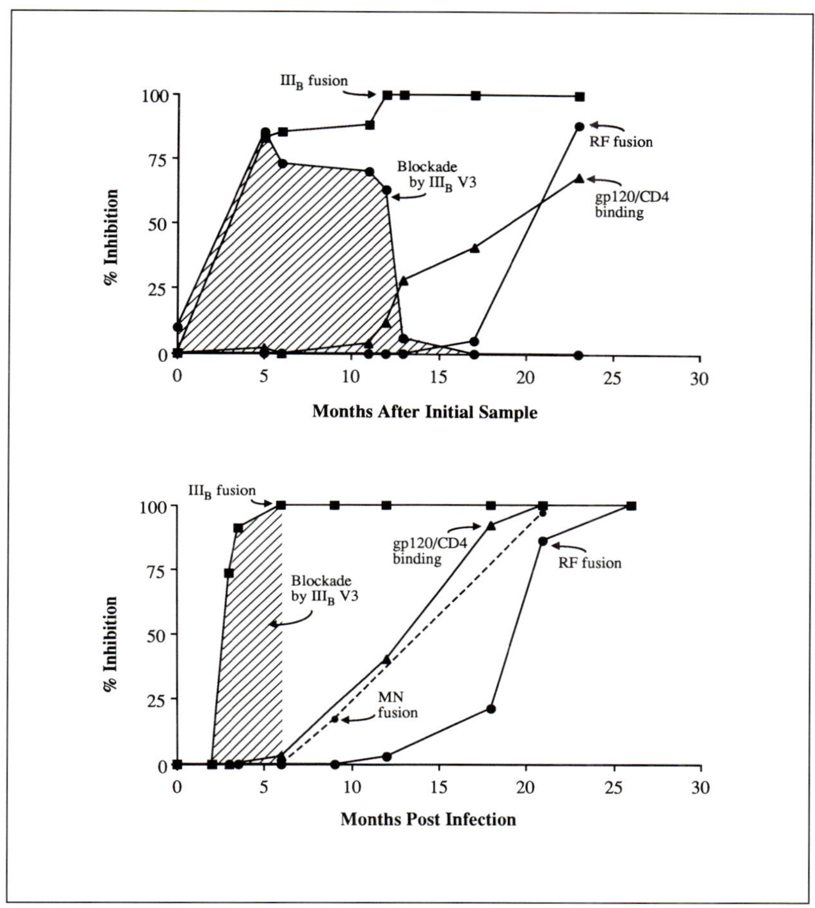

Fig. 2. Time course of immunobiologic responses in a laboratory worker infected with HIV (HIV IIIB) (top). The initial time points represent a phase of infection very soon after seroconversion. At that juncture, antibodies that mediate antibody-dependent cell cytotoxicity (ADCC) were already evident, although neutralizing antibodies or antibodies that blocked fusion were not detectable (not shown). The latter activities developed in parallel and were initially isolate restricted and remained so for approximately a year, as evidenced by inhibition using the IIIB V3 peptide. Reactivity to another HIV isolate, the highly divergent RF strain, was not detected until 23 months and coincided with a slowly increasing titer of antibodies that blocked binding of gp120 to CD4. Between 12 months and 23 months, it is likely that variants of the infecting strain emerged along with as well as recognition of conserved epitopes on the virus envelope. A similar pattern has been observed in a chimpanzee infected with HIV IIIB (bottom) were the MN isolate was used for comparison.

infected individual, while at the same time serving as a testing ground for vaccine immunogenicity and safety. With the exception of immunogens belonging to the category of inactivated or attenuated preparations of virus or infected cells, the candidate immunogens being evaluated in normal volunteers generally mirror those undergoing animal testing with both SIV and HIV. Thus, several envelope subunits, recombinant vaccinia bearing the HIV-1 envelope, recombinant particles with HIV-1 gene products prepared in yeast expression systems and peptides representing portions of the p17 shell protein of the virus have been introduced in man [45]. A virtual multitude of other virion components in various configurations are under development and consideration for clinical testing [47].

Until recently, the results of these trials have been unremarkable from the point of view of immunogenicity testing, but neither have the immunogen and adjuvant preparations generated untoward reactions. Both aspects, however, may change as doses are escalated. The most promising carried out thus far combines two separate trials in a prime/boost configuration: a limited number of individuals initially received a gp160 envelope recombinant vaccinia vector which was followed by boosting with a gp160 recombinant envelope subunit [48]. Strong responses were observed in both the cellular and humoral compartments even when the boosting with the recombinant subunit occurred as much as a year from the priming with the recombinant vector. This is indicative that long-term immunological memory was achieved. Sera from individuals receiving this combination showed, for the first time, high titers of neutralizing and fusion-inhibiting antibodies [49]. Furthermore, when peripheral blood mononuclear cells from vaccines receiving the individual immunogens alone and in combination were introduced into SCID mice, only animals receiving PBMC from the combination resisted challenge with infectious HIV-1 [50]. Evaluation of the presence of cytotoxic lymphocytes in these patients is currently underway.

Postexposure vaccination has also produced results which are of considerable interest. Three separate studies have been carried out employing killed virus [51], envelope subunits [52] and recombinant vaccinia vectors in combination with viral subunits and peptides [53]. Salk and colleagues performed studies with inactivated viral vaccines in patients already experiencing AIDS symptoms [51]. In spite of the fact that only limited immune responses specific to the vaccine were detected, there were benefits reported to the patients such as stabilization in their CD4 decline and weight loss. Clinical improvement along similar lines was reported by

Zagury and his colleagues [53] in AIDS patients who received their own fixed cells after infection with vaccinia recombinant vectors bearing envelope, *gag* and *pol* genes of HIV-1. Clinical benefits were observed by Redfield and colleagues [52] using a recombinant gp160 envelope subunit; but in this study, evidence was obtained that the immunogen used provoked novel immune responses not generally recognized in infected individuals.

To date, it has not been possible to determine the mechanism by which postexposure immunization brings about the clinical benefits observed. The hoped-for outcome is that these approaches improve upon the natural immune response resulting in better control of the virus and its pathogenic course [52]. On the other hand, the effect may not be due to novel immune responses to the virus, but instead result from a general immune stimulation which enhances natural host defense mechanisms [53]. Both are possible and not mutually exclusive. Clearly, more studies are warranted along these lines which should include appropriate control trials. These approaches should also be extended to the SIV model where to date the results with inactivated vaccines in the postexposure setting have, as yet, not produced comparable results.

Summary and Discussion

At the heart of the issue of how best to approach development of an effective vaccine is an understanding of what constitutes protective immunity and how to elicit it through vaccination. To date, as this review has illustrated, only fragmentary information has been obtained in either the SIV, chimp or human experiments as to what such immunity might entail. Moreover, since most of the experiments carried out in the animal models were in the nature of feasibility studies, even if the protective elements could be defined therein, they might not necessarily apply to the more complex situations which occur during natural transmission. Otherwise stated, immunity which is effective against experimental inocula with a given laboratory strain may not be effective against a highly divergent isolate, a mixture of natural isolates and even less likely against inocula consisting of both virus and infected cells. Likewise, the protective levels of immunity were achieved soon after a booster immunization and it is not apparent to what extent these need to be sustained in order for the vaccine to be effective in a practical sense. Hence, while one should view the initial protection trials in the animal models as an all important first step, the remaining issues of how to raise cross-protective immunity that will be effective against the existing swarm of HIV-1 isolates under natural modes of transmission will require that major gaps in the current knowledge base be filled. These include basic issues such as how to correlate virus variation with antigenicity, escape from immune defenses, viral transmission, tissue tropism and pathogenesis. Equally important is the necessity to develop new insights into vaccinology principles that would deal with the unprecedented variability of HIV-1 and how to elicit secretory immunity within the setting of comprehensive systemic immunity.

Such considerations dictate that animal models be available for extensive testing. This leads immediately to the question of relevancy and to what extent one can translate directly from one model to another and eventually to man. There will be some difficulties in this regard. For example, from the results thus far, it appears that there may well be significant differences in what constitutes protective immunity for SIV in monkeys versus HIV-1 in chimpanzees. Thus, correlation of protection against SIV with neutralizing activity is at best weak while neutralizing antibodies to the PND are currently thought to be the best correlate with protection of chimpanzees against HIV-1. This difference is punctuated by the apparent absence of a PND within the cognate region of SIV. Indeed, the corresponding SIV domain is relatively conserved between divergent isolates [54] and when used as an immunogen it induces little or no neutralizing antibody [55; Palker unpubl. observations]. Instead of a linear PND, current evidence suggests that anti-SIV neutralizing antibodies are directed at conformational determinants [55] which may nonetheless include the SIV V3 domain. It remains to be determined whether this region of SIV is involved in virus entry and which variable domains of the SIV envelope are important immune escape.

Quite aside from the issue of translation from one model to the other, one outstanding property that SIV possesses which is lacking in the chimpanzee HIV-1 model is the ability to induce disease which is very similar to human AIDS. The absence of this feature is considered a severe limitation of the chimp model and a great advantage of the SIV model in relation to development of a vaccine for man. On the other hand, a better understanding of why the chimp is better able to control the virus could provide fundamental clues toward vaccine approaches. The fact remains that the model can only be used to evaluate the capability of a vaccine to prevent infection, which is certainly a relevant issue, but at the same time, the most

Table 1. Possible applications of HIV vaccines

1	Vaccines that prevent infection
2	Vaccines that impact on disease without completely blocking infection
3	Vaccines that are able to curb virus transmission
4	Vaccines that are therapeutic in nature

difficult to achieve. Most of the viral vaccines in use protect against disease while allowing varying degrees of infection to occur. The immune response induced by the vaccine while not sufficient to block infection entirely, reduces the virus burden to a degree which allows the host to eventually clear the infection. To what extent vaccines can be designed to completely prevent infection remains unknown.

This argument raises the issue of how to set criteria for vaccines against HIV. First and foremost, should one point only toward the ideal vaccine which would guarantee blockade against infection or is it reasonable to consider more than on type of vaccine for different applications? For instance, a vaccine designed to prevent infection would demand the highest possible efficacy while ones designed to impact disease or transmission may require less rigid criteria. In this regard, the lessons from the SIV model are of particular significance in that they may provide information on what one can expect if varying amounts of virus are allowed to establish residence in a susceptible host. Likewise, animals that are better able to control virus replication may be less likely to transmit the virus to other animals or to their offspring. Understanding how to impact on transmission through vaccination would be of great importance for the growing problem of mother to infant infection in the human population.

It is implicit from the above discussions that both animal models will continue to guide vaccine development against HIV, each playing its respective role to answer a number of important questions. Already it is apparent that while preventive vaccines which are capable of inducing sterilizing immunity against all variants of HIV-1 may be very difficult to achieve, such may not be the case for vaccines which may have a significant impact on the course of infection, disease onset or disease severity. Likewise, vaccines might be more rapidly developed for situations where virus transmission could be reduced. In this regard, vaccines applied post-exposure for therapeutic purposes are already being evaluated, as noted above.

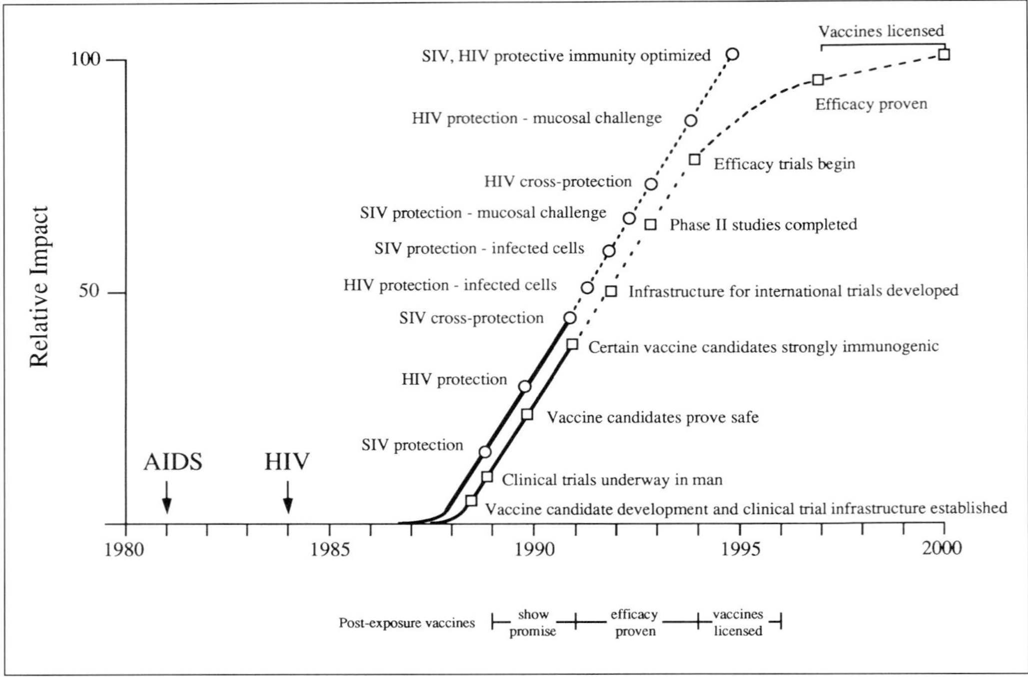

Fig. 3. Hypothetical pathway toward HIV vaccines: ○——○ = animal model studies achieved; ○- - -○ = animal model studies to be accomplished; □——□ = advances made in human trials; □- - -□ = future goals of human trials. Information below the abscissa reflects more rapid time course for postexposure vaccines.

It may therefore be worthwhile to consider developing several vaccines with varying criteria for efficacy and safety (table 1). Were this to take place, the design of clinical trials would meet several additional challenges. How many vaccines can reasonably be tested in humans, particularly in the efficacy phase? Given the low incidence of HIV infection in the general population with notable exceptions in highly endemic areas, coupled with the long and variable latent period between infection and the onset of disease, several parameters are expected to constrain the structure of clinical trials, such as size, geography, end-point measurements, etc. Fortunately, the infrastructure for conducting clinical trials with AIDS vaccines has been established and many of the initial obstacles have been overcome [45]. With this base, it is now possible to begin planning for international trials in developing countries [57].

In conclusion, while much remains to be accomplished before vaccines for AIDS becomes a reality, the significant progress made has stimulated considerable interest and commitment on the part of the federal government, academia and private industry toward this end. If this trend continues and research produces answers to the key scientific questions that remain, the prospects for efficacious vaccine by the year 2000 are not outside the realm of possibility (fig. 3).

Acknowledgments

The author is grateful to Miss Carla Blankenship for her expert assistance in the assembly of the manuscript. Supported by NIH Grant No. 2P01–CA43447–06.

References

1 Gardner M, Stott J: Personal commun.
2 Hu S-L, Abrams K, Barber G, et al: Protection of macaques against homologous SIV infection by immunization with live recombinant vaccinia virus followed by recombinant-made SIVmne gp160. 7th Int Conf AIDS, Florence, June 16–21, 1991, abstr THA12.
3 Putkonen P, Thorstensson R, Ghavamzadeh L, Albert J, Hild K, Biberfeld G, Norrby E: Prevention of HIV-2 and SIVsm infection by passive immunization in cynomolgus monkeys. Nature 1991;532:436–438.
4 Murphey-Corb M, Martin LN, Davison-Fairburn B, Ohkawa S, Baskin GB, Zhang J-Y, Montelaro RC, Miller M, West M, Allison AC, Eppatein D, Putney S: Induction of protective immune responses to viral infection by immunization of rhesus monkeys with formalin-inactivated whole SIV and glycoprotein-enriched SIV subunit vaccines. Vaccines 90, Cold Spring Harbor, Cold Spring Harbor Laboratory, 1990, pp 393–402.
5 Cranage MP, Stott EJ, Kent KA, et al: Inactivated SIV vaccine protects against heterologous SIV but not HIV-2 challenge. Nature 1991; submitted.
6 Marthas ML, Sutjipto S, Miller CJ, et al: Efficacy of live, virulence-attenuated and inactivated whole simian immunodeficiency virus vaccines against intravenous and genital routes of challenge. AIDS Res Hum Retroviruses 1991; submitted.
7 Stott J: Personal commun.
8 Desrosiers RC, Wyand MS, Kodama T, et al: Vaccine protection against simian immunodeficiency virus infection. Proc Natl Acad Sci USA 1989;86:6353–6357.
9 Marthas ML, Sutjipto S, Higgins J, et al: Immunization with a live, attenuated simian immunodeficiency virus (SIV) prevents early disease but not infection in rhesus macaques challenged with pathogenic SIV. J Virol 1990;64:3694–3700.
10 Schafferman A, Jahrling P, Benveniste R, et al: Inhibition of SIV infection in

macaques preimmunized with several SIV env peptides presented as B-galactosidase fusion proteins. VI Int Conf AIDS. Proc Natl Acad Sci USA, in press.

11 Putkonen P, Thorstensson R, Albert J, Hild K, Norrby E, Biberfeld P, Biberfeld G: Infection of cynomolgus monkeys with HIV-2 protects against pathogenic consequences of a subsequent simian immunodeficiency virus infection. AIDS 1990;4:783–789.

12 Murphey-Corb M, Davison-Fairburn B, Ohkawa S, et al: Immunization of healthy SIV-infected macaques with a formalin inactivated whole virus vaccine has no apparent effect on disease progression and survival (abstract). 8th Ann Symp Non-Human Primate Models for AIDS, New Orleans, 1990, p 55.

13 Gardner MB, Jennings M, Carlson JR, et al: Postexposure immunotherapy of simian immunodeficiency virus (SIV) infected rhesus with an SIV immunogen. J Med Primatol 1989a;18:321–328.

14 Girard MP, Eichberg JW: Progress in the development of HIV vaccines. AIDS 1990; 4(suppl 1):S143–S150.

15 Girard M, Kieny M-P, Pinter A, et al: Principal neutralizing domain of HIV-1 envelope protein. Proc Natl Acad Sci USA 1989;88:542–546.

16 Javaherian K, Langlois AJ, Silver S, et al: Principal neutralizing domain of HIV-1 envelope protein. Proc Natl Acad Sci USA 1989;86:6768–6772.

17 Berman PW, Gregory TJ, Riddle L, et al: Protection of chimpanzees from infection by HIV-1 after vaccination with recombinant glycoprotein gp120 but not gp160. Nature 1990;45:622–625.

18 Berman PW, Groopman JE, Gregory T, et al: Human Immunodeficiency virus type 1 challenge of chimpanzees immunized with recombinant envelope glycoprotein gp120. Proc Natl Acad Sci USA 1988;85:5200–5204.

19 Emini EA, Nara PL, Schleif WA: Antibody-mediated in vitro neutralization of human immunodeficiency virus type 1 abolishes infectivity for chimpanzees. J Virol 1990;64:3674–3678.

20 Emini EA, Schleif WA, Murthy K, Eda Y, Tokiyoshi S, Putney SD, Matsushita S, Nonberg JH, Eichberg JW: Passive immunization with a monoclonal antibody directed to the HIV-1 gp120 principal neutralization determinant confers protection against HIV-1 challenge in chimpanzees. 7th Int Conf AIDS, Florence, June 16–21, 1991.

21 Rusche JR, Javaherian K, McDanal C, et al: Antibodies that inhibit fusion of human immunodeficiency virus-infected cells bind a 24-amino acid sequence of the viral envelope, gp120. Proc Natl Acad Sci USA 1988;85:3198–3202.

22 Palker TJ, Clark ME, Langlois AJ, et al: Type-specific neutralization of the human immunodeficiency virus with antibodies to env-encoded synthetic peptides. Proc Natl Acad Sci USA 1988;85:1932–1936.

23 Kenealy WR, Matthews TJ, Ganfield M, et al: Antibodies from human immunodeficiency virus-infected individuals bind to a short amino acid sequence that elicits neutralizing antibodies in animals. AIDS Res Hum Retroviruses 1989;5:173–182.

24 Leonard CK, Spellman MW, Riddle L, et al: Assignment of intrachain disulfide bonds and characterization of potential glycosylation sites of the type 1 recombinant human immunodeficiency virus envelope glycoprotein (gp120) expressed in Chinese hamster ovary cells. J Biol Chem 1990;265:10373–10382.

25 Putney SD, McKeating JA: Antigenic variation in HIV. AIDS 1990;4(suppl 1): S129–S136.

26 LaRosa GJ, Davide JP, Weinhold KJ, et al: Conserved sequence and structural elements in the HIV-1 principal neutralizing determinant. Science 1990;249:932–935.

27 Simmonds P, Balfe P, Ludlam C, et al: Analysis of sequence diversity in hypervariable regions of the external glycoprotein of human immunodeficiency virus type 1. J Virol 1990;64:5840–5850.

28 Balfe P, Simmonds P, Ludlam C, et al: Concurrent evolution of human immunodeficiency virus type 1 in patients infected from the same source rate of sequence change and low frequency of mutations. J Virol 1990;64:6221–6233.

29 Javaherian K, Langlois AJ, LaRosa GJ, et al: Broadly neutralizing antibodies elicited by the hypervariable neutralizing determinant of HIV-1. Science 1990;250: 1590–1593.

30 Ivanoff LA, McDanal C, Morris J, et al: Biological analysis of HIV-1 proviruses containing mutations in the envelope V3 domain (abstract). AIDS Res Hum Retroviruses 1991;7:169–170.

31 Ivanoff LA, Looney DJ, McDanal C, et al: Alteration of HIV-1 infectivity and neutralization by a single amino acid replacement in the V3 loop domain. AIDS Res Hum Retroviruses in press.

32 Freed EO, Myers DJ, Risser R: Mutational analysis of the cleavage sequence of the human immunodeficiency virus type 1 envelope glycoprotein precursor gp160. J Virol 989;63:4670–4675.

33 Linsley PS, Ledbetter JA, Kinney-Thomas E, et al: Effects of anti-gp120 monoclonal antibodies on CD4 receptor binding by the env protein of human immunodeficiency virus type 1. J Virol 1988;62:3695–3702.

34 Skinner MA, Langlois AJ, McDanal CB, et al: Neutralizing antibodies to an immunodominant envelope sequence do not prevent gp120 binding to CD4. J Virol 1988; 62:4195–4200.

35 Hattori T, Koito A, Takatsuki K, et al: Involvement of tryptase-related cellular protease(s) in human immunodeficiency virus type 1 infection. FEBS Lett 1989;248: 48–52.

36 Clements GJ, Price-Jones MJ, Stephens PE, et al: The V3 loops of the HIV-1 and HIV-2 surface glycoproteins contain proteolytic cleavage sites: A possible function in viral fusion? AIDS Res Hum Retroviruses 1991;7:3–16.

37 O'Brien WA, Koyanagi Y, Namazie A, et al: HIV-1 tropism for mononuclear phagocytes can be determined by regions of gp120 outside the CD4-binding domain. Nature 1990;348:69–73.

38a Shioda T, Levy JA, Cheng-Mayer C: Macrophage and T cell-line tropisms of HIV-1 are determined by specific regions of the envelope gp120 gene. Nature 1991;349: 167–169.

38b Hwang SS, Boyle TJ, Lyerly HK, Cullen BR: Identification of the envelope V3 loop as the primary determinant of cell tropism in HIV-1. Science 1991;253:71–74.

39 Nara PL, Smit L, Dunlop N, et al: Emergence of viruses resistant to neutralization by V3-specific antibodies in experimental human immunodeficiency virus type 1 IIIB infection of chimpanzees. J Virol 1990;64:3779–3791.

40 Earl PL, Robert-Guroff M, Matthews TJ, et al: Isolate- and group-specific immune responses to the envelope protein of human immunodeficiency virus induced by a live recombinant vaccinia virus in macaques. AIDS Res Hum Retroviruses 1989;5: 23–32.

41 Bolognesi DP: Prospects for prevention of and early intervention against HIV. JAMA 1989;261:3007–3013.
42 Ho DD, McKeating JA, Li XL, et al: Conformational epitope on gp120 important in CD4 binding and human immunodeficiency virus type 1 neutralization identified by a human monoclonal antibody. J Virol 1991;65:489–493.
43 Thali M, Olshevsky U, Furman C, et al: Structure and function of a discontinuous gp120 epitope recognized by a broadly-reactive neutralizing human monoclonal antibody. 7th Int Conf AIDS, Florence, June 16–21, 1991.
44 Steimer KS, Nara P, Brooks E, et al: Immunization of primates with recombinant HIV-1 gp120 induces broadly neutralizing activity that cannot be explained by V3 sequence homologies. 7th Int Conf AIDS, Florence, June 16–21, 1991.
45 Koff WC, Schultz AM: AIDS vaccines 1990: A brief update. AIDS 1990;4(suppl 1): S179–S184.
46 Cohen J: AIDS vaccine conference: Is 'more' better? Science 1990;250:369–370.
47 Karn J, Almond JW, Tyrrell DAJ: Research strategies for AIDS vaccine development and evaluation: The MRC programme. AIDS 1990;4(suppl 1):S167–178.
48 Graham BS, Belshe R, Clements ML, et al: HIV-gp160 recombinant vaccinia vaccination of vaccinia-naive adults followed by rgp160 booster immunization. 7th Int Conf AIDS, Florence, June 16–21, 1991.
49 Matthews TJ: Unpubl. observations.
50 Mosier DE, Gulizia RJ, MacIsaac PD, et al: Evaluation of HIV-1 vaccines in hu-PBL-SCID mice (abstract). 7th Int Conf AIDS, Florence, June 16–21, 1991.
51 Levine A, Henderson BE, Grushea S, et al: Immunization of HIV-infected individuals with inactivated HIV immunogen: Significance of HIV-specific cell-mediated immune response (abstract). 6th Int Conf AIDS, San Francisco, June 1990, p 204.
52 Redfield R, Polonis BD, Davis V, et al: Active immunization of recombinant produced gp160 in patients with early HIV infections: Phase I trial immunogenicity and toxicity. AIDS Res Hum Retroviruses 1991;7:136–137.
53 Picard O, Giral P, Defer MC, et al: AIDS vaccine therapy: Phase I trial. Lancet 1990; 336:179.
54 Burns PW, Desrosiers RC: Selection of genetic variants of SIV in persistently infected rhesus monkeys. J Virol 1991;65:1843–1854.
55 Putney S, Langlois A, LaRosa G, et al: The principle neutralization determinant of SIV is different than HIV (abstract). 7th Int Conf AIDS, Florence, June 16–21, 1991.
56 Takahashi H, Cohen J, Hosmalin A, et al: An immunodominant epitope of the human immunodeficiency virus envelope glycoprotein gp160 recognized by class I major histocompatibility complex molecule-restricted murine cytotoxic T lymphocytes. Proc Natl Acad Sci USA 1988;85:3105–3109.
57 Statement from the consultation on criteria for international testing of candidate HIV vaccines. Global Programme on AIDS, WHO, Geneva, February 27–March 2, 1989.

Dr. D.P. Bolognesi, Surgical Virology Laboratory,
Duke University Medical Center, PO Box 2926, Durham, NC 27710 (USA)

Rossi GB, Beth-Giraldo E, Chieco-Bianchi L, Dianzani F, Giraldo G, Verani P (eds): Science Challenging AIDS. Basel, Karger, 1992, pp 159–174

The Psychosocial Aspects of AIDS

Marvellous M. Mhloyi

University of Zimbabwe, Harare, Zimbabwe

Introduction

A decade has passed. We have lived to see the AIDS pandemic spread like a carcinoma. The problem is becoming more intractable and complex as special high-risk groups are increasingly becoming societies at large. The devastating impact of AIDS is unfolding, as resources, both financial and human, are being spread even more thinly. The time has come for us to reflect on our efforts and reconsider new intervention approaches. At no time since the dawn of wonder drugs has it been so poignantly imperative that we go beyond the frontiers of our disciplinary academic formulations in their purity, for such purity has not resulted in the more aggressive, innovative and daring approaches needed to contain this multidimensional and most insidious problem. Still without a vaccine to prevent, or drugs to cure HIV infection, the only tool available to us is to change human behavior.

Several changes in the epidemiology of the epidemic during the 1980s are worth noting for their implications in shaping the nature of the psychosocial aspects of AIDS. First, the distribution of reported infections has shifted from the industrialized to developing countries. Of the 0.2% of the world's population infected with HIV-1, approximately 81% of these infections are in developing countries [1]. Africa alone contains 63% of the world infections, while Latin America and Asia contain 12 and 6%, respectively [1]. This geographic shift is accompanied by an increasing dominance of heterosexual and vertical transmission. For instance, up to 90% of transmissions in Africa are associated with sexuality [2–4]. Because of the increasing role of heterosexual transmission, Latin America is now classified as pattern I/II [5]. Parallelling these shifts is the increasing importance of women as victims and leaders in the epidemic. While both men and women can play a direct role in parenteral and sexual transmission, women have an exclusive role in vertical transmission. The number of women and children cases globally was 500,000 in the 1980s, by early 1990 more than 3 million women were HIV positive [6]. These shifts in geography, mode of transmission and population subgroups, pose challenges in resource allocation for intervention and research, in perceptions regarding risk factors and behavior, and, consequently, in intervention approaches.

Risk Factors and Intervention

In order to combat the spread of HIV infection, we need to change human behavior which expose people to the risk of parenteral, vertical or sexual infection. The intricate and dynamic interaction between a person's behavior, the social, economic and political contexts in which such behaviors occur, and the characteristics of the person [7] require in-depth understanding of both individuals, and their social and physical environments if change in human behavior is to be effected and sustained. The task of intervention programs is to orchestrate changes of those aspects of the individual and the environment relevant to determining health behavior [8]. In AIDS intervention programmes such orchestration requires that attention be paid to understanding the nature of high-risk behaviors, their distribution, the cultural and emotional symbolism attached to them, and the subjective and economic costs of the alternatives provided. Beliefs and perceptions about disease, and their effect on health-seeking behavior within specific socioeconomic, cultural and political contexts, remain poorly understood and nearly speculative. The relative absence of this qualitative knowledge is reflected in the impact intervention programs have had on HIV-1 transmission.

Parenteral Transmission

Significant progress has been made to reduce blood transmission. Most developed countries, through blood screening, have managed to virtually eliminate this form of transmission [9]. However, constrained resources in the developing world make complete blood screening a distant reality. For example, only 50 and 80% of the blood is screened in Zambia and Malawi, respectively. In Mozambique, 7 of 10 provinces have blood-screening facilities [10]. The progress made in reducing transmission through blood transfusion is consistent with health initiatives which place minimal psychosocial and economic constraints on the individual.

The difficulty of changing human behavior is illustrated by intravenous drug users (IVDUs), who share a subculture shaped by two powerful variables: the illegality of drug use, and the physiologic realities of dependency [11]. Because society stigmatizes IVDUs, they tend to withdraw and isolate themselves. IVDUs often lack a sense of community largely because they lack trust of each other [12]. AIDS intervention programs among IVDUs have been shaped more by subjective attitudes regarding drug use than by scientific knowledge. The scientific knowledge base is limited and progress has been slow. For instance, a survey of 1,500 IVDUs in the USA revealed that HIV infection remains a major public health problem in the country. Seroprevalence rates among this group ranged between 0 and 47% with a median of 5% [13]. Another study on IVDUs attending a chemical dependency treatment program in Minnesota showed seroprevalence rates of 1.3, 1.5 and 2.2% in 1987, 1988 and 1989, respectively [14]. Increasing HIV seroprevalence suggests a relapse to risky behavior. In a Miami study of 718 IVDUs, it was shown that denial of risk by those who engage in high-risk behavior was the highest impediment to intervention [15]. Only 10.8% of the clients perceived themselves at high risk of HIV infection. Another study in Switzerland revealed that drug users had a tendency to view AIDS as 'yet another problem to be dealt with' [16].

Fortunately, there is evidence that intervention programs can effect behavioral change. For example, needle sharing among 700 clients attending 18 syringe exchange programs in England declined from 27 to 36% in 1987/1988 to 21% in 1989 [17]. The program, however, could not attract younger clients, shorter term injectors and women during the intervention period. It was also noted that demand for syringes outstripped supply, and that the program lacked central co-ordination, supervision and direction [18]. In San Francisco an anonymous needle exchange program was able to create a broad base of community support pushing officials to establish needle exchange as public policy [19]. This community mobilization was facilitated by a coalition of strong AIDS advocacy groups. The need for broad-based national programs iwth varied modes of syringe distribution to facilitate the reaching of this heterogenous population has been demonstrated, and countries lagging behind may learn from those with such programs like Germany, Switzerland and Australia [16, 20, 21].

Studies have shown that exchanging sex for money is relatively common among some IVDUs. It is estimated that the proportion of female prostitutes who use drugs ranges between 40 and 85% depending on location and group. Drug use is more common among street sex workers than those working from homes [30]. Some studies have shown that drug users who engage in sex for money are more likely to engage in high-risk behavior with a consequent low rate of condom use [23]. This link between drug use and sex is an important pathway of HIV transmission from IVDUs to the larger population.

The relatively limited progress in effecting the necessary behavioral changes among IVDUs requires a reassessment of the existing intervention programs. It is possible that more benefits may accrue from intervention programs which combine efforts to deal with the underlying factors to drug abuse, and those which aim at making drug use safer. More research is needed to determine the nature and utility of such programs. In addition, there is a need to examine how sexual partners of drug users can be encouraged to demand condom use.

Although initial claims of extensive HIV contamination from contaminated instruments used in traditional scarification have not been substantiated [9], with the increase in infections, continuous effort should be made to make such practices, including male and female circumcision, safer than before.

Vertical Transmission

It is ironic that a child's biological link to its mother may also be a child's link to HIV infection. While changing high-risk behavior among potential mothers is currently the most effective way of stopping the spread of HIV infection to children [24], the achievement of this goal remains a battle. One option for seropositive women is not to have children. However, this option is difficult for many in societies where women derive status from children, and where infant mortality is also high. Increasingly, arguments against contraception for fertility regulation are growing in countries facing high mortality from AIDS [25]. Literally millions of women in Africa, Asia and Latin America face this dilemma. For HIV-positive women who get pregnant abortion may be yet another option. However, abortion is not legal in many countries. In addition, we know little about the psychosocial costs of abortion particularly to women in those societies where culture and religion oppose abortion, and even to some extent contraception – for example the Philippines, the Caribbean, and some parts of Central and South America [26]. Because of their low socioeconomic status, a large number of women cannot even make decisions regarding childbearing, let alone abortion. Renewe efforts are needed to legalize abortion, and to make it available and safe to all women.

Sexual Transmission

While sex is the least efficient mode of transmission, it is the most common and least amenable mode to intervention. HIV infection among the general heterosexual population is increasing in almost all countries. For instance, a study of women attending selected family planning clinics in the United States between 1988 and 1990 revealed seroprevalence rates of 0–2.3% with a median of 0.2% [27]. In Canada a study of 1,453 women undergoing abortions in Montreal clinics revealed a seroprevalence rate of 0.21% [28]. In Bahia, Brazil seropositivity among blood donors ranged between 0.29 and 2.2% [29].

In most of central sub-Saharan Africa, the perception that HIV infection is limited to high-risk groups like prostitutes, sexually transmitted disease patients, and truck drivers is no longer valid [9]. For instance, infection in the general population in Rwanda ranges from 1.3% in rural areas to 17.8% in the capital city. In rural Uganda, rates range from 2.7 to 14.7%; they range between 1.8 and 5.3 in Mozambique; in Zimbabwe, seroprevalence among blood donors was 5% in 1989 [10]. In Thailand, since the identification of the first AIDS cases in 1984 among homosexual prostitutes who were reported to have had sexual contact with foreigners, the infection rate has increased to 45% among drug abusers. Rates among low-paid Thai sex workers range between 45 and 72% [30, 31]. Seroprevalence among blood donors increased from 0.3% in June 1989 to 0.46% in December of 1990 at which point 0.19% of pregnant women also tested positive [30].

It is imperative therefore that programs should seek to change sexual behavior not only of the homosexual and prostitute populations, but also of the larger so-called mainstream population. It is a shift from the fixation on 'high-risk groups' to risky behavior in the general population, a shift which avoids stigmatization and consequent isolation of particular groups as responsible for the disease.

Evidence to date shows that there has been more progress in effecting changes in these stigmatized groups, particularly homosexuals, than in the general heterosexual population. In the United States and elsewhere, there has been a decrease in the prevalence of unprotected sex among homosexuals in recent years especially in the epicenters of the disease [32]. The changes in behavior have been facilitated, to a large extent, by broad-based community support with a wide variety of informal communication channels, enhanced by peer group support. Some of the gays, however, revert to risky behavior after an initial adoption – for example in a cohort study on homosexual and bisexual men in four United States cities [33]. Factors associated with relapse include low self-efficacy, poor perception of community support for sexual risk reduction, excessive alcohol consumption, and alcohol and/or drug use in conjunction with sex. An increase in infection rates among homosexuals was recently reported in San Fancisco [National Broadcast News, 2nd June, 1991]. A French study on the male homosexual population showed rapid changes in behavior between 1985 and 1987/1988 and a subsequent decline in 1989 [34]. Interestingly, the behavioral changes start with a reduction in the number of sexual partners before forsaking anal intercourse and/or adopting condom use.

Another high-risk sexual behavior bridging heterosexuality and homosexuality, but understood less than both, is bisexuality. In Latin America the 'machismo' pressure forces many homosexual men to marry but nonetheless continuing sex activities with other men [35]. Twenty five percent of homosexual men in the Netherlands also have sex with women but frequently hide it because they fear negative reactions from the gay and general society [35]. European conservative and religious communities which are hostile to homosexuality drive bisexual practice underground [35]. In the United States approximately 20.3% of adult men have had some sexual contact with other men [36].

The little that is known about bisexuality is, however, worrying. In Latin America studies have shown that men who do not identify themselves as bisexuals are less likely to use condoms especially with their wives; and bisexual men may engage both their male and female partners in high-risk behavior such as anal intercourse [35]. Because most bisexual men do not identify themselves as either homosexual or bisexual, they are least receptive to, and are sometimes bypassed by intervention programs. More research is needed to open channels of effective communication to this group.

Sexual behavior of the general heterosexual population is the most resistent to change. A study on 16,632 heterosexual women who attended the Pennsylvania Planned Parenthood clinics revealed that most women (67%) never used a condom with their regular partners [37]. An intervention study of heterosexual men and women with multiple sex partners in the Netherlands revealed no change in the frequency with which sexual techniques were practiced over a period of 20 months [38]. Some increase in frequency of condom use was reported only with commercial partners. Sexual behavior of this heterosexual population was found to be stable over time; only the number of partners was subject to change [38]. Changes in sexual behavior of the heterosexual population is particularly difficult to induce in many parts of Latin America, Asia and Africa where double standards regarding male and female sexuality exist [39–41]. For instance, some cultures encourage female virginity upon marriage [40, 42], and female marital fidelity is highly valued for it ensures legitimacy of children and thus the purity of the lineage [43]. In a number of African societies polygyny and levirate legitimize males having multiple sex partners. Yet even within marriage (including polygynous) women are not expected to raise issues on sexuality. For most women, breaking the silence in an attempt to practice safer sex may result in rejection, stigmatization, economic reprisals or even violence [26]. Yet most men do not want to use the condom because it is perceived to diminish sexual pleasure. Women's sexual subordinance is enhanced by their low socioeconomic status and their lack of education, property rights, and employment. A women's status in these societies is largely derived from children.

The economic dependency of women is the most important underlying factor to female prostitution. In his review, Carballo [44] noted that some women practiced prostitution as an economic 'right of passage'. He also noted that a common characteristic of female prostitutes in Africa and Britain is widowhood or separated/divorced status, another indication of the female economic dependence on males. For most prostitutes facing the dire need to survive and sometimes to feed a family, succumbing to their male patrons who refuse to use condoms is the only viable choice between two possible causes of death, starvation being the most immediate. There is great need to seriously and effectively put into place substitute income-generating projects for sex workers.

During the 1980s homosexuals and IVDUs were doubly unlucky as they bore the brunt of infection, and the political, psychological, and social burdens of the epidemic [45]. Women have taken over all the two plus the burden of caring for the sick, a triple jeopardy articulated in 1988 at the first conference of The Society of Women and AIDS in Africa in 1988. There is need for research to identify ways of effectively empowering women economically in the face of shrinking economies. There is also a need to identify the mechanisms of breaking the 'silent norm' regarding sexuality among women, and replace this silence with dialogue in equal sexual partnerships. Women are in critical need for a protective device which they can control. Research is also needed to find ways of assisting women to cope with the epidemic.

The Youth

A growing number of adolescents are acquiring HIV infection and developing AIDS [46, 47]. It is estimated that approximately one-fifth of global AIDS cases are people in their late twenties; the majority of these were infected in their adolescent years. A combination of the observed decline in age at first intercourse, and increasing mean ages at first marriage will increase opportunities for multiple premarital partnerships [44], and thus the risk of exposure to HIV infection. Unfortunately, condom use is also very low among the youth. Yet there is a pervasive denial of risk of infection by sexually active youth [47]. For most youngsters AIDS is just but another of the distant problems to be dealt with. For millions of Brazilian street children the fear of AIDS is crowded by other urgent concerns of homelessness, hunger, disease, violence and drugs. A British youngster makes this note about AIDS, 'Its like smoking, you look at a cigarette and you don't connect it with lung cancer ... I don't connect sex with disease' [47, p. 8].

Studies have shown that the involvement of people with AIDS (PWAs) in educational campaigns is effective in making AIDS a reality to the youth. Yet PWAs cannot be involved in public campaigns until fear and discrimination are replaced with understanding and companionship [47]. The challenge is for intervention programs to cultivate such qualities in the respective societies.

Sex education has been difficult in the face of limited knowledge about sexual practices of the youth. In addition, puritanical views regarding sex do not empower the youth with the knowledge needed to minimize risky behavior. For example, a study on teenage pregnancy in 37 developed countries revealed that the United States had higher rates of teenage pregnancy because sex education places emphasis on abstinence rather than on sexual decision-making, discussion of feelings of arousal and how to handle them, and safe sexual practices [48]. The presence of a large economically deprived class in the United States is also a contributing factor to the higher rates.

Research on cultural determinants and differences in early sexual experience, and their role in HIV transmission is needed. It is also important to assess the role of social factors like homelessness and family structure in determining risky behavior. Meanwhile, educational campaigns targeting the youth should be intensified. It has been shown that peer education is effective in influencing the adoption of risk reduction behaviors mainly because it fosters a sense of shared norms and identity [49]. Accelerating youth campaigns is therefore inevitable. Besides, it is easier to mould than to change behavior. A 'sex with condom' culture may save the youth, our window of hope.

Resistance against condom use transcends sexual orientation and age groupings. Condom use is less frequent with regular partners perhaps suggesting a reduction in motivation for condom use with increased intimacy. While intervention programs which assist women including prostitutes to take preventive measures against infection are needed, it is important to note that as long as the locus of condom control is among men, sustained and effective use of condoms can only be enhanced by the intensification of programs that also target men. For all the pleasure of parenthood universally cherished, it is likely that protecting the lives of women as mothers, and thus children, may be a worthwhile entry point for educational campaigns. The same men protesting condom use as inconsistent with manhood may forego the perceived pleasure of unprotected sex in the interest of healthy surviving offspring. It is a process of building on local values and empowering men to adopt less risky behaviors. The creation of a facilitative and supportive environment in which condoms are not only practical, but also embody a positive and/or erotic value to all sexually active persons is critical. Unfortunately, little headway has been made in this regard during the 30 years of family planning intervention. There is need to understand more about sexuality, and to fully utilize existing literature on sex from different disciplines and themes like gender and power relations, sexuality, and psychology. There is a need to generate new information for short-term intervention programs, and also for long-term basic research. Such research should endeavor to understand systems of meaning, knowledge, beliefs and practices that structure sexuality in different socioeconomic contexts [50]. It should differentiate cultural ideals from practice. Such work will require long-term financial support for both research and capacity building.

It is important to raise attention to the issue of morality which tends to be the inhibiting factor to most programs. Indeed, while these programs do reduce the risk of HIV infection, they do enhance behaviors which may in the long run ruin the health of program beneficiaries, a case in point is the provision of syringes to drug users. While programs which target the underlying factors to drug abuse may be complex and expensive, they would focus on the total health of the abusers, and thus may be more beneficial in the long run. Another issue is the incompatibility of condom use with procreation. Granted that condoms are not 100% effective, there is need for research to guide programs to encourage monogamous marital sex only. This borders on morality; yet the need for such morality cannot be denied. This need becomes even more apparent when one considers the cost and feasibility of supplying condoms to all sexually active persons who engage in extramarital sex. In addition, it should not be overlooked that use of a condom by an HIV-infected person is almost purely a moral consideration. The need for effective infusion of moral messages into intervention programs should not be ignored.

Coping

As we enter into the second decade of the AIDS epidemic towards the year 2000, attaining our health goals is increasingly becoming an illusion. The best that we can do is to cope at all levels in our societies. Regrettably, society attaches a stigma to AIDS patients, not because of the incurability of the disease, but because of negative and judgmental attitudes regarding how the disease is contracted. This destroys the emotional strength of those afflicted as they suffer from ostracism, guilt, anger and fear – all of which mitigate against health-seeking behavior. It is a struggle within a struggle. Family members who care for their sick in privacy in order to minimize shame run the risk of infection. Reports of abandonment of the sick by family members are not uncommon [26]. There is need to understand how those afflicted with AIDS can be empowered to meet their ordeals with courage, a sense of belonging, and dignity.

Medical staff, to some extent, may find it difficult to be useful partly because of their attitudes, lack of relevant training, and also fear, real and unreal, of contagion. For instance, 63% of the medical staff at a major New York City teaching hospital in 1989 expressed doubt over the fact that health workers who observe safety guidelines are at minimal risk of infection from patients [51]. Reports of medical staff burnout also exist. Mechanisms of relieving the medical staff of the increasing work load and the emotional stress, must be identified and implemented.

The devastating psychosocial impact of AIDS can now be observed on dependents who lose their most productive members and remain both economically and emotionally incapacitated. For instance, it is estimated that in sub-Saharan Africa alone there will be 10 million AIDS-related orphans by the end of this decade [8]. New issues also emerge. For instance, morbidity and mortality of these orphans and of aged dependents may increase as a consequence of the loss of financial support as AIDS victims die. It is also not clear how these children, surrounded by death of their most beloved, both young and old, will grow up to be normal people. But who can help, and how? [This issue is dealt with by Ankarah in this volume.]

A Flashback

What is the response of countries and organizations to this epidemic. As we look back over the past 10 years in pattern II countries, particularly Central African, an interesting pattern in the progression of the AIDS epidemic and the respective country responses emerge. Four stages can be identified (fig. 1). The duration of the stages depends on the rate at which the epidemic spreads in a given country.

The first stage is characterized by fairly low levels of infection often concentrated among 'high-risk' groups. At this stage (which can be described as the denial stage) national governments dismiss the disease as typical of those marginalized groups some of whom are in contact with foreigners. Some scientists endeavor to prove the foreign origin of the virus. The disease is thus considered inconsequential and unwarranting of investment. Publicizing the existence of the infection in a country is perceived as an unnecessary negative factor on the countries' tourist industry and foreign investment prospects. However, some governments may start initiating covert intervention programs like blood screening.

The second is the knowledge-based acceptance stage. As the disease spreads to the general population mostly demonstrated by its presence among blood donors and infants, most governments react by launching national AIDS control programs with educational campaigns to reduce risky behaviors. Intervention programs are mostly targeted at 'high-risk' groups. Although the general population becomes aware of the disease, it does little to reduce risky behavior because it assumes safety since it does not identify itself with the high-risk groups which includes foreigners. At this stage, there is a relapse in the momentum of the educational cam-

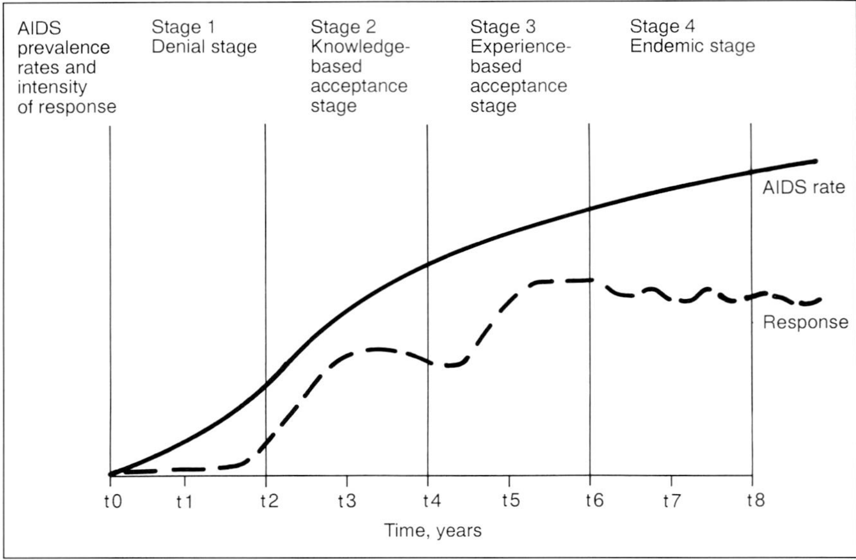

Fig. 1. The AIDS epidemic transition.

paigns. People feel bombarded and bored with the AIDS messages, the relevance of which they are skeptical of since they are not aware of individuals who died of AIDS.

The third stage, the experience-based stage, is characterized by an increasing number of people from the general population dying of AIDS. There is a boom of educational campaigns largely propelled by fear. Spontaneous formation of nongovernmental and voluntary organizations and clubs whose functions are mainly to educate people about risky behavior, and also to counsel those who are infected is characteristic of this stage. The momentum is maintained for a certain duration but subsequently declines as government and society become fatigued.

The fourth stage, the endemic stage, is characterized by problems of impact. The increasing number of AIDS cases imposes a significant impact on the health system. The number of orphans increases. Educational campaigns continue but the intensity is compromised as attention is diverted to coping with the consequences of the epidemic. As people observe more people dying of AIDS, some conclude that they too might be infected and thus find no reason to change their behavior. People learn to live with

AIDS and sometimes unwittingly tolerate death. Unless concerted efforts are made to instill a positive adaptive attitude, a 'silent conspiracy of complacency' may undermine efforts in changing risky behavior; it may also undermine the ability of governments to mobilize the necessary financial and human resources for combatting the disease.

A few lessons can be drawn. First, response lags behind the epidemic, and the rate of increase of the epidemic generally outpaces that of the response. This means that instead of being proactive, countries react to the epidemic. Second, once AIDS is identified in a country, whatever the population of entry is, it has the capacity to permeate the larger population like a carcinoma. Thus, every country has a potential to go through the four stages that some of the Central African countries have gone through. This is a time for global exchange of knowledge. As heterosexual transmission is increasing in every country lessons can be drawn from pattern II countries which have a 10-year experience with this mode of transmission. Countries currently reporting low levels of infection only among their high-risk groups, and linking such infection to external contacts, like Israel [52], Singapore [53], India [54], and Bolivia [55] must step up their educational campaigns to empower every individual regardless of sexual orientation, color or creed with knowledge and support to combat this disease. For those countries which are already in the fourth stage, this is the time to avoid the impact of AIDS from reducing their momentum in encouraging the adoption and sustenance of less risky behaviors, in order not only to reduce the continuous upward trend of the epidemic but also to eradicate the disease. They must not accept living with AIDS.

Conclusion

As we get into the second decade, the persistent trend of the epidemic after a decade of hard work is demoralizing, fatigue may increase, and consequently reduce the momentum in activities to combat the disease. The proclivity to escapism and denial may increase [56]. But fighting against AIDS is a war humanity cannot afford to lose. It is a war for human life. It transcends all wars. The highest priority is thus to support, strengthen and value the thousands of individuals working against, and living with AIDS. This conference should be a turning point at which every country reassesses its strategies and takes a proactive stance. The war is not over until science challenges AIDS.

Acknowledgements

The author is very grateful to Dr. A. Lucas for his comments, to Prof. L. Chen who edited the paper, and to Prof. D. Bell for his comments on the earlier version of the paper. Thanks are also due to two anonymous colleagues for their comments and encouragement.

References

1 Chen L: Aids and reproductive health. Workshop on AIDS and Reproductive Health, Bellagio, 1990.
2 Okware SI: Giving AIDS a New Face. Geneva, World Health Organization, 1989, October, p 18.
3 N'galy B: Epidemiology of HIV in Zaire; in Giraldo G, Beth-Giraldo E, Clumeck N, et al (eds): AIDS and Associated Cancers in Africa. Basel, Karger, 1988, p 37.
4 Mhalu FS, Dahoma A, Mbena E, et al: Some aspects of the epidemiology of AIDS and infections with the immunodeficiency virus in the United Republic of Tanzania; in Giraldo G, Beth-Giraldo E, Clumeck N, et al (eds): AIDS and Associated Cancers in Africa. Basel, Karger, 1988, pp 50–60.
5 Pan American Health Organization: AIDS surveillance in the Americas. AIDS Soc 1991;April/May:1.
6 Chin J: Current and future dimensions of HIV/AIDS pandemic in women and children. Lancet 1990;336:221–224.
7 Bandura A: The self-system in reciprocal determinism. Am Psychol 1978;33:344–360.
8 Baranowski T: Reciprocal determinism at the stages of behavior change: An integration of community, personal and behavioral perspectives. Int Q Hlth Educ 1989–90; 10:297–327.
9 Berkley Seth: AIDS in sub-Saharan Africa: An overview. Reports on Health and Development in Southern Africa 1990;Fall/Winter:5–9.
10 The Panos Institute: World AIDS 1990;991:5.
11 Valdiserri RO: Preventing AIDS, the Design of Effective Programs. New Brunswick, Rutgers University Press, 1989.
12 Des Jarlais DC, Friedman SR: Target groups for preventing AIDS among intravenous users. J Appl Soc Psychol 1987;17:251–268.
13 Allen DM, et al: Seroprevalence of HIV infections among drug user in the United States (abstract). VIth Int Conf on AIDS, San Francisco, 1990.
14 Thomas J, et al: A comparison of HIV-1 infection rates and behaviors in Minnesota of intravenous drug users over a three-year period: 1987–1989 (abstract). VIth Int Conf on AIDS, San Francisco, 1990.
15 McBride F, et al: The relationship between HIV risk behavior and perception of risk among intravenous drug users (abstract). VIth Int Conf on AIDS, San Francisco, 1990.
16 Hausser D, Lehmann P, Dubois-Aber F, Gutzwiller: Effectiveness of the AIDS prevention campaigns in Switzerland; in Fleming AF, et al (eds): The Global Impact of Aids. New York, Liss, 1988, pp 219–228.

17 Donoghoe M, et al: An evaluation of further development of syringe-exchanges in England (abstract). VIth Int Conf on AIDS, San Francisco, 1990.
18 Stimson G, Lart R: National survey of syringe-exchange program in England (abstract). VIth Int Conf on AIDS, San Francisco, 1990.
19 Downing M, et al: It is better to beg forgiveness, than to ask permission: A chronicle of prevention point needle exchange (abstract). VIth Int Conf on AIDS, San Francisco, 1990.
20 Friedrichs C, et al: The availability of sterile syringes in the Federal Republic of Germany and the offer by AIDS-Service organization and drug counselling centers for an exchange of syringes (abstract). VIth Int Conf on AIDS, San Francisco, 1990.
21 Ballard J: Regional profiles. World AIDS 1991;14:10.
22 Goldstein PJ: Prostitution and Drugs. Lexington, Lexington Books, 1979.
23 Plant ML (ed): AIDS, Drugs and Prostitution. London, Routledge, 1990.
24 UNFPA: Review of the impact of acquired immunodeficiency syndrome (AIDS) on women and children and the UNICEF response.
25 Perlez J: Toll of AIDS on Uganda's Women puts their roles and rights in question; in: Essential National Health Research Network Newsletter, vol II, No 1, 5th Jan 1991, p 5.
26 Gillespie MA: HIV: The global crisis. Ms January/February 1991;16–22.41.
27 Sweeney P, et al: HIV seroprevalence among women of reproductive age seeking clinic services, United States, 1988–1990 (abstract). VIth Int Conf on AIDS, San Francisco, 1990.
28 Remis R, et al: Prevalence and determinants of HIV infection among women undergoing an abortion in Quebec, Montreal (abstract). VIth Int Conf on AIDS, San Francisco, 1990.
29 Alves CB, et al: HIV seroprevalence and risk activities in blood donors, Bahia, Brazil (abstract). VIth Int Conf on AIDS, San Francisco, 1990.
30 Smith DG: Thailand: AIDS crisis looms. Lancet 1990;Mar 31 335:781–782.
31 Viravaidya Mechai: HIV/AIDS Situation and Surveillance in Thailand. Presented at The Havard School of Public Health.
32 Becker MH, Joseph JG: AIDS and behavioral change to reduce risk: A review. Am J Publ Health 1988;78:394–410.
33 O'Reilly KR, et al: Relapse from safer sex among homosexual men: Evidence from four cohorts in the AIDS community demonstration projects (abstract). VIth Int Conf on AIDS, San Francisco, 1990.
34 Pollack M, et al: Sexual behavior changes and HIV-seroprevalence in the French male homosexual population, 1985–1989 (abstract). VIth Int Conf on AIDS, San Francisco, 1990.
35 Tielman R: Educators need to understand bisexuality. World AIDS 1991;13:3.
36 Fay TM, Turner CF, Klassen AD, Gagnon JH: Prevalence and patterns of same-gender sexual contact among men. Science 1989;243:338–348.
37 Aral SO, et al: Condom use by women seeking family planning services (abstract). VIth Int Conf on AIDS, San Francisco, 1990.
38 Hooykaas C, et al: Changes in sexual behavior of men and women with multiple (private and commercial) partners (abstract). VIth Int Conf on AIDS, San Francisco, 1990.

39 Alonso AM, Koreck M, Silences: 'Hispanics', AIDS, and sexual practices. Differences 1989;1:101–119.
40 Sittirai W: Research on human sexuality in pattern II countries; in Chouinard A, Albert J (eds): Human Sexuality: Research Perspectives in a World Facing AIDS. Ottawa, IDRC, 1989.
41 Basset M, Mhloyi M: Women and AIDS in Zimbabwe: The making of an epidemic. Int J Soc Serv 1991;21:143–156.
42 Adeokun LA: Research on human sexuality in pattern II countries; in Chouinard A, Albert J (eds): Human Sexuality: Research Perspectives in a World Facing AIDS. Ottawa, IDRC, 1989.
43 Mhloyi M: Perceptions on communication and sexuality in marriage in Zimbabwe. Women Ther 1990;10:61–73.
44 Carballo M: Sexual behaviors. Temporal and cross-cultural trends. Forthcoming, King Homes.
45 MacDonald G, Helquist M: Reducing HIV transmission: Lessons from the past; in Chouinard A, Albert J (eds): Human sexuality: Research perspectives in a world facing AIDS. Ottawa, IDRC, 1989.
46 Kipke MD, Futterman D, Hein K: HIV infection during adolescence. Med Clin N Am 1990;75:1149–1167.
47 Kanene K, Kilalo G, Legion V, Terto V Jr, Panos: PWAs confront teen denial. World AIDS 1991;14:7–10.
48 Jones EF, et al: Teenage Pregnancy in Industrialized Countries. New Haven, Yale University Press, 1986.
49 Joseph JG, Montgomery SB, Emmons C, et al: Magnitudes and determinants of behavioral risk reduction: Longitudinal analysis of a cohort at risk of AIDS. Psychol Health 1987;1:73–95.
50 Parker H, Carballo M: Sexual culture, HIV transmission, and AIDS research. J Sex Res 1991;28:77–98.
51 Wallack JJ: AIDS anxiety among health care professionals. Hosp Community-Psychiatry 1989;40:501–510.
52 Bentwich Z: AIDS in Israel. Israel J Med Sci 1989;25:301–302.
53 Chew SK, Monterio EH: The acquired immunodeficiency syndrome in Singapore – epidemiological perspective. Singapore Med J 1989;30:28–31.
54 Seth P, Malaviya AN, Kiran U, Singh RR, Malaviya R, Khare SD, Bhargava NC, Chawla NP: Lack of evidence of endemicity of human immunodeficiency virus infection in Northern India. Ind J Med Res 1988;87:108–112.
55 Bartoloni A, Paradisi F, Aquilini D, Roselli M, Rivero R, Nunez LE, de Majo E, Parri F: Absence of HIV infection in low and high risk groups in Santa Cruz, Bolivia. AIDS 1989;March 3:184–185.
56 Mann J: Priorities of the second decade. World AIDS 1991;13:2.

Marvellous M. Mhloyi, PhD, University of Zimbabwe, P.O. Box MP167,
Mt. Pleasant, Harare (Zimbabwe)

Rossi GB, Beth-Giraldo E, Chieco-Bianchi L, Dianzani F, Giraldo G, Verani P (eds): Science Challenging AIDS. Basel, Karger, 1992, pp 175–187

The Impact of AIDS on Social, Economic, Health and Welfare Systems

AIDS: Preserving the Family of Mankind

E. Maxine Ankrah

Department of Social Work and Social Administration, Faculty of Social Sciences, Makerere University, Kampala, Uganda

Introduction

AIDS is a killer disease which is relentlessly spreading suffering and grief throughout our social order. My task is to direct our discussion to a consideration of its impact on our social, economic, health and welfare systems. I take the view that the potential of this devastating pandemic to effect adverse consequences is most fully appreciated when applied to that basic unit of caring [1], namely the family system. I have chosen a synoptic and descriptive more than an empirical approach to examine some of the ramifications of the disease at that level. I contend that if we are to withstand the impact of AIDS and eliminate it from our midst, then a family perspective which heightens global solidarity and global learning is needed. I conclude that a global ethic becomes an imperative.

Changing Nature of Families and Family Life

Sarah, as a single female i.v. drug user, could not find a treatment facility at the time she resolved to get off drugs. Then, she met Mike. As they shared needles, she could not risk losing him, so she no longer tried to stop. The baby came. But the welfare worker would not recognize the three as a family. They had not married, officially. All being HIV positive, Mike died; the baby died. As Sarah left her parents and siblings to come to New York City, Sarah is dying ... alone. Mr. Mallya of Kagera Region in Eastern Africa had four grown children. He has now buried all but one of them because of AIDS. They left him the responsibility of caring for 19 orphans. He lives on in the village among others of his generation who must also cope with a similar loss.

Families and family life are being drastically changed by AIDS. Yet, we have marginalized the family and made it an invisible factor in our AIDS equation. Let me tell you why.

We, as scientists, have individualized the AIDS problem. First, we placed the gay white male 'at the epicenter of the epidemic'. Currently, the woman and the orphan are being brought to center stage. But, the individual, whether a gay man, a heterosexual woman, or an orphaned child cannot be taken out of the family system, the context in which most of mankind is located, and which provides ultimate support. Such individualization of HIV infection and AIDS unwittingly permitted us to define the problem as a 'person with AIDS' rather than as 'families, whose member, or members, have AIDS'. The other social and economic, health and welfare systems are being overwhelmed, therefore, not merely by the numbers of individuals who are HIV infected or have AIDS, but also by the varied, unanticipated and unprecedented problems afflicting families as they attempt to address the needs of a member.

There is the need to focus on the individual person when we are concerned mainly with the question of transmission. Nor are we totally unmindful of the family. But to see that the family dimension of AIDS has been largely overlooked is evident, first, by the paucity of policies, programs, financing and research that are directed toward this unit in its entirety. Consider research alone: out of about 5,500 papers selected for the Montreal Conference in 1989, a mere 35 studies focused on the family impact of AIDS. At San Francisco in 1990, the number was only 18. Only in 15 of the 53 studies were the issues of family resource needs, support services, treatment approaches and the social and economic impact addressed holistically. Thus, whether of the traditional, biological family of origin [2], the affiliated family of 'choice' [3], the larger extended family [4] – common in the developing countries – our knowledge of its impact, overall, and the family's response to AIDS is grossly insufficient.

This narrow view of the 'person with AIDS' can leave unattended the other multiple problems besetting this unit. Family strengths, as opposed to individual strengths, do not figure automatically as a major resource in problem solving [3]. Survival needs and future planning may well be peripheral issues. Moreover, a person-centered approach is basically alien to many cultures of developing countries. There is a danger that the ensuing strategies will not encourage scientists and program implementers to involve families proactively, to be aligned with them in finding ways to prevention and change. Rather, they could be called in as a reaction, to be used to manage a crisis situation where other systems have failed.

It is not surprising that the most dramatic impact of AIDS is being felt by the family. The most basic institution, responsible for procreation, for the development and socialization of the young, and for the care and welfare of the old, the family is the mainstay of most societies. Certainly, it is the key to social development no less than demographic continuity in most of Africa, the other developing regions, and throughout the so-called developed world.

AIDS is affecting families for a variety of reasons. The most important is that once HIV and AIDS are present in an individual, irrespective of whether the person who is infected is single or married, has children or not, is male or female, the biological problem of AIDS is almost always, and immediately, compounded by the social and economic ties that person has with others. In some way or another they are almost always cared for, or are being cared for by someone else, be this in terms of emotional or economic support [1]. Men and women assume responsibility for each other, for offspring, for older parents, for friends, for loved ones.

In many of our African communities the mortality rates among bread winners or heads of households is now so high that large numbers of families are being left without the type of economic base that would otherwise provide for a healthy survival by those who are left. In a context where the most prevalent transmission of AIDS is heterosexual, the presentation of AIDS in one spouse is often indicative if infection in the other, even if the process to AIDS itself does not follow the same timing.

But in a social environment where fear and misunderstanding about AIDS still dominate so much about individual relationships, AIDS brings with it other difficulties. Communication problems arise between parents and their children when they are made aware of, for example, homosexuality; between a wife, or a husband who was not aware of the other's sexual partners; and problems among young children, bewildered and confused by the pain and turmoil around them.

In many communities, families are becoming characterized by relationships in which spouses are being rejected because the wife or husband knows or suspects or fears that the other is infected. And in societies where the social, economic and personal status of both women and men is intimately linked to the framework of family – be it nuclear or extended – the social opportunities for that rejected person are immediately reduced. All too often rejection by the family is tantamount to rejection by society. For, in many communities there are few, if any, other resources to support individuals who are ill or destitute. The numbers of sick people who are 'on the streets', are, therefore, increasing, especially in the industrialized countries.

Cates et al. [5] refer to these various consequences as an 'exceptional crisis' in which a continuum of reactions are evoked from the family toward its member with AIDS. These range from complete support to total abandonment; from fear to stigma and denial, from anger to blame, withdrawal and isolation of those for which integration and care are most needed.

Our already high adult morbidity figures, then, are being further exacerbated by illnesses which might otherwise have been prevented, and whose severity would surely have been reduced, had there been a family environment able and willing to support individual coping. Particularly, the welfare of persons without families is deteriorating: their health is being most acutely threatened. And they will go on to add to the burden of health and social services that are already stretched to the breaking point. Their contribution to the economic well-being of the societies in which they live is meanwhile more and more rapidly diminishing.

Adults, of course, are not the only ones to be caught up in this vicious situation. Children – neglected in the disruption of adult relationships, or as a result of death – already present a major public health and ethical dilemma. Contrary to what many would like to believe, they are not always caught by supportive nets of extended families, by welfare systems, or charitable institutions.

In the 1990s it is estimated that between 1.5 and 2.9 million women of child-bearing age will die due to HIV/AIDS; and as a result of HIV, between 3.1 and 5.5 million children will be orphaned in the eastern and central African regions alone [6]. In New York, 60,000–70,000 children had one or no parent in 1988 [6]. These orphaned survivors, both seropositive and seronegative, needed to be placed in someone's care. But where institutions are ill-equipped for the responsibility, and where the biological relatives are too old, poor or overwhelmed, such as in Mallya's case, a family will have to be created for them. The spread of AIDS among young children and adolescents who find themselves on the streets after being orphaned, or due to poverty and broken homes, is also of great concern. From Brazil [7] to Nairobi [8] and New York [6] children scavenge on the streets for survival. As these street children become adolescents, they experience profound physical and psychological changes. Experimenting with a wide range of behavioral styles in the city from which to choose, they are exposed to sex, HIV infection, AIDS and death. The plight of the children represent the ultimate tragedy and the ultimate loss of the society's future intellectual and economic potential.

Attention should also be given briefly to the near familyless vulnerable 'others' who must count in what Levine [2] calls an expanded definition of family. HIV and AIDS have mostly afflicted groups that were already stigmatized before its advent [6]. In a number of industrialized countries the addiction of some women to i.v. drug use is used to criminalize the disease, to further stigmatize and to deny them the health care that is due. With respect to these women, the authorities are often content to perceive AIDS as a small problem. Minority male i.v. drug abusers that are HIV infected face the prospects, too, of a bleak future.

There are still other perturbing trends and shifts. The impacting force of AIDS on the nontraditional and nonfamilial arrangements, such as among homosexuals and the Sarah's of the inner cities, is extremely severe. The presence of HIV infection among young adolescents is also a cause for growing alarm in Europe, the Americas and Asia because this group has shown a reluctance to change the lifestyles that put them at risk. The workforce is being adversely affected, too. In Zaire, for example, HIV infection is higher among managers than workers [9]. These demographic patterns are worrisome; for, wherever such are prevalent, they make the task of coping evermore problematic.

Coping: Dissipating Individual and Community Strengths

Our preoccupation with the individual persons with AIDS may have blurred our perception of the degree to which families which remain relatively intact and communities are able to manage where there is a high incidence of AIDS cases. With the exception of the middle- and upper-income families in western countries, few households have been secure enough to bear the financial expenses connected with the treatment of AIDS. Rather, many families face a process of 'poveritization', a situation in which they must 'spend down to poverty', if they take responsibility for care. In the developing countries, security nets against AIDS outside the extended family hardly exist [10]. Even the affluent find that their assets of money and property are dissipated as they attempt to meet the costs of AIDS.

Beyond individual families, communities are being devastated by the impact of AIDS. The prospect of accelerated mass poverty is real in rural areas of developing countries where there is a high prevalence of HIV infection among those who are producers of food and upon whom the survival of millions depend now and in the future [11]. Community strengths are being ravaged, too, by a lack of transitional structures, such as support networks and nongovernmental, as well as government, organizations that enable the local people to mobilize for maximal collective management of the AIDS problem. In the inner cities of North America and Europe that capacity is often impeded by massive deterioration and social decay [12].

Communities will differ in their will to prevent the further transmission of the disease among their inhabitants. Cultivating that will is likely to be a slow process. Now is the time, therefore, for those nations whose peoples have hardly experienced the epidemic, and whose resources are not yet severely strained to act with the utmost haste to develop structures within communities through which to combat AIDS. These multiple dimensions of the problems of families – and beyond this unit to the community – call into question the current trend in patient care where the burden is shifted to the home and family, and the locus of treatment and action is the community, where we have not adequately assessed the capacity of the family or the community to cope.

Whether the family in the context of AIDS can successfully readjust and cope with the nature of the problem still remains to be seen. We do not believe it can do so without a major support initiative from national and international bodies.

Resource Constraints: The Health, Welfare and Economic Systems under Threat

Against this background then, we now examine more directly the impact of AIDS on our health, welfare and economic systems. By the mid-1980s, many of our governments were still reluctant to admit that they had an AIDS problem. Many of those who did were optimistic that they could adequately respond to the requirements of the few targeted groups through existing medical facilities and welfare services. But quickly, the picture has changed. In 1988, the US Surgeon General [13] stated that 'if the number of AIDS cases continued to double every 13 months, the United States would run out of hospital beds in 1991'.

We know that AIDS cases are now doubling every 9–12 months, and that in some hospitals in New York City [14] it is as difficult to find a bed as in Mama Yemo Hospital in Kinshasa, Zaire [9], or Mulago in Kampala! In Africa and many other parts of the nonindustrialized world we must continue to remind ourselves that AIDS has emerged in an environment already beset by health problems that have been neglected, or which have defied our collective ability to handle. It is an environment in which 100 million people suffer from and are constantly incapacitated by malaria. It is a world in which 750 million children a year suffer from diarrheal diseases and from which 4 million die as a result. It is a world in which approximately another 4 million children die from acute respiratory diseases every 12 months; and in which another 3 million die from vaccine-preventable diseases. It is a world in which over 3 million poeple still die annually from tuberculosis and in which countless many more millions are afflicted and die from gross, and what we should consider in 1991, unacceptable, malnutrition. It is a world in which many countries are still trying to meet the basic needs of their people for food and shelter, when their health budgets scarcely surpass approximately US$ 10 per person, annually.

The diminished economic productivity that is virtually inevitable in the context of a disease that primarily affects those who are most able to work and earn, in turn, undercuts the financing of services. Many developing countries have still not been able to extend most critical health and welfare programs beyond the relatively narrow confines of their urban centers. And even in the cities the equitable distribution of these has been hampered by a host of factors, not least of which has been the rapid rural-urban migrations that have outmatched the planning and financing resources of governments.

The economic impact is challenging. The figures differ as to the direct costs and the percentage increases in any given country. Indirect costs greatly exceed direct costs at present. What appears applicable worldwide is this: that rich nations like the poor nations are having to spend ever larger sums of their gross national product on HIV/AIDS management;

that the medical facilities and welfare services are being overwhelmed by the demands of providing for the needs of the infected; that there is heightening tension with respect to who is entitled to care, who should pay and for what services [15]. The epidemic is revealing that there is a divide: nowhere do all citizens have automatic access to care according to need. The costs in terms of stress and burn-out on the health workforce are incalculable, as health care professionals opt out of medical services. And as the financial load on society grows, private sources of funding are drying up [16].

These several conditions have extremely grave consequences for the peoples of Africa, Latin America, the Caribbean – regions whose governments are having to cut health care and social services as part of structural adjustment policies, whose economies are in decline due to natural disasters, crippling debts, low commodity prices, and, in some countries, military conflict as well. An accumulation of such adverse circumstances is likely to do great harm to family systems already under threat. If this is the case, there must be global solidarity. For, no people should have to stand alone to face AIDS.

Global Solidarity

WHO's constant call to global solidarity seems most appropriate as we think about the pandemic from a family perspective. The disease cannot be stopped without a world united against AIDS. Yet, we may have a power gap, wherein those members of the broader family most at risk are at the same time most powerless to change the circumstances of their lives.

We see this gap in power between the sexes, between classes and between nations. Women are being infected at a faster rate than men. In Thailand the sex ratios of infected persons had dropped from 17 males to 1 female in 1986 to 5 males to 1 female in 1990 [17]. Three studies which covered all but the northern and eastern districts of Uganda showed that the infection rate in females was 1.4 times higher than in males [18]. This phenomenal rise of infection in women is predictable when poverty and powerlessness are perceived as cofactors of HIV transmission [19]. Low levels of education, skills and relevant experience for the marketplace render the women – especially in developing countries – highly dependent on males, economically. These factors also reduce the woman's bargaining power to insist on protective sexual behavior. The technology that women could control and that would protect them from HIV transmission, currently claims very little of national or international research budgets, either at the level of biology or clinical or social science [20]. Yet, public policy has not raised this issue to a high level of debate, not even in developed countries. Women – some powerless, unequal and exploited – can be excused when they ask: 'But, with whom are we to be in solidarity?'

Is AIDS emerging as a class disease, and, thus, does its prevalence alone signal a division between classes within societies? The question is prompted by increased references to AIDS as the 'disease of the poor'. In the western industrialized countries while estimates of HIV infection rates are going down for white gay males, they are being revised upward for i.v. drug users [18]. Persons in this latter category are predominantly 'Blacks' and 'Hispanic'. In the United States while these groups make up 20% of the population, they constitute 42% of the AIDS cases [2]. There is believed to be a close correlation in New York City between the proportion of people below the poverty level and the number of AIDS cases [2].

But, a distinction must be made between the developed and developing countries as well as between classes within countries with respect to the issue of poverty. Some research out of Africa [9] reveals that AIDS is most visible among the educated and productive elite, especially in the urban communities. Nevertheless, as the infection spreads to rural areas in the developing countries, more and more of the poor in each community will be affected.

Within these nations, classes of persons – localized geographically, and by their socioeconomic circumstances and behavior – can be identified. Five of the estimated 8–10 million infected persons are in Africa, which also contains 28 of the world's 36 poorest nations.

The problem, then, is one beyond class differences; it is one of underdevelopment and of maldistribution of resources and power. It is a matter of poverty of our rural areas in most developing countries. In many others, it is poverty in the enormous shanty towns in the mushrooming urban cities of the Third World. These are increasing much more rapidly there than in the industrialized nations, where the disease is adding to the poverty of the inner cities. AIDS makes it quite clear that the concept, 'global solidarity', can be a hollow term to those at risk, but lacking power.

Realization of global solidarity, thus, requires that we redefine 'family' even at the global level. There are members of that global family, who, because of their sex, or class, or nation, are powerless and trapped by circumstances that render them highly susceptible to HIV infection and AIDS. Half of the world enslaved, half free of the disease: could we accept such a world?

Global Learning

Because of a global experience of AIDS, we now have much to learn from each other. Coping can be defined as staged learning, and learning as new adaptations [11]. Implicit in these definitions is the idea that we must become students of the lessons AIDS is teaching mankind, since its impact is demanding rapid adaptations in our social, economic, health and welfare systems.

Yet, we appear not to have taken a broad perception of the learning task. For example, it is taken as a truism that in the absence of an effective vaccine, we must educate people to change their sexual behavior. Thus, in essence, we have reduced the task of coping largerly to the behaviors of individuals – and except for blood transfusions and injections – to the matter of sexual behavior, alone.

Governments and policymakers could therefore be forgiven if they have assumed that AIDS has nothing to do with the problems of underdevelopment and poverty. They could conceivably become impatient with drug users, prostitutes, and populations of developing countries who refuse to change their sexual behavior or to take the lessons of health educators seriously.

When we consider the need for global learning, we may find that there are more questions than answers, and that we are in danger of oversimplifying deep cultural, economic and social traditions. With respect to educating, for example, what do we really know? Who determines what must be learned? How do we sustain and translate the lessons learned into viable strategies and structures for combatting AIDS? With respect to change, what is that range of behaviors that must be changed? As we raise these questions for consideration at the societal level, and not only as they apply at the family level, it is clear how urgent is our need to learn, empirically, what is meaningful education and meaningful behavioral change in each cultural context.

None of us, therefore, should presume that we have all the answers. Among the global family, one member will help because of direct experience with AIDS; another will contribute through advanced science. Together, we all will learn. None of us should have the illusion or the dread that we are alone.

Global Ethic

Let this be our conclusion: underlying our thoughts about the advancing pandemic, about the common issues, particularly the resource constraints, and the reflections on power and solidarity and learning, is the desire to find a common base for collective action. This search is prompted by awareness that there could be a reluctance on the part of governments, public policymakers and donors to make substantial and long-term commitment to a systematic provision of help for the most needy of the global family. Yet, recent expenditures of one billion dollars a day on the Gulf War show that we have the means, if the political will exists, to make policies which ensure no one is denied the right to be free of AIDS. Another basis than means, therefore, is clearly indicated.

We contend that that basis is ethics. The idea of the family applied here is of social bonding. It is conceived in relational terms and connotes an emotional content and concern. In this conception the world can qualify as a family. AIDS, just as war sometimes does, could serve to bring about something good. For in its destructiveness, it calls for a global commitment to an ethical imperative that transcends science. That ethical imperative is that we care.

Ethics have always been at the base of science, typified by science's scrupulous search for truth. But AIDS has raised ethics to a new and high level of public consciousness and debate. Drug firms are now being challenged on ethical grounds; medical specialists are called to give an account for their behavior vis-à-vis the public; whole communities get involved around the issue of vaccine trials and human rights; and the media are monitoring the compliance of politicians with respect to ethical procedures.

The matter of stopping AIDS is ethics more than science. Science has a limit. It can be limited by selective application. For example, science has produced AZT. It works – but only for those who can pay. Caring is limitless.

From the standpoint of AIDS control, we do not adopt a family perspective because we want to preserve our individual nations, or families or just our own 'persons with AIDS'. Rather, this we must do because we cannot save ourselves unless we join with others to fight the epidemic that is attacking us all. Linking hands against the onslaught of AIDS challenges the whole family of mankind: the Latin American, but also the African as well as the Eastern European. More so, the poor challenge us to an ethic of caring; the powerful is called on to preserve the powerless. Helping the larger family is the way ahead.

Science, indeed, challenges AIDS. But AIDS challenges our ethical and moral foundations as no disease has ever done. If we can build an ethic of caring into all of our scientific thinking, our scientific work, our scientific enterprise, we will go that extra mile because we are one family. And we will ensure that Sarah is protected as a member of that global family. We will ensure that Sarah does not stand alone to fight AIDS.

Acknowledgements

Acknowledgement with profound gratitude is accorded to Dr. S. Lwanga, Dr. D. Mulder and Dr. Rumashuana of Uganda who reviewed the paper. Dr. M. Carballo is thanked for his encouragement and technical contributions. Mr. S. Wangalwa, research assistant, and Miss Rose Kiggundu and Miss Bonnie Rose who prepared the manuscript also rendered valuable assistance.

References

1 Macklin ED (ed): AIDS and Families: Report of the AIDS Task Force, Groves Conference on Marriage and the Family. New York, Haworth, 1989, pp 1–284.
2 Levine C: AIDS and changing concepts of family. Milbank Q 1990;68(suppl 1):33–58.
3 Lovejoy NC: AIDS: Impact on the gay man's homosexual and heterosexual families. Homosex Fam Relat 1989, pp 285–316.
4 Lloyd GA: HIV-Infection, AIDS and family disruption; in Fleming AF (ed): The Global Impact on AIDS. New York, Liss, 1988, pp 183–190.
5 Cates JA, Graham LL, Boeglin D, Tielker S: The effect of AIDS on the family system. Families in society. J Contemp Hum Serv 1990;195–201.
6 Children and AIDS: An Impending Calamity. The Growing Impact of HIV Infection on Women and Family Life in the Developing World. New York, UNICEF, 1990, pp 1–24.
7 Logie DE: Brazil: Neither angels nor devils. Lancet 1990;335:1332.
8 Manguyu F: The Challenge of HIV/AIDS in Women, Adolescents and Children: The Women's Perspective. Nairobi, African Fertility Society, 1991, pp 1–22.

9 Hassig SE, Perriens JOS, Baende EK, Kahotwa M, Bishagara K, Kinkela N, Kupita B: An analysis of the economic impact of HIV infection among patients at Mama Yemo Hospital, Kinshasa, Zaire: Current Science. AIDS 1990;4:883–887.
10 Ankrah EM, Lubega M, Nkumbi S: The family and care-giving in Uganda. Vth Int Conf AIDS, Montreal, June 4–9, 1989, abstr WGP 32.
11 Barnett T, Blaikie P: Aids and food production in East and Central Africa: A research outline. Food Policy 1989;142–146.
12 Drucker E: Epidemic in the war zone: AIDS and community survival in New York City. Int J Health Serv 1990;20:601.
13 AIDS: Hospital care. Health Soc WK 1988;13:10.
14 Edmondson B: AIDS and aging. Am Demograph 1990;28–34.
15 Fox DM: Chronic disease and disadvantage: The new politics of HIV infection. J Health Polit Policy Law 1990;15:341–355.
16 Lokken D: AIDS marches on: State of the state of AIDS by gay men's health crisis. Daily Nation (Kenya), April 7, 1991.
17 Nkowane BN: The Status of the HIV/AIDS Epidemic in Africa: The Challenge, Constraints and Opportunities. Nairobi, African Fertility Society, 1991, in press.
18 Berkley SF, Namaara W, Okware S, Downing R, Konde-Lule J, Wawer M, Musagaara M, Musgrave S: AIDS and HIV infection in Uganda – are more women infected than men? AIDS 1990;4:1237–1242.
19 Radlett M, Foreman M, Mariasy J: Social conditions heighten vulnerability to HIV for many women. Les implications du SIDA pour la mer et l'enfant. Conf Int Paris, Nov 27/30, 1989, abstr D20.
20 Stein ZA: HIV prevention: The need for methods women can use. Am J Publ Health 1990;80:460–462.

E. Maxine Ankrah, PhD, Department of Social Work and Social Administration, Faculty of Social Sciences, Makerere University, c/o Church of Uganda, PO Box 14123, Kampala (Uganda)

Rossi GB, Beth-Giraldo E, Chieco-Bianchi L, Dianzani F, Giraldo G, Verani P (eds): Science Challenging AIDS. Basel, Karger, 1992, pp 188–195

Small Regions of the *env* and *tat* Genes Control Cellular Tropism, Cytopathology and Replicative Properties of HIV-1

C. Cheng-Mayer, T. Shioda, J.A. Levy

Cancer Research Institute, University of California, School of Medicine, San Francisco, Calif., USA

Introduction

Strains of the human immunodeficiency virus type 1 (HIV-1) display a high degree of biological heterogeneity which may be linked to certain clinical manifestations of AIDS. They vary in their cellular host range or tropism, kinetics of replication, and susceptibility to serum neutralization [1, 2, 5, 8]. Differences among isolates in their ability to down-modulate the cell surface CD4 receptor molecule and to induce cytopathology in infected cells have also been observed [4, 11]. Some of these biological properties in vitro correlate with virus pathogenicity in vivo. We and others have shown that HIV-1 isolates recovered from patients with advanced disease are more cytopathic, replicate with faster kinetics and display a wider cellular host range than isolates obtained from the same individual in a previously healthy state [3, 12]. To identify the genes or domains within genes that control these in vitro pathogenic properties, recombinant viruses were generated between two closely related HIV-1 isolates that show distinct biological properties.

Table 1. Comparison of biological properties of HIV-1_{SF2} and HIV-1_{SF13} molecular clones

		HIV-1_{SF2}	HIV-1_{SF13}
Initial source		PMC	PMC
Clinical state		oral candiasis	KS, PCP
Biological properties	replication in		
	PMC	+	++
	HUT78	+	++
	MT-4	+	++
	CEM	–	+
	U937	–	+
	cytopathology		
	syncytia formation	+	++
	plaque formation	–	+

PMCs, T cell and monocytic cell lines were infected with 1,000 $TCID_{50}$ (tissue culture infectious dose) of virus stocks derived from molecularly cloned genomes. Virus replication was monitored by the presence of reverse transcriptase (RT activity) in culture supernatant. For replication in PMC, HUT 78, and MT-4 cell, results represent RT activity at days 7–10 postinfection. ++ = RT activity > 500,000 cpm/ml; + = RT activity of 20,000–50,000 cpm/ml. For replication in CEM and U937 cells, results represent RT activity at days 25–30 postinfection; + = RT activity > 100,000 cpm/ml; – = RT activity < 5,000 cpm/ml. Background RT was generally < 2,000 cpm/ml. Plaque assay in the MT-4 cell line was conducted as described [11]; + indicates presence of plaque at day 6 postinfection. Cytopathology was assessed by the presence of giant cells and balloon degeneration in infected cells. ++ indicates > 90% infected cells showing cytopathology; – indicates < 20% infected cells. KS = Kaposi's sarcoma; PCP = *Pneumocystis carinii* pneumonia.

Comparison of the Biological Properties of HIV-1_{SF2mc} and HIV-1_{SF13mc}

The two isolates studied, HIV-1_{SF2} and HIV-1_{SF13}, were sequentially recovered from the peripheral mononuclear cells of a patient who had progressed from a relatively asymptomatic state to severe disease [3] (table 1). Both isolates were molecularly cloned (mc) and sequenced [7, 9], and the genomes were found to display > 97% nucleotide sequence homology. Comparison of their biological properties revealed that both isolates replicated to comparable titers in PMC from seronegative donors as well as in the HUT 78 and MT-4 T cell lines. HIV-1_{SF13mc}, however, replicated with faster kinetics in these cell types, and also productively infected the CEM T-cell line and the monocytic cell line U937. Furthermore, HIV-1_{SF13mc} is highly cytopathic; it readily induces syncytia formation in infected PMC and T-cell lines, and forms plaques in the MT-4 cell line [11].

Functional Mapping of Viral Determinants for Replication Kinetics and Plaque-Forming Ability of HIV-1

To identify the genetic determinants that control the differences in replicative rate and cytopathic properties of HIV-1_{SF2mc} and HIV-1_{SF13mc}, recombinant viruses were generated exchanging smaller and smaller portions of sequences in the 3′ region of the genome using a method that had previously been described [6, 13]. The kinetics of replication of these recombinant viruses in PMC, HUT 78, and MT-4 cells were examined, and their ability to form plaques in the MT-4 cells was determined. Results are summarized in figure 1. Recombinant viruses R5, R7, R9, R11, and R13 were found to replicate with faster kinetics in PMC and HUT 78 cells than their reciprocal recombinant viruses. The finding that R13, an HIV-1_{SF2mc} isolate that had acquired a 0.29-kb EcoRI/HindIII fragment from HIV-1_{SF13mc}, replicated with faster kinetics indicates that this portion of the genome determines this biological property. Faster kinetics of replication in MT-4 cells, however, was observed for recombinant viruses R5, R7, R10, R12, and R14. The genomic region shared between R7 and R10, a 0.49-kb StuI/MstII fragment of HIV-1_{SF13mc}, therefore, contains a major determinant for faster growth in this cell line. This same region also confers the plaque-forming ability of HIV-1 in the same cell line.

Genetic Determinants for Host Range and Syncytia Formation of HIV-1

Infection of the CEM and U937 cell lines by recombinant viruses R5-R14 described above indicate that a relatively large portion of the envelope gene, a 1.3-kb MstII/PstI fragment of HIV-1_{SF13mc}, contains a major determinant for infection of these cell types (data not shown). Additional recombinant viruses were generated in an attempt to identify the minimum genetic sequences that confer infectivity in CEM and U937 cells. Results are summarized in figure 2. Similar to parental HIV-1_{SF13mc}

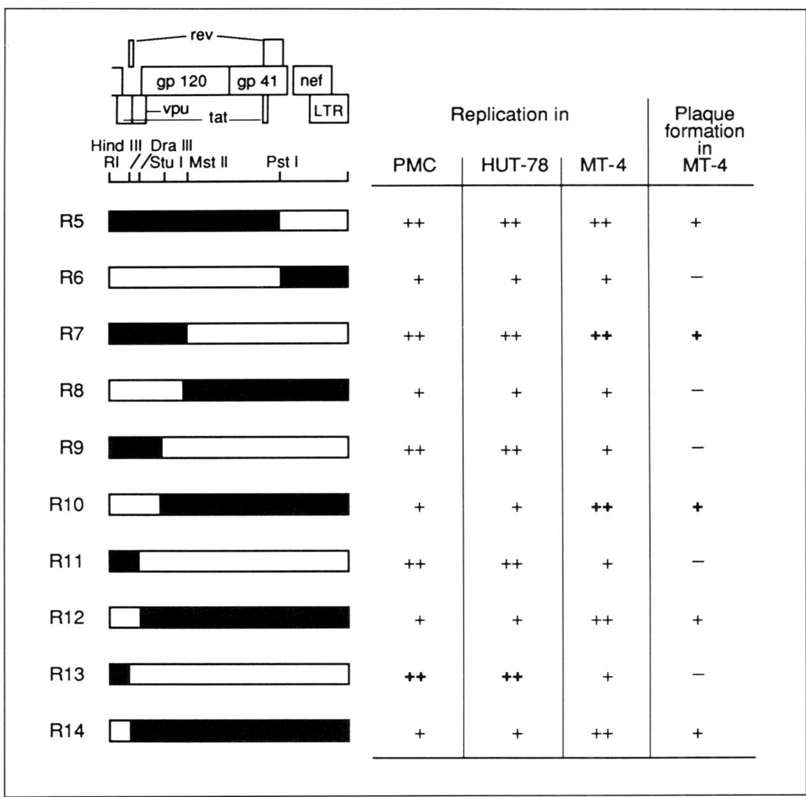

	Replication in PMC	Replication in HUT-78	Replication in MT-4	Plaque formation in MT-4
R5	++	++	++	+
R6	+	+	+	–
R7	++	++	++	+
R8	+	+	+	–
R9	++	++	+	–
R10	+	+	++	+
R11	++	++	+	–
R12	+	+	++	+
R13	++	++	+	–
R14	+	+	++	+

Fig. 1. Viral determinants for replication kinetics and plaque-forming ability of HIV-1. Recombinant viruses were assayed for their infection of PMC, HUT 78, and MT-4 cells, and their ability to induce plaques in the MT-4 cell line as described in table 1. For virus replication, results represent RT activity at days 7–10 postinfection. ++ = RT activity > 500,000 cpm/ml; + = RT activity of 20,000–50,000 cpm/ml. For plaque formation in the MT-4 cells, + indicates presence of plaques at day 6 postinfection. Odd number recombinant viruses have HIV-1_{SF2mc} 5′ portion; even numbers have HIV-1_{SF13mc} 5′ region [7].

(R2), recombinant viruses R15, R17, and R20 productively infected the CEM and U937 cell lines. This finding, therefore, further defines the region containing the determinant for infection of these cell types; it is a 0.39-kb MstII/StyI fragment of the gp120 of HIV-1_{SF13mc} shared between R17 and R20. Extensive syncytia formation in the HUT 78 cells were

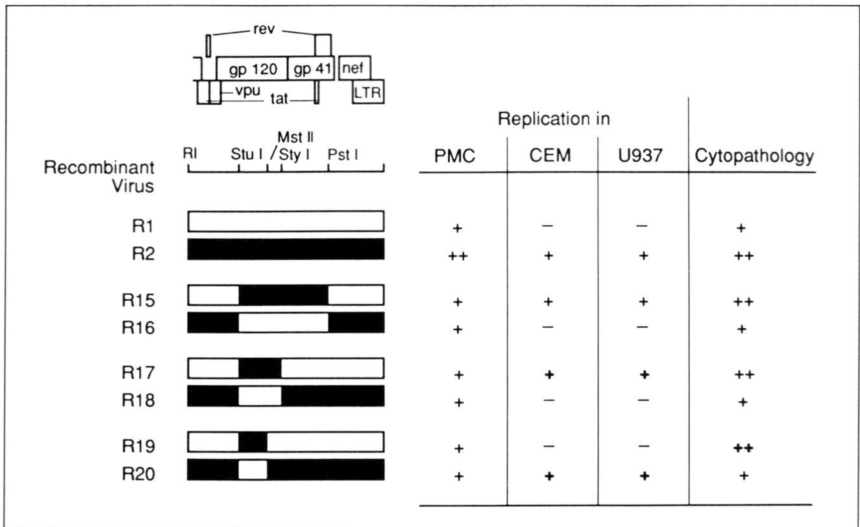

Fig. 2. Genetic determinants of host range and cytopathic properties of HIV-1. Recombinant viruses were assayed for their infection of the CEM and U937 cell lines, and their ability to induce cytopathology in the HUT 78 cell line as described in table 1. For replication, + indicates RT activity > 100,000 cpm/ml; – indicates RT activity < 5,000 cpm/ml. The odd number recombinants have the 5′ portion of HIV-1_{SF2mc}, the even number recombinants have the 5′ portion of HIV-1_{SF13mc}.

observed for recombinant viruses R15, R17, and R19. These results confirm that the 0.49 kb StuI/MstII fragment of gp120 contains a major determinant for cytopathology as assessed by syncytia formation (fig. 2) or plaque-forming ability (fig. 1).

Small Regions of the env and tat Genes Determine Cellular Host Range, Replicative, and Cytopathic Properties of HIV-1

Table 2 summarizes the data obtained with the recombinant viruses generated between two closely related, but biologically distinct HIV-1 isolates. Figure 3 schematically presents the regions identified. The results indicate that a 0.29-kb EcoRI/HindIII fragment of 97 amino acids (aa), encompassing the first coding exons of *tat* and *rev,* determines the rate of replication in PMC and the HUT 78 T cell line. A comparison of the

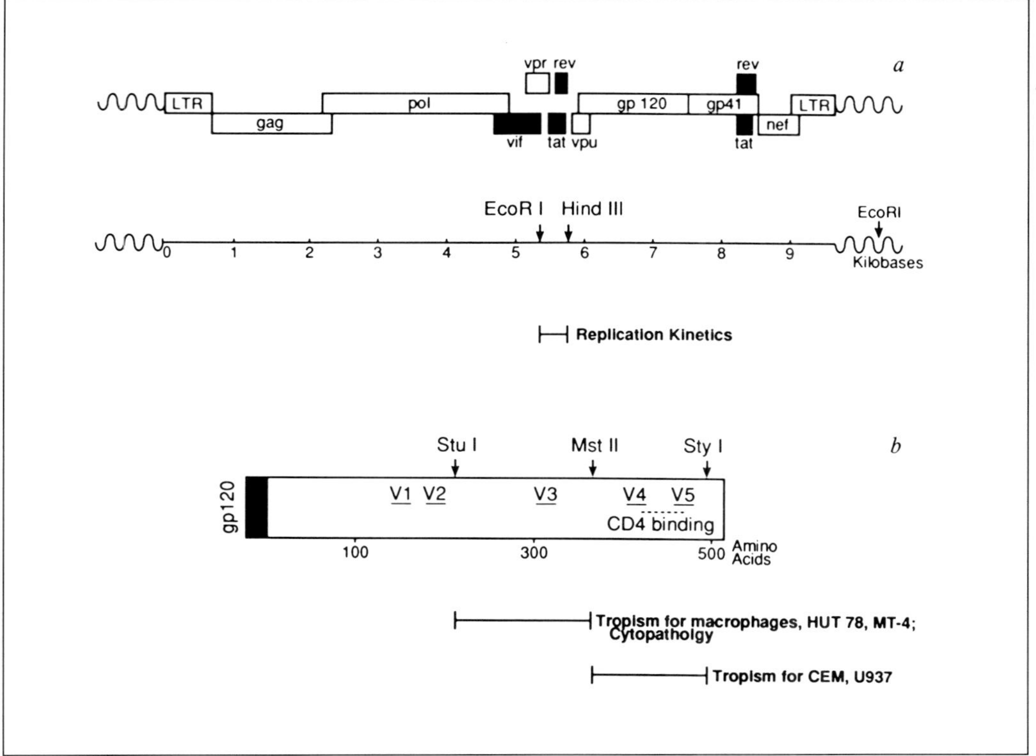

Fig. 3. Schematic representation of the genomic regions identified as containing determinants for HIV-1 host range, replicative, and cytopathic properties. Structure of the provirus genome is shown, and the structural and regulatory genes of HIV-1 are marked. *a* The EcoRI/HindIII viral region identified as determining HIV-1 replication rate in PMC and HUT 78 cells. *b* Structure of envelope gp120. Solid lines mark the five hypervariable regions and dotted lines define the CD4 binding region. Regions indentified as containing major determinants for infection of T cell and monocytic cell lines are delineated. Amino acid numbering is according to the coding sequence of HIV-1$_{SF2mc}$ [8].

amino acid sequences between HIV-1$_{SF2mc}$ and HIV-1$_{SF13mc}$ in this region reveal that there are only two amino acid differences, both of which are in the *tat* gene. A 0.49 kb StuI/MstII fragment of 160 aa, that encompasses the V3 hypervariable domain of gp120, contains a major determinant for infection of MT-4 cells and induction of cytopathology. This same region has previously been shown to determine tropisms for primary macro-

Table 2. Summary of the viral genetic regions responsible for host range, replicative, and cytopathic properties of HIV-1

Genomic	Fragment	Domain	Property	Amino acid change
0.29 kb	R1/HindIII (97aa)	first coding exons of *tat* and *rev*	replication efficiency in PMC, HUT 78	2 (in *tat*)
0.49 kb	StuI/MstII (160aa)	V3 of gp120	cytopathology infection of φ, HUT 78, MT-4 cells	10 (4aa changes in V3)
0.39 kb	MstII/StyI (130aa)	V4, V5, CD4 binding of gp120	infection of CEM, U937 cells	12

R1 = EcoRI; aa = amino acid.

phages and the HUT 78 T cell line [10]. Ten amino acid differences are present in this portion of the genome, four of which are in the V3 domain. Finally, a 0.39 kb MstII/StyI fragment of 130 aa, consisting of the V4, V5 and CD4-binding domains of gp120, contains a major determinant for infection of the CEM and U937 cell lines. Twelve amino acid differences were found within this region of the viral genome; the majority were clustered in the V4 domain. Moreover, some of these amino acid differences would result in the generation of unique glycosylation sites for both HIV-1_{SF2mc} and HIV-1_{SF13mc}.

Conclusions

Our studies with recombinant viruses generated between two closely related HIV-1 isolates that show differences in host range, replicative and cytopathic properties have revealed that very few amino acid changes within small regions of the *env* and *tat* genes determine these biological properties. The results also suggest regulation of HIV-1 infection and replication at the entry (i.e. envelope) and postentry (i.e. *tat*) levels in various cell types. Functional analyses with site-directed mutant viruses in both the *tat* and *env* genes should provide additional valuable information on the molecular mechanism of HIV-1 infection and pathogenesis.

References

1 Castro BA, Cheng-Mayer C, Evans LA, Levy JA: HIV heterogeneity and viral pathogenesis. AIDS 1988;2(suppl 1):s17–s27.
2 Cheng-Mayer C, Homsy J, Evans LA, Levy JA: Identification of human immunodeficiency virus subtypes with distinct patterns of sensitivity to serum neutralization. Proc Natl Acad Sci USA 1988;85:2815–2819.
3 Cheng-Mayer C, Seto D, Tateno M, Levy JA: Biologic features of HIV that correlate with virulence in the host. Science 1988;240:80–82.
4 Cheng-Mayer C, Weiss C, Seto D, Levy JA: Isolates of human immunodeficiency virus type 1 from the brain may constitute a special group of the AIDS virus. Proc Natl Acad Sci USA 1989;86:8575–8579.
5 Cheng-Mayer C: Biological and molecular features of HIV-1 related to tissue tropism. AIDS 1990;4(suppl 1):s49–s56.
6 Cheng-Mayer C, Quiroga C, Tung JW, Dina D, Levy JA: Viral determinants of human immunodeficiency virus type 1 T-cell or macrophage tropism, cytopathogenicity, and CD4 antigen modulation. J Virol 1990;64:4390–4398.
7 Cheng-Mayer C, Shioda T, Levy JA: In vitro pathogenic properties of HIV-1 are determined by few amino acid changes in *tat* and gp120. Submitted.
8 Fenyo EM, Albert J, Asjo B: Replicative capacity, cytopathic effect, and cell tropism of HIV. AIDS 1989;3(suppl 1):s5–s12.
9 Sanchez-Pescador R, Power MD, Barr PJ, Steimer KS, Stempien MM, Brown-Shimer SL, Gee WW, Ranard A, Randolph A, Levy JA, Dina D, Luciw PA: Nucleotide sequence and expression of the AIDS-associated retrovirus (ARV-2). Science 1985;227:484–492.
10 Shioda T, Levy JA, Cheng-Mayer C: Macrophage and T-cell line tropisms of HIV-1 are determined by specific regions of the envelope gp120. Nature 1991;234:167–169.
11 Tateno T, Levy JA: MT-4 plaque formation can distinguish cytopathic subtypes of the human immunodeficiency virus (HIV). Virology 1988;176:299–301.
12 Tersmette M, Gruters RA, de Wolf F, de Groede REY, Lange JMA, Schellekens PTA, Goudsmit J, Huisman HG, Miedema F: Evidence for a role of virulent human immunodeficiency virus (HIV) variants in the pathogenesis of acquired immunodeficiency syndrome: Studies on sequential HIV isolates. J Virol 1989;63:2118–2125.
13 York-Higgins D, Cheng-Mayer C, Bauer D, Levy JA, Dina D: Human immunodeficiency virus type 1 cellular host range, replication, and cytopathicity are linked to the envelope region of the genome. J Virol 1990;64:4016–4020.

Cecilia Cheng-Mayer, PhD, University of California, School of Medicine,
Cancer Research Institute, 3rd & Parnassus Avenue S1280,
San Francisco, CA 94143–0128 (USA)

Rossi GB, Beth-Giraldo E, Chieco-Bianchi L, Dianzani F, Giraldo G, Verani P (eds): Science Challenging AIDS. Basel, Karger, 1992, pp 196–206

Development of Opportunistic Non-Hodgkin's Lymphomas in Severely Immunosuppressed HIV-Infected Patients Receiving Long-Term Antiretroviral Therapy

James M. Pluda[a], *Jill Lietzau*[a], *Giovanna Tosato*[b], *David Venzon*[a], *Deborah L. Birx*[c], *Samuel Broder*[a], *Robert Yarchoan*[a]

[a] National Cancer Institute, Bethesda, Md.;
[b] Center for Biologics Evaluation and Research, Bethesda, Md.;
[c] Walter Reed Army Medical Center, Washington, D.C., USA

Introduction

The association of non-Hodgkin's lymphomas (NHL) with human immunodeficiency virus (HIV) infection and acquired immunodeficiency syndrome (AIDS) has been recognized since 1982 [1, 2]. HIV-associated NHL tend to be aggressive, high-grade, B cell tumors, frequently occurring in extranodal sites, particularly in the central nervous system (CNS) [3–5]. Lymphomas arising in the setting of HIV infection have been reported to the United States Centers for Disease Control (CDC) since their initial recognition, and in 1985 the CDC revised the case surveillance definition of AIDS to include certain high-grade NHL [6]. These lymphomas account for approximately 3% of all AIDS cases in the United States [7]. However, this figure is certainly an underestimate of the actual incidence of HIV-associated NHL since tumors developing after the initial AIDS-defining illness are not usually reported and, in fact, there are relatively little data on the incidence of these types of lymphomas. The life expectancy of patients with AIDS has been substantially prolonged in recent years by the advent of improved therapies for the treatment of HIV and its associated complications. In this chapter, we have examined one of the first cohorts of patients to ever receive antiretroviral therapy for the development of NHL and for factors which may be related to the development of these tumors.

Non-Hodgkin's Lymphomas in Patients on Long-Term Antiretroviral Therapy

The National Cancer Institute (NCI) of the National Institutes of Health has followed a cohort of 55 patients with AIDS or severe AIDS-related complex (ARC) receiving zidovudine (AZT)-based antiretroviral therapy for up to 4.5 years [8]. These patients were participants in 3 separate trials, and were among the first patients ever to receive targeted antiretroviral therapy. At present, 8 of the 55 patients (14.5%) have developed an NHL. The NHL were diagnosed a median of 21.8 months (range 2.7–76.6 months) after their AIDS diagnosis and a median of 23.6 months (range 15.4–34.8 months) after initiation of antiretroviral therapy. We originally reported the development of NHL in this cohort in 1990 [8], and none of the surviving patients have developed an NHL since then. As of May, 1991, the estimated probability of developing an NHL within 24 months of initiating antiretroviral therapy was 12% (95% CI 5–27%), increasing to 29% (95% CI 15–49%) after 36 months, with a hazard rate of 0.077 per patient-year of follow-up.

The median CD4 cell count for the entire cohort at the initiation of therapy was 71 cells/mm^3 (range 0–953 cells/mm^3). However, the lymphomas did not develop until the patients had a substantial decline in their CD4 cells. Indeed, the 8 patients developing NHL had a median of 6 CD4 cells/mm^3 (range 4–30 cells/mm^3) at the time of their lymphoma diagnosis. There was a statistically significant difference in the rate of development of NHL during the time that the patients had more than 50 CD4 cells/mm^3 (0/57.3 patient-years of follow-up) compared to the time they had less than 50 CD4 cells/mm^3 (8/46.3 patients-years of follow-up; $p = 0.002$ by the test for a difference in two Poisson rates [9]). This relationship was independent of the time that they received antiretroviral therapy. Thus, it appears that prolonged survival in a severely immunosuppressed state, in particular with less than 50 CD4 cells/mm^3, may result in a significantly increased incidence of developing NHL.

The lymphomas that developed in this cohort were typical of those previously described as occurring in association with HIV infection [3–5]. All of the lymphomas occurred at extranodal sites, with 5 presenting only within the CNS. Histologically, 4 of the tumors were small noncleaved cell (this designation includes Burkitt's, non-Burkitt's and diffuse undifferentiated lymphoma) and 4 were large-cell immunoblastic lymphomas. Seven of the 8 NHL were B cell in origin, while 1 was a null cell type. (Interest-

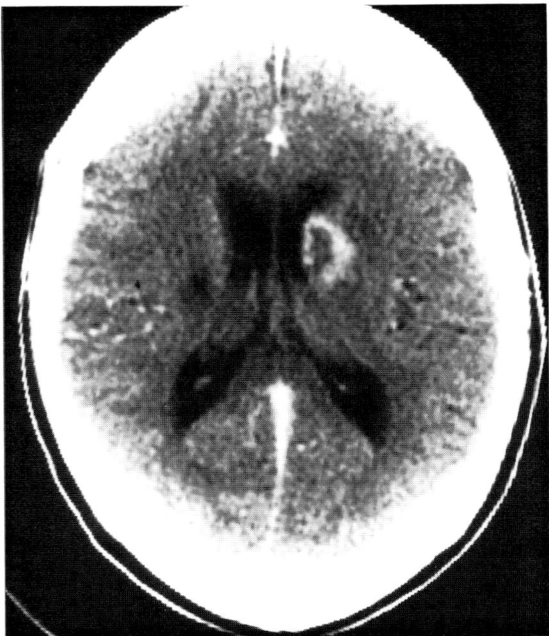

Fig. 1. Contrast-enhanced computerized tomographic scan of a patient with biopsy-proven cerebral toxoplasmosis who subsequently developed NHL. After initial successful treatment of his toxoplasmosis, the patient developed increasing headaches, and a computerized tomographic scan showed enhancing lesions. Stereotactic biopsy of the lesion shown revealed small noncleaved cell lymphoma, B cell type.

ingly, this latter patient subsequently developed a second, histologically distinct B cell lymphoma.) Response to therapy was poor, with a median survival of 1.8 months (range 0.6–3.2 months) for the 5 patients with primary CNS lymphomas, and 7 months (range 0.5–18 months) for the 3 patients with other extranodal presentations.

It should be stressed that NHL can develop in patients with higher CD4 counts; indeed, at the NCI, we have observed HIV-related NHL in patients with as many as 451 CD4 cells/mm^3. However, the results here suggest that NHL are much more common in advanced AIDS patients, and in this setting can be called 'opportunistic'. Such patients often have multiple signs and symptoms of HIV infection as well as opportunistic infections. Indeed, NHL may be extremely difficult to diagnose in such patients. The patients in the NCI cohort were followed in a research hospital on an AIDS/oncology service where the level of suspicion for tumors may have been especially high, and several of these tumors could have gone undiagnosed in other settings. For instance, 2 of the lymphomas developed in patients with known cerebral toxoplasmosis (fig. 1), and another was found at autopsy to involve the leptomeninges only. This experience mimics that reported in other institutions. For example, in an autopsy study of 101 HIV-related deaths in teaching hospitals in New York City, 20 patients were found to have NHL, and 8 of these (40%) were unsuspected antemortem [10].

Recently, Moore et al. [11], reviewing the hospital and office records of 1,030 patients with advanced HIV infection receiving AZT, reported a somewhat lower incidence of NHL than that found in the NCI cohort. Over a 2-year period of follow-up, 24 NHL were diagnosed, yielding an estimated probability of developing lymphoma during that time of 3.2%. It is not completely clear why they saw a lower incidence of lymphomas at 2 years in their study than was seen in the NCI cohort. One possible reason is that the patients in the NCI cohort were sicker and had a greater degree of immunosuppression at entry onto the study (median entry CD4 count 71 cells/mm^3 in the NCI cohort vs. 104 cells/mm^3 in the Moore study). Also, as discussed above, many of these tumors may be missed unless physicians maintain a high index of suspicion and utilize aggressive diagnostic modalities, and there may have been a difference in follow-up of the patients in the two cohorts. Finally, it should be stressed that the incidence of lymphoma in the NCI cohort increased after 2 years, and it will be of interest to determine whether there is a substantial increase in the estimated probability of NHL in the cohort reported by Moore et al. by year 3 of follow-up. Other defined data bases should also be examined for the incidence of lymphoma.

Pathogenesis of HIV-Related Non-Hodgkin's Lymphomas

The pathogenesis of HIV-related lymphomas is complex and incompletely understood at present. However, most researchers would agree that the polyclonal B cell activation seen in patients infected with HIV plays a central role in lymphoma genesis. Patients with AIDS or ARC have been shown to have a polyclonal B cell lymphoproliferative lymph node expansion [12], the cause of which may be multifactorial. HIV itself can induce polyclonal B cell hyperactivation, either through mitogenic or antigenic stimulation by HIV, or via HIV-induced T-cell-dependent B cell activation [13–16]. There is also evidence that interluekin-6 (IL-6) production, a B-cell-stimulating cytokine, may be increased in the setting of infection with HIV [17–19]. Thus, induction of IL-6 may be another mechanism by which HIV infection results in B cell hyperactivity, which, in turn, may increase the likelihood of some transforming event or events. In most cases where this has been examined, NHL tumor cells have not been found to be infected by HIV [12], and the role of HIV in lymphoma genesis is believed to be indirect.

The role of Epstein-Barr virus (EBV) in the development of these HIV-associated lymphomas is not well defined. Patients with HIV infection have increased numbers of circulating EBV-infected B cells, which contribute to the pool of hyperactive B cells [13, 20]. The increased number of EBV-infected B cells is related to the profound immunodeficiency with defective T cell regulation of EBV infection. The exact role of EBV in the pathogenesis of HIV-related NHL is still under investigation. EBV sequences have been found in only 40–50% of HIV-related lymphomas [21, 22], and in those cases that are EBV-positive there is evidence that the EBV infection precedes the clonal B cell expansion leading to lymphoma [23]. Also, there is recent evidence from Shibata et al. [24] that the development of NHL is more likely to occur in patients in whom the EBV genome can be detected in benign lymph node tissue. However, the mechanism by which EBV may contribute to the development of these lymphomas remains unclear.

The molecular event or events involved in the malignant transformation leading to a lymphoma are also incompletely understood, although there is evidence that translocations involving the *c-myc* gene are involved in as many as 75% of these tumors [12, 21]. The typical t(8:14) translocation seen in non-HIV-related Burkitt's lymphoma (BL) is present in most if not all of the HIV-associated BL and BL-like lymphomas (small noncleaved cell lymphomas). Translocations involving the *c-myc* locus t(8q24) are not limited to small noncleaved cell lymphomas. Ladanyi et al. [25] have recently demonstrated the presence of t(8q24) in 71% of HIV-related large-cell lymphomas, including immunoblastic types. There are known to be molecular differences between endemic BL (that typically found in Africa) and sporadic BL, and the question has been raised as to which type HIV-associated NHL most closely resemble. For example, HIV-associated NHL have been identified as having patterns of chromosomal breakpoints associated with both endemic BL, where the breakpoint occurs far outside the *c-myc* gene locus, as well as sporadic BL, with the breakpoint being near or within the *c-myc* locus [21, 26, 27]. In non-HIV associated lymphomas, it has been suggested that there may be an association between the location of the chromosome breakpoints and the presence of EBV, with 95% of tumors having breakpoints outside the *c-myc* locus (similar to endemic BL) being EBV-positive [28]. It is possible that a similar association might be found in HIV-related NHL. Thus, EBV may play a direct role in the development of a certain subset of these tumors. Additional research will be needed to further characterize these tumors at a molecular level.

A common feature of the t(8q24) translocations in NHL is an increased expression of *c-myc*. The exact mechanism by which this activation of *c-myc* results in the development of a malignancy remains unknown. It has been shown in vitro that increased *c-myc* expression in EBV-transformed or BL cells results in decreased expression of EBV surface antigens, decreased or absent expression of certain class I major histocompatibility complex antigens, and a down-regulation of lymphocyte-function-associated antigen-1 adhesion molecules [29, 30]. Thus, in vitro *c-myc* overexpression may endow a transformed cell with a selection advantage in escaping normal immunologic control mechanisms. This selection advantage conferred upon tumors having *c-myc* translocation and activation, in conjunction with profound immunologic defects, particularly in T cell function, may allow for the substantially increased occurrence of these tumors in HIV-infected patients.

Regardless of the molecular events leading to malignant transformation, the immunosuppression induced by HIV appears to play a major role in the development of these lymphomas. This is similar to the occurrence of high-grade B cell NHL that is seen in association with primary immunodeficiency disorders such as the Wiskott-Aldrich syndrome (WAS), severe combined immunodeficiency, combined variable hypogammaglobulinemia, Chediak-Higashi syndrome and ataxia-telangiectasia [31, 32]. The improvements in supportive measures resulting in prolonged survival of patients with WAS are believed to have resulted in an increase in the incidence of lymphomas [33]. This is analogous to the finding in the NCI cohort that prolonged survival with less than 50 CD4 cells/mm^3 is a major risk factor for the development of NHL. Recently, Feichtinger et al. [34] have reported the development of high-grade, B cell NHL in 9 of 24 (38%) severely immunosuppressed cynomolgus monkeys infected with simian immunodeficiency virus [34]. Interestingly, the majority of these lymphomas occurred in monkeys with less than 100 CD4 cells/mm^3. The occurrence of NHL in this setting may provide a useful animal model for the development of HIV-associated lymphomas.

As noted above, however, HIV-associated NHL can also occur in patients with higher CD4 cells and, indeed, there is a suggestion that there may be differences in the lymphomas which occur in patients with higher CD4 cells as compared to those which occur in patients with advanced AIDS. For example, Roithmann et al. [35], after reviewing the French Registry of HIV-associated tumors, have suggested that the CD4 cell count may be related to the histologic type of lymphoma that develops. They found that small noncleaved cell lymphomas developed more frequently in patients with relatively high CD4 cells, while large-cell immunoblastic lymphomas occurred more frequently in patients who were more severely immunosuppressed as indicated by the presence of low CD4 cells. This latter situation may be analogous to the development of NHL in transplant patients, particularly those receiving treatment with relatively higher doses of OKT3 anti-T cell antibodies [36]. Also, Levine [37] has recently reported a significant difference in CD4 cells in patients presenting with primary CNS lymphoma versus those with systemic lymphomas: patients with tumors presenting in the brain had a median CD4 cell count of 30 cells/mm^3 while those with systemic lymphomas had a median CD4 cell count of 189 cells/mm^3. However, at this time, more data are necessary before any of these generalizations can be considered firm.

Since the patients in the NCI cohort were all on long-term AZT-based therapy, the question occurs as to whether AZT is directly causing the lymphomas. It has been observed that rats and mice treated with lifelong high-dose AZT have been reported to develop vaginal malignancies. However, these rodents metabolize AZT differently from humans, concentrating it in the urine, and AZT has not been shown to induce lymphomas per se in any animal models [38]. Thus, this observation would not likely be related to the development of NHL in patients. Also, the lymphomas that developed in the patients in our cohort were typical of those previously reported in HIV-infected patients before the development of antiretroviral therapy [3–5]. Finally, in a placebo-controlled study of asymptomatic HIV-infected patients with greater than 200 CD4 cells/mm^3, there was no increased incidence of lymphomas seen in the AZT-treated group compared to placebo [39]. Putting these facts together, we believe it unlikely that AZT is directly causing the lymphomas. However, we cannot absolutely rule out this possibility with the information at hand.

Future Incidence Trends of HIV-Related Lymphomas

In the period between 1973 and 1987, the Surveillance, Epidemiology and End Results (SEER) Program of the NCI documented an increase of greater than 50% in the overall incidence of NHL [40]. This trend includes a dramatic increase in NHL in never-married males aged 20–49 beginning in 1983, presumably due to HIV-associated lymphomas. The SEER data base includes both AIDS-defining lymphomas as well as those arising in patients with a prior AIDS diagnosis. A projection of this trend, taking into account the regional variation in the incidence of HIV infection, would predict that in the United States there will be about 4,700 cases of AIDS-associated NHL in 1992 [41]. However, projections can also be made based on estimates of future AIDS incidence, as well as on estimates of the incidence of lymphomas that are AIDS-defining and that occur in patients with a prior AIDS diagnosis. Using the results of the NCI cohort, projections made by Gail et al. [41] suggest that there may be between 2,900 and 9,800 AIDS-related NHL in 1992 (with an intermediate estimate of about 5,000 cases). This would involve 8–27% of all NHL occurring in the United States during that year [41]. Nearly 80% of these lymphomas are expected to develop in patients who were diagnosed with AIDS prior to the development of their lymphoma. Thus, it would appear that HIV-related NHL will have a significant impact on the overall incidence of NHL in the very near future.

Conclusion

A growing body of evidence would seem to suggest that NHL will be an increasing source of morbidity and mortality in HIV-infected patients in the near future. Prolonged survival in a severely immunosuppressed state appears to be associated with a significantly increased risk of developing lymphoma. Research into the pathogenesis of these tumors may yield valuable information leading to the development of newer, more effective, and less toxic therapies, as well as the possibility of preventing their development altogether. Newer antiretroviral therapies and strategies aimed at preventing the development of severe immunosuppression, particularly maintaining CD4 cells above 50/mm^3, may delay or possibly prevent the development of this most formidable complication of infection with HIV.

References

1 Doll DC, List AF: Burkitt's lymphoma in a homosexual. Lancet 1982;i:1026–1027.
2 Ziegler JL, Miner RC, Rosenbaum E, Lennette ET, Shillitoe E, Casavant C, Drew WL, Mintz L, Gershow J, Greenspan J, Becksted J, Yamamoto K: Outbreak of Burkitt's-like lymphoma in homosexual men. Lancet 1982;ii:631–633.
3 Ziegler JL, Becksted JA, Volberding PA, Abrams DI, Levine AM, Lukes RJ, Gill PS, Burkes RL, Meyer PR, Metroka CE, Mouradian J, Moore A, Riggs SA; Butler JJ, Cabanillas FC, Hersh E, Newell GR, Laubenstein LJ, Knowles D, Odajnyk C, Raphael B, Koziner B, Urmacher C, Clarkson BD: Non-Hodgkin's lymphoma in 90 homosexual men: Relation to generalized lymphadenopathy and the acquired immunodeficiency syndrome. N Engl J Med 1984;311:565–570.
4 Lowenthal DA; Straus DJ, Campbell SW, Gold JWW, Clarkson BD, Koziner B: AIDS-related lymphoid neoplasia: The Memorial Hospital experience. Cancer 1988; 61:2325–2337.
5 Knowles DM; Chamulak GA, Subar M, Burke JS, Dugan M, Wernz J, Slywotzky C, Pelicci P-G, Dalla-Favera R, Raphael B: Lymphoid neoplasia associated with the acquired immunodeficiency syndrome (AIDS): The New York University Medical Center experience with 105 patients (1981–1086). Ann Intern Med 1988;108:744–753.
6 Centers for Disease Control: Revision of the case definition of acquired immunodeficiency syndrome for national reporting – United States. MMWR 1985;34:373–375.
7 Beral V, Peterman T, Berkelma R, Jaffe H: AIDS-associated non-Hodgkin lymphoma. Lancet 1991;337:805–809.
8 Pluda JM, Yarchoan R, Jaffe ES, Feuerstein IM, Solomon D, Steinberg S, Wyvill KM, Raubitschek A, Katz D, Broder S: Development of non-Hodgkin lymphoma in a cohort of patients with severe human immunodeficiency virus (HIV) infection on long-term antiretroviral therapy. Ann Intern Med 1990;113:276–282.
9 Cox DR, Hinckley DV: Theoretical Statistics. London, Chapman & Hall, 1974; p 136.
10 Wilkes M, Fortin AH, Felix JC, Godwin TA, Thompson WG: Value of necropsy in acquired immunodeficiency syndrome. Lancet 1988;ii:85–88.
11 Moore RD, Kessler H, Richman DD, Flexner C, Chaisson RE: Non-Hodgkin's lymphoma patientsc with advanced HIV infection treated with zidovudine. JAMA 1991;265:2208–2211.
12 Pelicci P-G, Knowles DM II, Arlin ZA, Wieczorek R, Luciw P, Dina D, Basilico C, Dalla-Favera R: Multiple monoclonal B cell expansions and c-myc oncogene rearrangements in acquired immune deficiency syndrome-related lymphoproliferative disorders. J Exp Med 1986;164:2049–2076.
13 Yarchoan R, Redfield RR, Broder S: Mechanisms of B cell activation in patients with acquired immunodeficiency syndrome and related disorders: Contribution of antibody-producing B cells, of Epstein-Barr virus-infected B cells, and of immunoglobulin production induced by human T cell lymphotropic virus, type III/lymphadenopathy-associated virus. J Clin Invest 1986;78:439–447.
14 Schnittman SM, Lane HC, Higgins SE, Folks T, Fauci AS: Direct polyclonal activation of human B lymphocytes by the acquired immune deficiency syndrome virus. Science 1986;233:1084–1086.

15 Pahwa S, Pahwa R, Saxinger C, Gallo RC, Good RA: Influence of the human T-lymphotropic virus/lymphadenopathy-associated virus on functions of human lymphocytes: Evidence for immunosuppressive effects and polyclonal B-cell activation by banded viral preparations. Proc Natl Acad Sci USA 1985;82:8198–8202.

16 Amadori A; Zamarchi R, Ciminale V, Del Mistro A, Siervo S, Alberti A, Colombatti M, Cheico-Bianchi L: HIV-1-specific B cell activation: A major constituent of spontaneous B cell activation during HIV-1 infection. J Immunol 1989;143:2146–2152.

17 Breen EC, Rezai AR, Nakajima K, Beall GN, Mitsuyasu RT, Hirano T, Kishimoto T, Martinez-Maza O: Infection with HIV is associated with elevated IL-6 levels and production. J Immunol 1990;144:480–484.

18 Birx DL; Redfield RR, Tencer K, Fowler A, Burke DS, Tosato G: Induction of IL-6 during human immunodeficiency virus infection. Blood 1990;76:2303–2310.

19 Amadori A, Zamarchi R, Veronese ML, Panozzo M, Barelli A, Borri A, Sironi M, Colotta F, Mantovani A, Chieco-Bianchi L: B cell activation during HIV-1 infection. II. Cell-to-cell interactions and cytokine requirement. J Immunol 1991;146:57–62.

20 Birx DL, Redfield RR, Tosato G: Defective regulation of Epstein-Barr virus infection in patients with acquired immunodeficiency syndrome (AIDS) or AIDS-related disorders. N Engl J Med 1986;314:874–879.

21 Subar M, Neri A, Inghirami G, Knowles DM, Dalla-Favera R: Frequent c-myc oncogene activation and infrequent presence of Epstein-Barr virus genome in AIDS-associated lymphoma. Blood 1988;72:667–671.

22 Hamilton-Dutoit SJ, Pallesen G, Franzman MB, Karkov J, Black F, Skinhoj P, Pedersen C: AIDS-related lymphoma: Histopathology, immunophenotype, and association with Epstein-Barr virus as demonstrated by in situ nucleic acid hybridization. Am J Pathol 1991;138:149–163.

23 Neri A, Barriga F, Inghirami G, Knowles DM, Neequaye J, Magrath IT, Dalla-Favera R: Epstein-Barr virus infection precedes clonal expansion in Burkitt's and acquired immunodeficiency syndrome-associated lymphoma. Blood 1991;77:1092–1095.

24 Shibata D, Weiss LM, Nathwani BN, Brynes RK, Levine AM: Epstein-Barr virus in benign lymph node biopsies from individuals infected with the human immunodeficiency virus is associated with concurrent or subsequent development of non-Hodgkin's lymphoma. Blood 1991;77:1527–1533.

25 Ladanyi M, Offit K, Khanwar SC, Filippa DA, Chaganti RSK: *MYC* rearrangement and translocations involving band 8q24 in diffuse large cell lymphomas. Blood 1991;77:1057–1063.

26 Neri A, Barriga F, Knowles DM, Magrath IT, Dalla-Favera R: Different regions of the immunoglobulin heavy-chain locus are involved in chromosomal translocations in distinct pathogenetic forms of Burkitt lymphoma. Proc Natl Acad Sci USA 1988; 85:2748–2752.

27 Haluska FG, Russo G, Kant J, Andreef M, Croce CM: Molecular resemblance of an AIDS-associated lymphoma and endemic Burkitt lymphomas: Implications for their pathogenesis. Proc Natl Acad Sci USA 1989;86:8907–8911.

28 Shiramizu B, Barriga F, Neequaye J, Jafri A, Dalla-Favera R, Neri A, Guttierez M, Levine P, Magrath I: Patterns of chromosomal breakpoint locations in Burkitt's lymphoma: Relevance to geography and Epstein-Barr virus association. Blood 1991; 77:1516–1526.

29 Klein G: Multiple phenotypic consequences of the *Ig*/Myc translocation in B-cell-derived tumors. Genes Chrom Cancer 1989;1:3–8.
30 Inghirami G, Grignani F Sternas, Lars, Lombardi L, Knowles DM; Dalla-Favera R: Down regulation of LFA-1 adhesion receptors by *c-myc* oncogene in human B lymphoblastoid cells. Science 1990;250:682–686.
31 Page AR, Berendes H, Warner J, Good RA: The Chediák-Higashi syndrome. Blood 1962;20:330–343.
32 Filipovich AH, Heinitz KJ, Robison LL, Frizzera G: The immunodeficiency cancer registry: A research resource. Am J Pediatr Hematol Oncol 1987;9:183–184.
33 Cotelingam JD, Witebsky FG, Hsu SM, Blaese RM, Jaffe ES: Malignant lymphoma in patients with the Wiskott-Aldrich syndrome. Cancer Invest 1985;3:515–522.
34 Feichtinger H, Putkonen P, Parravicini C, Li S-L, Kaaya EE, Böttiger D, Biberfeld G, Biberfeld P: Malignant lymphomas in cynomolgus monkeys infected with simian immunodeficiency virus. Am J Pathol 1990;137:1311–1315.
35 Roithmann S, Toledano M, Tourani JM, Raphael M, Gentilini M, Gastaut JA, Armengaud M, Morlat P, Tilly H, Dupont B, Taillan B, Theodore C, Donadio D, Andrieu J-M: HIV-associated non-Hodgkin's lymphomas: Clinical characteristics and outcome. The experience of the French registry of HIV-associated tumors. Ann Oncol 1991;9:289–295.
36 Swinnen LJ, Costanzo-Nordin MR, Fisher SG, O'Sullivan EJ, Johnson MR, Heroux AL, Dizikes GJ, Pifarre R, Fisher RI: Increased incidence of lymphoproliferative disorders after immunosuppression with monoclonal antibody OKT3 in cardiac-transplant recipients. N Engl J Med 1990;323:1723–1728.
37 Levine MA: AIDS-related lymphoma: Clinical aspects and biology of disease. Adv Oncol 1991;7:18–25.
38 Burroughs Wellcome Company: Lifetime Bioassay Studies (eds): Comprehensive Information for Investigators, RETROVIR. Research Triangle Park, Burroughs Wellcome, 1989, p 48.
39 Volberding PA, Lagakos SW, Koch MA, Pettinelli C, Myers MW, Booth DK, Balfour HH, Reichman RC, Barlett JA, Hirsch MS, Murphy RL, Hardy D, Soeiro R, Fischl MA, Bartlett JG, Merigan TC, Hyslop NH, Richman DD, Valentine FT, Corey L, and the AIDS Clinical Trial Group of the National Institute of Allergy and Infectious Diseases: Zidovudine in asymptomatic human immunodeficiency virus infection: A controlled trial in persons with fewer than 500 CD4-positive cells per cubic millimeter. N Engl J Med 1990;322:941–949.
40 The Surveillance Program Division of Cancer Prevention and Control National Cancer Institute: Table II–1. Summary of 15 year trends: Age adjusted cancer incidence rates; in Ries LAG, Barrett MJ, Labbe RR (eds): Cancer Statistics Review: 1973–1987. Bethesda, US Department of Health and Human Services, 1990:II.4.
41 Gail MH, Pluda JM, Rabkin CS, Biggar RJ, Goedert JJ; Horm JW, Sondik EJ, Yarchoan R, Broder S: Projections of the incidence of non-Hodgkin's lymphoma related to acquired immunodeficiency syndrome. JNCI 1991;83:695–701.

James M. Pluda, MD, National Institutes of Health, National Cancer Institute, 9000 Rockville Pike, Bldg. 10, Rm 13N248, Bethesda, MD 20892 (USA)

Rossi GB, Beth-Giraldo E, Chieco-Bianchi L, Dianzani F, Giraldo G, Verani P (eds): Science Challenging AIDS. Basel, Karger, 1992, pp 207–213

Risk Reduction and Stabilization of HIV Seroprevalence among Drug Injectors in New York City and Bangkok, Thailand[1]

Don C. Des Jarlais[a], *K. Choopanya*[b], *J. Wenston*[c], *S. Vanichseni*[b], *J.L. Sotheran*[c], *K. Plangsringarm*[b], *P. Friedmann*[a], *W. Sonchai*[a], *M. Carballo*[d], *S.R. Friedman*[c]

[a] Beth Israel Medical Center, New York, N.Y., USA;
[b] Health Department, Bangkok Metropolitan Administration, Bangkok, Thailand;
[c] Narcotic and Drug Research, Inc., New York, N.Y., USA;
[d] World Health Organization, Geneva, Switzerland

HIV infection among persons who inject ilicit drugs has now been reported in over 30 different countries. The variation in current rates of HIV infection among injecting drug users (IDUs) – as well as in the responses to the threat of HIV epidemics among IDUs – make this a topic where international collaborative studies may be particularly fruitful. Two such international studies have been undertaken, one sponsored by the European Economic Community, the other by the World Health Organization. The data presented here were collected as part of the WHO study.

There are at least two critical questions in the epidemiology of HIV infection among IDUs where international data can be of particular importance. First, we need to better understand the behavioral dynamics during periods of rapid transmission of HIV among IDUs. Epidemics where HIV seroprevalence among IDUs has increased by 10% or more per year have occurred in a number of cities in many different countries [1]. Preventing such periods of rapid transmission is of the highest priority. If a period of rapid transmission does occur in a city, those IDUs will then serve as a large reservoir for heterosexual and perinatal transmission of HIV.

[1] This research was supported by the US National Institute on Drug Abuse (grant 03574) and the World Health Organization.

Second, we need to better understand the behavioral dynamics of periods of stable HIV seroprevalence among IDUs. Such stable seroprevalence includes continuing low-to-moderate rates of new HIV infections among IDUs, and should therefore be seen as an endemic problem rather than as a solved problem.

Both New York City and Bangkok have undergone periods of rapid transmission and are currently in situations of stable seroprevalence [2, 3]. Comparative data from the two cities provides some preliminary insight into the dynamics of both rapid transmission and stable seroprevalence. (Time and space limitations do not permit full analyses of the data here; these will be presented elsewhere.)

Methods

A working group of World Health Organization collaborators has developed a standard questionnaire for comparative studies of HIV risk behavior among IDUs. This questionnaire was administered to 601 subjects in Bangkok (after translation into Thai) in 1989, and to 810 IDUs in New York City in 1990. The questionnaire is sufficiently similar to the questionnaire used in a study of 314 IDUs recruited from short-term detoxification and methadone maintenance treatment programs in New York in 1984 so that comparisons can also be made with those data. Comparisons of the 1984 New York data with the 1989 Bangkok data can be particularly useful in that both cities were in the early phase of an HIV stabilization situation in those years.

Recruitment in 1989 and 1990 was primarily from drug abuse treatment programs: in Bangkok from long-term detoxification and aftercare programs (n = 342), and in New York from short-term detoxification and methadone maintenance programs (n = 446). Additional subjects were recruited from persons who were new to the treatment system in Bangkok (n = 259) [3] and from a street outreach research storefront in New York (n = 366). Informed consent was obtained from all subjects for participation in the study. HIV pretest counseling was conducted, and a blood sample was collected. HIV testing was done with double ELISA tests and Western blot confirmation.

Results

Selected demographic and drug injection characteristics are presented in table 1. The IDUs in New York were older, more likely to be females, and had much greater ethnic/racial diversity. Employment (full or part-time) was much higher among the 1989 Bangkok subjects than among the 1990 New York City subjects. (Employment data were not collected for the 1984 New York subjects.) The IDUs from New York also reported slightly

Table 1. Demographics and drug injection

	New York City 1984 (n = 314)	New York City 1990 (n = 813)	Bangkok 1989 (n = 601)
Male, %	75	75	95
Mean age, years	34	36	30
Mean injections/month	87	88	75
Race, %			
White/Thai	29	20	99
Black	30	45	
Hispanic	39	34	
Other	2	1	1
Employed, %		23	70

Table 2. Seroprevalence

	New York	Bangkok
Early stabilization	50% treatment sample	40% treatment sample 27% new-to-treatment sample
Continued stabilization	50% treatment sample 41% street sample	

higher frequencies of recent drug injection; this was primarily due to their injection of cocaine, either alone or in combination with heroin (a 'speedball'). Cocaine injection was almost totally absent among the Bangkok IDUs, with less than 1% reporting injecting it.

HIV seroprevalence in these samples is presented in table 2. For the in-treatment samples, seroprevalence has stabilized at approximately 50% in New York over a 6- to 7-year period [1], and at approximately 40% in Bangkok [4]. The 1989 new-to-treatment sample in Bangkok had a significantly lower seroprevalence than the in-treatment sample ($p < 0.01$ by chi square), and the 1990 street-recruited sample in New York also had a significantly lower seroprevalence ($p < 0.01$ by chi square).

Risk factor analyses for HIV exposure have been performed on the 1984 New York in-treatment sample [5] and the 1989 Bangkok total sample [3]. The most important risk factors for the 1984 New York sample were frequency of injection and injecting in shooting galleries, while the most important factors for the 1989 Bangkok sample were incarceration within the previous 5 years and sharing injection equipment with 2 or more other persons in the previous 6 months. Injecting in shooting galleries, incarceration, and sharing with 2 or more other persons all would permit transmission of HIV across friendship groups. Large percentages of these samples reported engaging in these specific risk behaviors: 58% of the 1984 New York sample reported injecting in shooting galleries, while 70% of the 1989 Bangkok sample reported having been incarcerated and 30% reported sharing with 2 or more persons.

Large-scale risk reduction has occurred among IDUs in both cities. In response to the question 'Have you changed your behavior to protect yourself against AIDS?', 59% of the 1984 New York sample, 78% of the 1990 New York sample, and 92% of the 1989 Bangkok sample reported that they had changed their behavior in some way. Fifty-four percent of the 1984 New York sample, 57% of the 1990 New York sample, and 78% of the 1989 Bangkok sample reported that they had started to practice some form of 'safer' injection. (Safer injection included cessation or reduction of sharing, increased use of illicitly obtained sterile injection equipment and, to a lesser extent, using bleach to disinfect potentially contaminated injection equipment.) The 1990 New York data indicate that the AIDS risk reduction has been maintained at the group level, though there undoubtedly has been relapse at the individual level.

The large-scale risk reduction and stabilization of HIV seroprevalence in Bangkok and New York does not mean an absence of new HIV infections. The seroconversion rate for the Bangkok sample was estimated from the 175 persons who reported having been tested for HIV prior to the 1989 data collection [3]. The seroconversion rate for the year preceding the data collection was approximately 10 per 100 person-years at risk. Seroconversion rates for New York have been estimated from a cohort follow-up of the 1984 sample and through modelling of the epidemic using turnover rates from the 1984 and 1990 data. The seroconversion rate for 1984–1985 was 7 per 100 person-years at risk [6], declining to approximately 2 per 100 person-years at risk [Marmor et al., unpubl. data; Des Jarlais et al., unpubl. data].

Risk factor analyses have been conducted for HIV seroprevalence in the 1990 New York sample [Wenston et al., unpubl. data] and for seroconversion in the cohort follow-up of the 1984 New York sample [Marmor et al., unpubl. data]. Frequency of injection, particularly cocaine injection, continues to be a risk factor for HIV exposure, but a new set of risk factors

Table 3. Phase of HIV epidemic among drug injectors

	New York	Bangkok
Introduction of HIV	mid-1970s	1986
Rapid spread	1978–1983	1978–1988
Early stabilization	1984–1987	late 1988 on
Continued stabilization	1987 on	

has emerged that were not statistically significant independent predictors in the 1984 sample. These new risk factors – age, ethnicity and gender – suggest that recent HIV transmissions among IDUs in New York are occurring primarily within close social relationships such as friendship groups or heterosexual relationships.

A long-term problem for both cities will be the new persons starting to inject illicit drugs. 31% of the 1989 Bangkok sample and 11% of the 1990 New York sample had begun injecting in the previous 5 years. In New York, these persons who began injecting did so after knowledge about how AIDS is transmitted was widespread among IDUs in the city [7].

Discussion

Combined with previous data [2, 4], the 1989–1990 seroprevalence data permit a 'best estimate' reconstruction of the HIV epidemics among IDUs in New York and Bangkok. A structural representation of these epidemics is presented in table 3. Despite the multiple cultural and economic differences between the two cities, both epidemics appear to have the same structure, with New York currently being one phase ahead of Bangkok. The biggest difference between the two epidemics appears to be the relatively compressed time frame in which the Bangkok epidemic occurred.

Both New York and Bangkok experienced periods of rapid transmission of HIV: over a 4-year period in New York and a 2-year period in Bangkok. The data from both cities indicate that large numbers of IDUs were engaging in behavior that would transmit HIV across friendship groups – using shooting galleries, being incarcerated and sharing with multiple partners. Sharing injection equipment while incarcerated may be a particularly effective means for rapid transmission of HIV. Incarceration would closely approximate the random mixing that would produce the most rapid transmission of a communicable agent through a population. Transmission caused by sharing injection equipment while incarcerated would also probably be missed by HIV screening at drug treatment programs in Bangkok, which would create the impression of more rapid transmission after the IDUs had left prison and later were tested at treatment programs. While almost all countries with large numbers of IDUs imprison many of them, almost no countries have developed programs that would prevent HIV transmission among IDUs in prison. The Bangkok data suggest that this may be a critical mistake.

The high percentage of IDUs practicing safer injection has also occurred within a relatively short period of time in Bangkok. Moreover, it occurred without the example of many IDUs with AIDS as had happened in New York City. Additional research is clearly needed to fully understand how such large-scale risk reduction occurs, but we would offer as a working hypothesis that the relatively close integration of IDUs in Thai society facilitated the risk reduction. As compared to New York, IDUs in Bangkok are more closely integrated into the larger society, both economically (they have a 70% employment rate) and ethnically (almost all belong to the majority ethnic group in the country). The fact that ethnic minority IDUs in New York have higher HIV seroprevalence rates also suggests that a relative lack of social integration poses a higher risk of HIV exposure among IDUs.

Comparison of the New York and Bangkok data raises interesting questions about the role of cocaine injection in the spread of HIV. Cocaine and speedball injections contribute the greatest proportion of the variance in drug injection in New York, and have been associated with HIV throughout the epidemic in New York [8, 9; Friedman et al., unpubl. data] and in other American cities [10]. Yet the Bangkok data clearly illustrate that it is possible to have very rapid spread of HIV among IDUs in the absence of any meaningful cocaine injection. Study of the spread of HIV among IDUs in cities where cocaine injection predominates (e.g. Rio de Janeiro or Buenos Aires) would provide a useful additional comparison to New York and Bangkok.

In both New York and Bangkok, the additional sample had a lower HIV seroprevalence rate than the standard in-treatment sample. The in-treatment samples in both cities contained many drug injectors who have repeatedly cycled in and out of treatment without being able to permanently reduce their levels of illicit drug use. These drug users may constitute a subgroup of IDUs who are particularly at risk for HIV exposure.

Differences among behavioral subgroups of IDUs should be considered when estimating the seroprevalence among IDUs in a city.

Finally, while large-scale risk reduction has occurred in both New York and Bangkok, the recent data from both cities show a substantial percentage of new drug injectors. Transmission of HIV among drug injectors over the long term will depend heavily both on the number of new persons who enter the IDU population and the extent to which safer injection practices develop among these new injectors. Additional research in both cities will need to focus on this problem.

References

1 Friedman SR, Des Jarlais DC: HIV among drug injectors: The epidemic and the response. AIDS Care 1991; in press.

2 Des Jarlais DC, Friedman SR, Novick DM, Sotheran JL, Thomas P, Yancovitz SR, Mildvan D, Weber J, Kreek MJ, Maslansky R, Bartleme S, Spira T, Marmor M: HIV-1 infection among intravenous drug users in Manhattan, New York City, from 1977 through 1987. JAMA 1989;261:1008–1012.

3 Choopanya K, Vanichseni S, Plangsringarm K, Sonchai W, Carballo M, Des Jarlais DC: Prevalence and risk factors of HIV infection among injecting drug users in Bangkok. Abstr 7th Int Conf AIDS, Florence, 1991, in press.

4 Vanichseni S, Sakuntanaga P: Results of three seroprevalence studies for HIV in IVDU in Bangkok. Abstr 6th Int Conf AIDS, San Francisco, 1990, p 116.

5 Marmor M, Des Jarlais DC, Cohen H, Friedman SR, Beatrice ST, Dubin N, El-Sadr W, Mildvan D, Yancovitz S, Mathur U, Holzman R: Risk factors for infection with human immunodeficiency virus among intravenous drug abusers in New York City. AIDS 1987;1:39–44.

6 Des Jarlais DC, Friedman SR: HIV infection among intravenous drug users: Epidemiology and risk reduction. AIDS 1987;1:67–76.

7 Friedman SR, Des Jarlais DC, Sotheran JL, Garber J, Cohen H, Smith D: AIDS and self-organization among intravenous drug users. Int J Addict 1987;22:201–219.

8 Novick DM, Trigg HL, Des Jarlais DC, Friedman SR, Vlahov D, Kreek MJ: Cocaine injection and ethnicity in parenteral drug users during the early years of the human immunodeficiency virus (HIV) epidemic in New York City. J Med Virol 1989;29:181–185.

9 Schoenbaum EE, Hartel D, Selwyn PA, Klein RS, Davenny K, Rogers M, Feiner C, Friedland G: Risk factors for human immunodeficiency virus infection in intravenous drug users. N Engl J Med 1989;321:874–879.

10 Chaisson RE, Baccheti P, Osmond D, Brodie B, Sande MA, Moss AR: Cocaine use and HIV infection in intravenous drug users in San Francisco. JAMA 1989;261: 561–565.

Don C. Des Jarlais, MD, Beth Israel Medical Center, 11 Beach Street,
New York, NY 10013 (USA)

Rossi GB, Beth-Giraldo E, Chieco-Bianchi L, Dianzani F, Giraldo G, Verani P (eds): Science Challenging AIDS. Basel, Karger, 1992, pp 214–225

Change in AIDS Risk Behaviors from Adolescence to Adulthood[1]

Arlene Rubin Stiffman[a], *Renee Cunningham*[a], *Felton Earls*[b], *Peter Dore*[a]

[a] George Warren Brown School of Social Work, Washington University, St. Louis, Mo.; [b] Harvard School of Public Health and Harvard Medical School, Boston, Mass., USA

Introduction

Extensive interventions directed to adolescents and young adults are being mounted on local, regional, national, and international levels to prevent the spread of the HIV virus. They largely rely on educational methods that would impact knowledge. But we have evidence that these interventions do not effectively change behavior [1–4], or achieve inadequate or temporary levels of change [5–9]. This is very discouraging for those attempting to intervene with this population.

A review of the current state of knowledge about predictors of risk behaviors in youths prompted our interest in using a multivariate model to examine the determinants of change in AIDS-related risk behavior among young adults. Our model postulates that risk behavior is determined by social norms (peer behaviors), intervention effects (knowledge about AIDS, knowledge about prevention, counseling, estimate of personal risk), mental health problems (depression, substance misuse, anxiety, suicidality, posttraumatic stress), stressors (stressful life events, trauma, mistreat-

[1] Research for this project was funded by Robert Wood Johnson Foundation, and NIMH Grant No. 1RO 1 MH 45118-01.

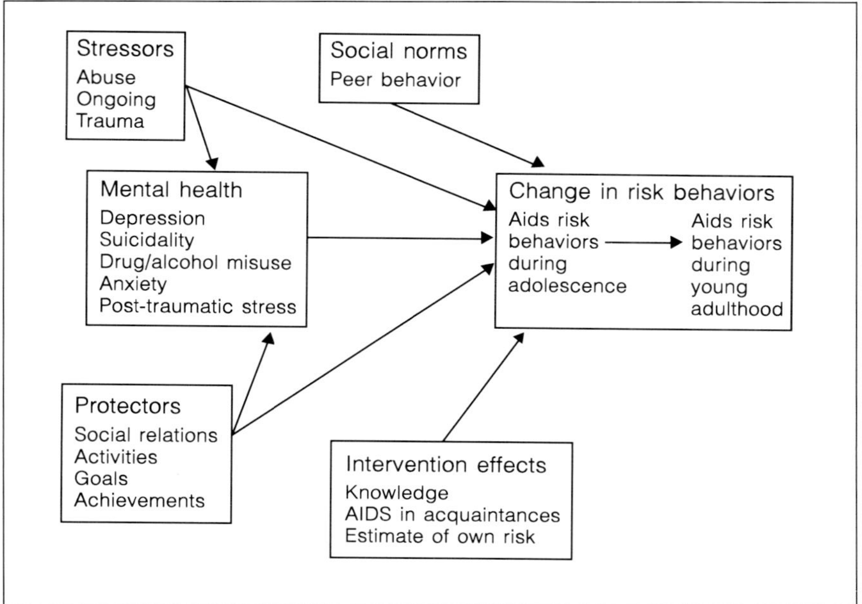

Fig. 1. Multivariate model of change in AIDS risk behaviors.

ment) and protectors (personal goals, social relationships, achievements, social activities) (fig. 1).

Prior research supports the existence of each univariate element of the model; however, our study is the first to examine them from a longitudinal multivariate perspective.

Mental Health Problems

We know that certain mental health indicators may put youths at heightened risk: low levels of impulse control in conduct disorder, lack of ability to assess riskiness or act on that assessment, disinhibitory effects of alcohol and drug misuse, and helplessness due to depression [10]. Research on risk-taking behavior in general has found a connection between psychological distress and risky behavior, such as depression and the use of illicit drugs [11–13], stress-related depression and sexual behavior [11] and low self-esteem and poor contraceptive use [14]. We also have some indication that there may be a connection between depression and AIDS [15, 16].

Social Norms

Adolescents are known to be particularly subject to influence by peer behavior, and cultural and social norms which may inhibit the adoption of protective behaviors [17].

Stressors and Protectors

Research findings suggest that risk status (for behavior problems) may be directly associated with environmental stressors and the lack of protectors or resiliency factors [18, 19]. Earlier studies by two of the authors (Stiffman and Earls) found that the joint and interactive effect of stressors and protectors explained approximately one-third of the variance in behavioral outcomes for children [18, 19] and for adolescents [20]. If there is an association between general behavior problems and risk taking for HIV infection, then environmental stressors and social support are, at least, indirectly, if not directly, associated with risk taking.

Methods

Design

Repeated interviews with 602 youths in 1984, 1985, and 1989 provide a history of change in AIDS-related high-risk behaviors from adolescence to young adulthood in relation to the youths' knowledge, experiences, and mental health.

Sample

The 602 young adults were selected through a stratified random sampling procedure (based on risk status) from 2,787 inner-city youths who were part of a larger evaluation of a multisite program by the Robert Wood Johnson Foundation [21]. There were no significant demographic differences between the stratified random sample and the original full sample. In both, the subjects were predominantly female (77%), black (70%), and age 15 years or older when first interviewed (85%). Hollingshead's schema [22] classified the occupations of the majority of the parents as skilled blue-collar (58%).

Instruments

The current interview consisted of information concerning subjects' protectors, stressors, social norms, intervention effects, and mental health. Other questions concerned sex and drug behaviors, and knowledge about, attitudes toward, and experiences with AIDS/HIV infection.

AIDS Risk Behaviors

We calculated the number of AIDS-related risk behaviors engaged in during young adulthood from the following list: male homosexual or bisexual behavior, prostitution, injectable drug use, 6 or more sexual partners in 1 year, irregular use of condoms, sharing of i.v. drug paraphernalia without cleaning it, and choice of a partner who was known to be at risk (male homosexual activity, i.v. drug user, HIV positive, prostitute) [23]. We used parallel information to calculate the number of AIDS-related risk behaviors engaged in during adolescence.

Intervention Effects

We posited the possible effects of interventions to be greater knowledge about AIDS or HIV infection, greater awareness of one's own risk, and/or greater awareness of others who have died of AIDS. The questions about AIDS came from the questionnaire used by the National Center of Health Statistics [24].

Mental Health

We used the Diagnostic Interview Schedule (DIS) [25–27] to derive symptom counts for symptoms occurring in the last year for, respectively, depression, anxiety, post-traumatic stress, and substance misuse (alcohol or drug).

Protective/Resiliency Factors

Protective factors included social activities, personal goals, achievements, and social relationships. We measured social activities by counting the number of different types of activities participated in by the respondent. We assessed goals by questions asking about high school graduation, and intention to continue in school or obtain special training. We counted the number of awards, promotions, or goals reached in the last year to measure achievement. To measure relationships, the adolescents rated their agreement with several statements indicating the extent to which they felt they could not rely on parents or peers for support, and to which they fought or argued with them.

Stress Factors

Three different stressors were measured: post-traumatic stressors, stressful life events, and mistreatment. The Diagnostic Interview Schedule section on posttraumatic stress disorder yields counts of number of experiences which induced traumatic reactions. We also developed a summary score for stressful life events from a list enumerating experiences of illness, poverty, household violence, family death, homelessness, parental separation/divorce, and unemployment. Mistreatment incidents were the number of each of the following types of events: physical abuse by guardian/parent, sexual abuse, or rape.

Social Norms

To measure peer behaviors, the youths rated how many of their friends (none, a few, about half, most, or all of their friends) had trouble with the police, used drugs or marijuana, or drank alcohol almost every day. A summary score from these three questions indicates peer misbehavior.

Analyses

A multivariate model is used to predict young adult AIDS risk behaviors based on risk behaviors during adolescence, mental health problems, presumed intervention effects from AIDS prevention programs, peer behavior, and on interim stressors and protectors. We included in this multivariate risk model those above-cited variables that demonstrated a significant association in the preliminary univariate analyses. Simultaneous multiple regressions, controlling first for adolescent risk behaviors, were used to analyze the effect of mental health, knowledge, social climate, stress and protective factors on young adult risk behaviors. The ultimate effect of a multivariate longitudinal model that forces in earlier behavior as a control variable on the dependent variable of later behavior is to examine the effects of the other independent variables on change in that behavior over time [28].

Results

Univariate Analyses

Univariate analyses of the various elements of the model revealed consistently significant relationships as predicted. All variables from the stress factor – posttraumatic events ($f = 25.3$, d.f. = 2, 601, $p < 0.0001$, $b = 0.11$), mistreatment ($f = 23.4$, d.f. = 2, 601, $p < 0.0001$, $b = 0.18$), and stressful events ($f = 53.1$, d.f. = 2, 601, $p < 0.0001$, $b = 0.09$) – were univariately significant predictors of change in AIDS risk behaviors.

Similarly, all variables from the mental health factor – depressive symptoms ($f = 41.4$, d.f. = 2, 601, $p < 0.0001$, $b = 0.03$), anxiety symptoms ($f = 27.9$, d.f. = 1, 601, $p < 0.0001$, $b = 0.04$), posttraumatic stress symptoms ($f = 41.4$, d.f. = 2, 601, $p < 0.0001$, $b = 0.04$), substance abuse ($f = 62.1$, d.f. = 2, 601, $p < 0.0001$, $b = 0.14$), and suicidality ($f = 53.9$, d.f. = 2,601, $p < 0.0001$, $b = 0.20$) – are significant univariate predictors of change in AIDS risk behaviors.

Of the variables that comprise the protector factor, only social relationships ($f = 22.8$, d.f. = 2, 601, $p < 0.0001$, $b = 0.03$) is a significant univariate predictor of change in AIDS-related risk behaviors (social activities, goals and achievements are not significant).

The social norm variable of proportion of friends who engage in a variety of risk behaviors is a significant univariate predictor ($f = 33.5$, d.f. = 2, 601, $p < 0.0001$, $b = 0.03$). Of the intervention effects variables, only awareness of those who had died of AIDS ($f = 22.9$, d.f. = 2, 601, $p < 0.0001$, $b = 0.08$), and estimate of current personal risk level ($f = 31.9$, d.f. = 2, 577, $p < 0.0001$, $b = 0.20$) and future personal risk level ($f = 24.9$, d.f. = 2, 577, $p < 0.0001$, $b = 0.20$) were significant univariate predictors of change in AIDS risk behaviors. Knowledge about AIDS was not significant. Note that, for variables from the intervention factor, the direction of effects was opposite that of the goal of treatment: those who knew more individuals who had died of AIDS, and viewed their own risk level as higher changed their risk behaviors in the direction of engaging in more rather than fewer risk behaviors.

Univariate associations are not enough to test the model, as there are high rates of comorbidity in the various mental health problems as well as correlations between the various elements of the model. If we are looking to develop a model that will direct intervention approaches parsimoniously to those areas that predict unique variance in AIDS-related behavior change, we must rely on multivariate analyses.

Multivariate Model Building

We tested the multivariate effects on change in AIDS risk behaviors of the independent variables (measured between adolescence and young adulthood). To do this, we first forced into the equation the individual's risk behaviors during adolescence as an independent variable predicting status at young adulthood.

The multiple regression analysis reveals a highly significant relationship between change in adolescents' AIDS-related risk behaviors and stressors (stressful life events), mental health problems (substance misuse symptoms and suicidality), and intervention effects (estimate of personal risk). These variables explain 22% of the variance in the number of risk behaviors at young adulthood (table 1).

Although there were theoretical reasons for including each of the variables in this analysis, we had no a priori basis for assuming any specific causal ordering among the independent variables. Therefore, after controlling for differences in the number of risk behaviors during adolescence (in order to force the equation to show change in behavior), each main effect was incorporated into the equation after all of the other main effects were entered. This form of multiple regression yields a test of the salient variables' unique contribution in predicting change in risk behaviors.

Prior Risk Behavior

As expected, a proportion (3%) of the variance in risk behaviors in young adulthood is predicted by the youth's risk behaviors at adolescence. Once risk behaviors during young adulthood are controlled in this manner, the effects of other independent variables indicate their contribution to change in risk behaviors between adolescence and young adulthood. The question of clinical significance concerns the effect of subsequent stressors and protectors on future or continued risk behaviors.

Table 1. Multiple regression analysis, predicting young adult risk behaviors while controlling for adolescent behaviors

Source	Degrees of freedom	Sum of squares	f value	Probability	r^2
Model	5	31.2	31.6	0.0001	0.22
Error	572	112.8			
Total	577	144.0			

Independent variables	Sequential r^2	p	b
Adolescent AIDS risk behaviors	0.03	0.0001	0.33
	Unique r^2	p	b
Mental health problems			
Drug/alcohol misuse symptoms	0.05	0.0001	0.10
Suicide symptoms	0.01	0.02	0.06
Stressors			
Interim stressful events	0.01	0.003	0.03
Intervention effects			
Estimate of personal risk	0.03	0.0001	0.13

Mental Health

Of all the mental health problems, only symptoms of alcohol and drug misuse and suicidality contributed unique variance (5% and 1%, respectively) to change in AIDS risk behaviors. This is not surprising since drug or alcohol misuse is associated with disinhibition effects on both sexual behavior and protective behaviors as well as increased likelihood of i.v. drug use or prostitution to obtain drugs. Similarly, the effect of suicidality on continuance or increase in AIDS-related risk behaviors is not surprising, since those risk behaviors might ultimately be a form of attempting suicide, and, conversely, desisting from risk behaviors may indicate a desire to prolong one's life.

Stressors

The number of interim stressful life events uniquely contributed 1% of the variance to change in AIDS risk behaviors. Youths who increased AIDS risk behaviors had experienced higher numbers of stressful events than youths who had not experienced an increase. Posttraumatic events and mistreatment had no unique contribution to change in AIDS risk behaviors.

Protectors/Resiliency Factors

None of the protectors (social relationships, social activities, goals, or achievements) contributed unique variance to an understanding of change in AIDS risk behaviors.

Social Norms

Once personal mental health and stress are entered into the equation, the effect of peer behavior drops out as a significant predictor. This indicates that youths probably choose, as friends, individuals with similar problems who are likely to engage in similar behaviors. Therefore, once the underlying similarity in personal and environmental precipitates of high risk behaviors are controlled, the similarity in risk behaviors themselves drop out.

Intervention Effects

The youths' estimate of their personal risk uniquely contributed 3% of the variance to change in AIDS risk behaviors. However, the direction of effects was not toward greater protection. Those youths with higher estimates of their personal risk for HIV infection actually were those increasing the most in risk behaviors. They were aware of their risk, but did nothing. If one assumes a causal direction of effects, and invervention programs increase one's estimate of personal future risk for becoming HIV positive, then intervention effects could be having exactly the opposite effect from that desired. The higher youths feel their future chance of becoming HIV positive is, the less they do to decrease risk behaviors.

Discussion

Our results are extremely important to those who fund, design and implement preventive programs for HIV infection, particularly programs designed for high-risk teenagers and young adults. They reveal that a variety of mental health problems and stressful events directly influence change in the risk behaviors of these youths. Thus, preventive programs should focus on influencing these factors, if they wish to close the knowledge-behavior gap that is known to exist among young people in relation to AIDS-related risk behaviors.

Despite our clear findings, we offer some caveats before elaborating on the meaning or making generalizations to broader populations. The young adults participating in this study do not represent a general sample of young adults in the United States. In their teen years they were all health clinic users from cities with high rates of homicide, suicide, drug abuse, and unwed pregnancy. Furthermore, all our data is through self-report. Also, due to the multiwave nature of the data, those who were not able to be located upon subsequent interviews were more likely to come from the highest risk groups. Finally, our sample was overweighted by the high-risk subjects compared to a normal representative sample of inner city adolescents.

Despite these methodological shortcomings, we have data that add important information to the current state of knowledge about what influences change in AIDS-related risk behaviours among inner city youths. We have found that two factors, mental health and life stress, neither usually included in HIV infection prevention programs, actually are more influential of change in AIDS-related risk behaviors than is increased knowledge, the one factor on which most prevention programs are based.

The impact of substance misuse and suicidality on change in risk behaviors reveals that those who have advocated mental health services as a part of preventive interventions are correct [11]. The results of our study confirmed others' statements that risky behaviors are related to mental health problems [10–13], and that they are specifically related to AIDS risk behaviors [15, 16]. Due to the longitudinal nature of our data, our study found that *changes* in AIDS risk behaviors are due, in part, to these mental health factors. Obviously, the findings indicate that treatment for those symptoms should be a part of preventive interventions with high-risk young people.

Further, our data also confirm others' research showing that stressors are important determinants of risk behaviors [18–20]. Of major import, however, is the multivariate model utilized on longitudinal data. Our data goes beyond those sources to show the importance of stressful life events (such as financial worries, ill health or death of family members, family violence, unemployment) in influencing change in AIDS-related risk behaviors.

In terms of treatment implications, the findings show that various elements of the youths' environment that might be manipulated by treatment agents are important in determining change in AIDS-related risk behaviors. The current study, showing that change in AIDS risk behaviors is predicted by a combination of stressors and mental health, tells us a great deal about the possibility of effectuating change by reducing stressors and by offering mental health treatment. Given that we know that knowledge alone is not sufficient to prevent young people from engaging in risk behaviors [29], and that making youths aware of their own risk might have the opposite effect from that desired [4], preventive programs should provide mental health treatment and services to ameliorate stressful life events.

Policy implications are made apparent in the vicious cycle that is identified. Many of the stresses associated with mental health problems are a direct result of poverty (money worries, housing problems and unemployment) and some mental health problems (such as drug and alcohol misuse) may increase unemployment. Therefore, our findings identify a cycle with obvious implications for prevention and treatment on both a societal and an individual level.

Although our findings point to the need for mental health treatment and environmental intervention as a part of preventive interventions, we cannot say definitively that such programs will be more effective than purely educational ones in reducing AIDS risk behaviors. The findings simply show that programs and research should incorporate this knowledge about the effects of mental health symptoms and stress on change in AIDS risk behaviors. Future evaluations of such programs will serve to test two hypotheses: that interventions with high-risk youths will result in reductions of mental health problems, and that such reductions will be associated with lower rates of AIDS-related risk behaviors.

References

1 DiClemente RJ: Adolescents and AIDS: An update. Multicult Inquiry Res AIDS 1989;3:3–7.

2 Weisman CS, Nathanson CA, Ensminger M, Teitelbaum MA, Robinson JC, Plichta S: AIDS knowledge, perceived risk and prevention among adolescent clients of a family planning clinic. Fam Plann Perspect 1989;21:213–217.

3 Brown LK, Fritz GK, Barone VJ: The impact of AIDS education on junior and senior high school students. J Adolesc Health Care 1989;10:386–392.

4 Stiffman AR, Earls F, Dore P, Cunningham R: Change in AIDS related risk behavior after adolescence: Relationship to knowledge and experience concerning HIV infection. Under review.

5 Chitwood DD, Comerford M: Drugs, sex, and AIDS risk. Am Behav Sci 1990;33: 465–477.
6 McCoy CB, Khoury E: Drug use and the risk of AIDS. Am Behav Sci 1990;33: 419–431.
7 Fineberg HV: Education to prevent AIDS: Prospects and obstacles. Science 1988; 239:592–596.
8 Sonenstein FL, Pleck JH, Ku LC: Sexual activity, condom use and AIDS awareness among adolescent males. Fam Plann Perspect 1989;21:152–158.
9 DiClemente RJ, Pies CA, Stoller EJ, Straits C, Olivia GE, Haskin J, Rutherford GW: Evaluation of school based AIDS education curricula in San Francisco. J Sex Res 1989;26:188–198.
10 Hein K, Hurst M: Human immunodeficiency virus infection in adolescence: A rationale for concern. Adolesc Pediatr Gynecol 1988;1:73–82.
11 Rotheram-Borus MJ, Koopman C, Bradley JS: Barriers to successful AIDS prevention programs with runaway youth; in Woodruff JO, Doherty D, Athey JG (eds): Troubled Adolescents and HIV Infection. Georgetown, CASSP Technical Assistance Center: Georgetown University Child Development Center, 1989.
12 Kandel DB, Davies M: Epidemiology of depressive mood in adolescents. Arch Gen Psychiatry 1982;39:1205–1212.
13 Deykin EY, Levy JC, Wells V: Adolescent depression, alcohol, and drug abuse. Am J Publ Health 1987;77:178–180.
14 Hayes CD: Risking the Future: Adolescent Sexuality. Pregnancy, and Childbearing. Washington, National Academy Press, 1987.
15 Diley J, Ochitill H, Perl M, Volberding P: Findings in psychiatric consultations with patients with AIDS. Am J Psychiatry 1985;421:82–86.
16 Perry CL, Jessor R: The concept of health promotion and the prevention of adolescent drug use. Health Educ Q 1985;12:169–184.
17 Rolf J, Nanda J, Thompson L, Mamon J, Chandra A, Baldwin J, Delahunt M: Issues in AIDS prevention among juvenile offenders. Troubled Adolesc HIV Infect 1989; 56–69.
18 Feldman RA, Stiffman AR, Jung K: Children at Risk: The Web of Parental Mental Illness. New Brunswick, Rutgers, 1987.
19 Rutter M: Psychosocial resilience and protective mechanisms; in Rolf J, Masten A, Cicchetti D, Nuechterlein KH, Weintraub S (eds): Risk and Protective Factors in the Development of Psychopathology. New York, Cambridge University Press, 1990, pp 181–214.
20 Stiffman AR, Jung K, Feldman RA: A multivariate risk model for childhood behavior problems. Am J Orthopsychiatry 1986;56:204–211.
21 Earls F, Robins LN, Stiffman AR, Powell J: Comprehensive health care for high-risk adolescents: An evaluation study. Am J Publ Health 1989;79:99–106.
22 Hollingshead AG: Four-factor index of social status. Working paper, Yale University, 1974.
23 Stiffman AR, Earls F: Behavioral risks for human immunodeficiency virus infection in adolescent medical patients. Pediatrics 1990;85:303–310.
24 Dawson D, Cynamon M, Fitti JE: AIDS knowledge and attitudes for September 1987. NCHS Advance Data. Washington, US Department of Health and Human Services, 1988, p 148.

25 Helzer JE, Robins LN, McEvoy LT, Spitznagel EL, Stoltzman RK, Farmer A, Brockington IF: A comparison of clinical and diagnostic interview schedule diagnoses. Arch Gen Psychiatry 1985;42:657–666.
26 Robins LN: Epidemiology: Reflections on testing the validity of psychiatric interview. Arch Gen Psychiatry 1985;42:918–924.
27 Robins LN, Helzer JE, Croughan J, Ratcliff KS: The NIMH Diagnostic Interview Schedule: Its history, characteristics and validity. Arch Gen Psychiatry 1981;38: 381–389.
28 Bohrnstedt GW: Observations on the measurement of change; in Borgatta E (ed): Sociological Methodology: 1969, San Francisco, Jossey-Bass, 1968, pp 113–136.
29 Stanton B, Black M, Keane V, Feigelbaum S: HIV risk behaviors in young people: Can we benefit from 30 years of research experience? AIDS Publ Pol J 1990;5: 17–23.

Arlene Rubin Stiffman, PhD, George Warren Brown School of Social Work, Washington University, St. Louis, MO 63130 (USA)

Summary Reports

Rossi GB, Beth-Giraldo E, Chieco-Bianchi L, Dianzani F, Giraldo G, Verani P (eds): Science Challenging AIDS. Basel, Karger, 1992, pp 226–237

Basic Science Challenging AIDS

Jean-Claude Gluckman

CNRS URA1463, CERVI, Hôpital de la Pitié-Salpêtrière, Paris, France

AIDS was first described in the literature only 10 years ago [1]. Since then, much effort and much money have been spent, but HIV infection is still facing us. Knowledge has rapidly accumulated, yet! After only 2 years, in 1983, isolation of AIDS virus, presently known as HIV, was first reported by Barré-Sinoussi et al. [2], HIV tropism for $CD4^+$ cells was demonstrated, the first serology data were obtained, and hypotheses as to the relationship between HIV and AIDS were presented by Montagnier et al. [3]. Enormous progress has been accomplished since then, the most recent highlights of which have been presented at the VII International Conference on AIDS.

Among hundreds of communications one can distinguish a pattern that ranges from the molecular interactions between HIV and its target cells to the development of a vaccine against HIV. This inevitably leaves aside many important results in important fields, which are nonetheless covered elsewhere in this book.

HIV Receptor(s) and Membrane-Virus Interactions

Till now, HIV binding to target cells was thought to merely involve a well-defined region of the CD4 first domain and as yet less well-known regions of gp120. In fact, the process appears rather more complex. The initial interaction of gp120 with the CDR2 domain of CD4 induces conformational changes of both gp120 and CD4, leading to additional binding of gp120 to CD4 outside of CDR2 [4–6], and to the exposure of cryptic domains of gp41 [5]. This results then in 'receptor-activated' virus-cell fusion [7]. Such process is a possible reason why, under some conditions, soluble CD4 (sCD4) can enhance SIV infectivity [8, 9], and also why it can promote HIV-2 infection even of CD4⁻ cells [10]. In contrast, at least for laboratory strains, binding of sCD4 to gp120 can strip gp120 from gp41 and block infection [11]. Additional noncompetitive mechanisms of viral inactivation by sCD4 are also possible [12]. These effects of sCD4 should be considered when contemplating its therapeutic use.

The question as to whether there are other membrane receptors for HIV than CD4, and how the virus can infect CD4⁻ cells, has long been in debate. Harouse et al. [13] have brought the first clear demonstration of the possibility that galactocerebroside or an antigenically related molecule can function as a cellular receptor for HIV in neural cells.

Another point of interest was examined by Fauci's group [14, 15]. It is the possibility that besides CD4 membrane adhesion molecules of the integrin family – such as LFA-1 – are necessary, not for virus infection, but to reinforce the interactions that are necessary for cell-to-cell infection and fusion. Different molecules might well be involved for different target cells. Such type of adhesion molecules could, thus, contribute to HIV infectivity and to its tropism, as they contribute to intercellular communication [16].

HIV Phenotype and Tropism

Many groups have now demonstrated that, during progression to AIDS, the virus can switch from a low to a high syncytium-inducing and replication rate phenotype, which is often asssociated with either primary macrophage or T cell line tropism, respectively. Such phenotypic modifications rest on subtle genomic differences that are determined in the *env* gene – especially in the region that codes for the V3 loop of HIV-1 gp120 – and at the regulatory genes level, especially *tat,* as discussed by several reports [17–19]. How and why such kind of 'linkage disequilibrium' is selective for HIV variants remains to be examined. It has been proposed to utilize the virus phenotype as a new prognostic marker, based on this strong relationship between phenotypic switch and progression to AIDS [20].

Macrophages, however, do not epitomize HIV target cells among antigen-presenting cells. Knight's [21–23], and more recently Haseltine's [24] groups have shown that other widely distributed bone-marrow-derived cells, namely dendritic cells, known as 'veiled' cells in the blood, may in fact represent a major virus reservoir. These cells are presently considered to be the primary antigen-presenting cells to $CD4^+$ helper T lymphocytes, and HIV infection alters their function [25]. In addition, they may then readily transmit the virus to helper T lymphocytes upon antigen presentation. From this, one can easily elaborate to account for most of the immunologic defects observed in HIV infection.

One should note, here, the possibility that besides tropsim as determined at the cell entry level, HIV pathogenesis may also rest on differences of the regulation of virus replication in different susceptible cells and on the possible pleiomorphic properties of HIV proteins on cell functions, as abundantly dicussed at the meeting.

Neutralizing Antibodies to HIV and the V3 Loop of gp120

So far, no known in vitro parameters of the immune response against HIV correlate with protection against disease progression in infected individuals. Neutralizing antibodies have been largely investigated in this respect [26, 27].

These antibodies essentially appear to be directed against the V3 loop [27–29], the amino acid sequence variability of which around a conserved tip accounts for neutralization type specificity. Functionally, this region certainly plays an important role in post-CD4 binding events leading to cell infection, and V3 may also be important in determining virus tropism and pathogenesis [17]. Data have been presented here by Schulz et al. [30] that support the hypothesis that V3 proteolytic cleavage may be necessary for viral infectivity, neutralizing antibodies acting, thus, by hindering access of a putative membrane protease.

There is now strong evidence that antibodies to V3 are sufficient to protect against HIV-1 infection, as indicated by Emini et al. [31], who showed that passive in vivo immunization with a 'humanized' monoclonal antibody to V3 has protected a chimpanzee against a homologous HIV-1 challenge. Besides its theoretical importance regarding antibody neutralization and V3, such finding provides a hint as to the basis for immune prophylaxis.

However, V3 sequence variability represents a way for the virus to escape neutralization. This can especially arise through the mechanism that Goudsmit et al. [28] and Nara et al. [29] have called 'the original antigenic sin', which can be summarized as follows: after mutation in the V3 sequence and the occurrence of antigen drift, current V3 epitopes stimulate better the memory antibody response to the V3 sequence of the original infecting variant than the primary response to themselves, as if the antibody response were definitely 'locked-in' once triggered by the first V3 sequence [32]. Conversely, observations from Herlihy et al. [33] have confirmed that short synthetic peptides, which comprise the tip-of-the-loop GPGRAF sequence that is conserved in the majority of the strains analyzed, can elicit broadly neutralizing antibodies to divergent isolates. This suggests that a peptide-based vaccine can, indeed, be contemplated to protect against the majority of HIV-1 isolates.

Currently, much attention is focused on the antibody response to V3, but other neutralizing epitopes should not be neglected. In fact, if type-specific, V3-directed neutralizing antibodies seem to predominate in the early phase of HIV infection, lower-titered and broader neutralizing antibodies develop later that recognize divergent strains [34]. Reports indicate that the latter antibodies are mainly directed at discontinuous conformational epitopes of gp120. Mapping of these epitopes has been realized with mouse or human monoclonal antibodies, some of which have been shown to react against gp120 putative CD4 binding site [34, 35]. Both types of antibodies, those directed to V3 and those that inhibit the binding to CD4, may synergize to neutralize virus infectivity [35].

Specific T Lymphocyte Responses against HIV

T cell reactivity is an important parameter with respect to the immune response against HIV: helper T cell activation is considered as the requisite for the appropriate development of any effective immunity, while induction of cytotoxic T lymphocytes is thought to be essential for protection against virus infection. Mapping the epitopes recognized by T lymphocytes on different viral proteins, which is necessary for identifying HIV essential protective components, has retained much attention. Actually, T lymphocyte epitope recognition is a three-partner process, involving a processed peptide presented by a polymorphic major histocompatibility complex (MHC) molecule on the one hand, and the polymorphic T cell receptor that specifically recognizes both of these molecules on the other; each of these molecules has to exactly fit the others [36]. This results in so-called MHC restriction of recognition, which theoretically represents a difficult problem to overcome in the prospect of a synthetic peptide-based vaccine. Therefore, the observation that the same *env* peptides can be recognized both by cytotoxic T lymphocytes and by helper T lymphocytes, and that they can be promiscuously presented by different MHC class I and class II molecules [37], appears intriguing, though auspicious for vaccination purposes.

Many investigators have given a closer look at the role cytotoxic T lymphocytes might play in controlling HIV infection in vitro [38, 39] or in vivo [40], and some results raise the possibility that HIV may escape cytotoxic T lymphocyte recognition [41] in the same manner as it escapes neutralizing antibodies.

Immune Intervention against HIV

Among the many therapeutic approaches that have been considered in the last years, one could stress the new strategies to design specific inhibitors of HIV protease directly based on the knowledge of its cristallographic structure [42], and in vitro attempts at so-called 'intracellular immunization' [43, 44]. However, attempts at increasing immune responses to the virus, either in the form of preinfection vaccination or as postinfection immunization, have mostly been in the limelight.

Immunotherapy has been advocated in already HIV-infected individuals to enhance protective immunity and, thus, to delay disease progression [45]. This is not without potential pitfalls [46]. As a caveat, one should be aware of Fultz et al.'s observation regarding a chimpanzee which, despite vaccination, had not been protected from challenge [47]. To test the hypothesis that stimulation of the immune system might lead to increased HIV replication, this chimpanzee, which had previously been submitted to a hepatitis B vaccination trial and whose vaccination protocol comprised recombinant vaccinia virus expressing HIV proteins [47], was successively inoculated with these recall antigens, with MDP-based adjuvant, and with anti-HIV IgGs. Meanwhile, numbers of HIV-infected blood mononuclear cells were monitored by serial end-point dilution and cocultivation with indicator normal human phytohemagglutinin-stimulated blood mononuclear cells [48]. Most inoculations resulted, within a 2-week period, in increased numbers of infected blood mononuclear cells, even after the injection of inactivated HIV or anti-HIV IgGs! Transient

Table 1. Primate animal models and vaccination against HIV

PRO	CON
Chimpanzee/HIV-1	
Can be infected by HIV-1	No disease; often undetectable viremia
Human-like antibody response	Endangered species
V3 is the principal neutralizing epitope	Expensive
MHC (T cell responses)	MHC typing is difficult
Macaques/SIV	
Disease and pathology	SIV differences with HIV-1
Possible short time course of disease	V3 is not the principal neutralizing epitope
Availability	Neutralizing antibodies do not correlate with protection
Relatively 'cheap' model	
Macaques/HIV-1	
Immunogenicity model	

increases in virus load were usually accompanied by enhanced, HIV-specific, immune responses [P. Fultz et al., in preparation]. Can the expected enhanced immunity always override the possible increase in viral expression? This question should be kept in mind when trying immunotherapy in man.

Many reports dealt with vaccination trials. Indeed, how to confer lasting protection against an extremely variable virus represents a formidable challenge. Current efficacy trials are essentially conducted in the chimpanzee and in different macaque species [49]. None of these models being perfect (table 1), which one to favor raises difficult policy problems. Briefly, the chimpanzee is susceptible to HIV-1 and its immune response to the virus resembles that of humans, especially the recognition of V3 neutralizing epitope; but it is a poor model for pathogenesis as compared with human infection because of a different clinical outcome and evolution of viremia [48]. In contrast, SIV infection in macaques closely mimics human HIV disease and pathology, with a time course that can be very short for some SIV-macaque species combinations. Therefore, this model can be used to examine the pathogenesis of immunodeficiency virus-related diseases [50, 51] as well as vaccines and therapy. But SIV differs from HIV-1 in many respects: in particular, SIV's principal neutralizing determinant is not the V3 loop [52, 53] and, after vaccination, neutralizing antibodies do not appear to correlate with protection [54], in contrast with the first

Table 2. Problems to solve for HIV vaccine development

Need for surrogate protection markers
Neutralization assay: define a 'gold standard'
New adjuvants
Protect against cell-free *and* infected cell challenge
Route of immunization and of infection (local immunity?)
Define the best injection schedule
Determine the duration of protective immunity
Homologous versus heterologous strains
Protection against infection or disease?
Human effficacy trials?

chimpanzee results [55, 56]. Therefore, both models should presently be considered as necessary and not to be too hastily extrapolated from to the human situation.

Most results regarding vaccination presented at the Conference have been obtained in macaques, immunization and challenge being usually performed with homologous strains, which is obviously far different from the real situation one would have to face. Homologous and the first hints of heterologous protection have been reported, especially using inactivated SIV [57, 58]. Homologous protection has also been obtained after live recombinant vaccinia virus priming and subunit recombinant envelope boosting [59].

Also, new immunogens have been tested, with interesting prospects, among which one can distinguish BCG recombinant live vectors [60–62], or recombinant HBs particles [63].

As already discussed, passive immunization can been successful in the chimpanzee [31]. In addition, it has also been shown that preinfection passive immunization of *Cynomolgus* macaques can protect against a low dose HIV-2 or SIV homologous challenge [64]. Such results pave the way for potential early immunotherapy, but studies on postexposure prophylaxis and protection against heterologous viruses have still to be performed.

However, before contemplating to enter into large-scale human vaccination efficacy trials and to reach the goal of protecting humans against HIV, many questions remain still to be answered (table 2). First is the obvious need to dispose of surrogate biologic markers that truthfully predict protection. Presently, neutralizing antibody titers represent the better of a bad choice in this respect [56], and standardizing the assays through a real international cooperation is urgently needed. Second, the only adjuvant authorized in humans, alum, is not the most effective and it is, thus, necessary to find a safe substitute. Third, in most current animal trials, immunization and challenge are performed with homologous strains, and challenge uses the intravenous route, while the final vaccines should provide group-specific protection against free virus as well as virus-infected cells entering the body through the mucosal as well as the intravenous route. Finally, should and effective vaccine protect against infection or against disease only?

Thus, despite all the recent progress, more insight into HIV and the pathogenesis of HIV-related disease has yet to be gained before finally solving AIDS problems. This we have to face in the years to come.

References

1 Anonymous: Pneumocystis pneumonia – Los Angeles. MMWR 1981;30:250–251.

2 Barré-Sinoussi F, Chermann JC, Rey F, Nugeyre MT, Chamaret S, Gruest J, Dauguet C, Axler-Blin C, Vezinet-Brun F, Rouzioux C, Rozenbaum W, Montagnier L: Isolation of a T-lymphotropic retrovirus from a patient at risk for acquired immune deficiency syndrome (AIDS). Science 1983;220:868–871.

3 Montagnier L, Barré-Sinoussi F, Dauguet C, Rozenbaum W, Brun-Vezinet F, Rouzioux C, Klatzmann D, Gluckman JC, Chermann JC: A human T-lymphotropic retrovirus: Characterization and possible role in A.I.D.S. (abstract); in Essex M, Gallo RC (eds): Meeting on Human T-cell Leukemia Viruses, Sept, 1983. Cold Spring Harbor. Cold Spring Harbor Laboratory, 1983, p 41.

4 Truneh A, Buck D, Cassat D, Juszczak R, Kassis S, Ryu SE, Healey D, Sweet R, Sattentau QJ: A region in domain 1 of CD4 distinct from the primary gp120 binding site is involved in HIV infection and virus-mediated fusion. J Biol Chem 1991;266: 5942–5948.

5 Sattentau QJ, Moore JP: CD4 binding to gp120 on HIV infected cells initiates membrane fusion (abstract). VII Int Conf on AIDS, Florence, June, 1991, vol 2, p 54.

6 Repke H, Jaekyoon S, Haseltine WA: Quantitative analysis of the binding kinetics and thermodynamics of gp120-CD4 interaction reveals a three step process and a new mode of allosteric inhibition of gp120 binding (abstract). VII Int Conf on AIDS, Florence, June, 1991, vol 1, p 24.

7 Moore JP, McKeating JA, Weiss RA, Clapham PR, Sattentau QJ: Receptor-mediated activation of immunodeficiency viruses in viral fusion. Science 1991;252: 1322–1323.

8 Allan JS, Strauss J, Buck DW: Enhancement of SIV infection with soluble receptor molecules. Science 1990;247:1084–1088.

9 Werner A, Winskwosky G, Kurth R: Soluble CD4 enhances simian immunodeficiency virus SIVagm infection. J Virol 1990;64:6252–6256.

10 Clapham PR, McKnight A, Weiss RA: CD4 independent fusion and infection by HIV-2 strains: Enhancement by soluble CD4 (abstract). VII Int Conf on AIDS, Florence, June, 1991, vol 1, p 41.

11 Moore JP, McKeating JA, Norton WA, Sattentau QJ: Direct measurement of soluble CD4 binding to human immunodeficiency virus type 1 virions: gp120 dissociation and its implication for virus-cell binding and fusion reactions and their neutralization by soluble CD4. J Virol 1991;65:1133–1140.

12 Berkower I, Murphy D, Smith GE: Noncompetitive inactivation of HIV-1 by rCD4 and CD4-Ig (abstract). VII Int Conf on AIDS, Florence, June, 1991, vol 1, p 25.

13 Harouse JM, Bhat S, Laughlin M, Stefano K, Spitalnik S, Silberberg DH, Gonzalez-Scarano F: Galactocerebroside mediates entry of HIV-1 in neural cell lines (abstract). VII Int Conf on AIDS, Florence, June, 1991, vol 2, p 54.

14 Butini L, Pantaleo G, Poli G, Dayton AI, Fauci AS: Role of the LFA-1/ICAM-1/2 pathway of cell adhesion in HIV infection (abstract). VII Int Conf on AIDS, Florence, June, 1991, vol 1, p 26.

15 Pantaleo G, Butini L, Grazisi C, Schnittman SM, Greenhouse JJ, Poli G, Gallin JI, Fauci AS: LFA-1 molecule regulates HIV-mediated cell fusion and syncytia fromation in CD4+ T lymphocytes (abstract). VII Int Conf on AIDS, Florence, June, 1991, vol 2, p 41.

16 Springer TA: Adhesion receptors of the immune system. Nature 1990;346:425–434.

17 Cheng-Mayer C, Shioda T, Levy JA: Small regions of the *env* and *tat* genes control cellular tropism, cytopathology and replicative properties of HIV-1 (abstract). VII Int Conf on AIDS, Florence, June, 1991, vol 1, p 21.

18 Westervelt P, Trowbridge DB, Gendelman HE, Ratner L: Regulation of silent and productive HIV-1 infection in primary monocytes by two distinct genetic determinants (abstract). VII Int Conf on AIDS, Florence, June 1991, vol 1, p 42.

19 Groenink M, Fouchier RAM, van der Jagt RCM, de Goede REY, Huisman JG, Tersmette M: Genetic mapping of regions in the HIV-1 genome determining syncytium induction and cytotropism (abstract). VII Int Conf on AIDS, Florence, June, 1991, vol 1, p 58.

20 Koot M, Keet IPM, Vos AHV, de Goede REY, Roos MTL, Schellekens PTA, Couthino RA, Miedema F, Tersmette M: HIV biological phenotype as a prognostic marker for AIDS in a large population of seropositive individuals (abstract). VII Int Conf on AIDS, Florence, June, 1991, vol 2, p 43.

21 Patterson S, Gross J, Bedford P, Knight SC: Morphology and phenotype of dendritic cells from peripheral blood and their productive and non-productive infection with human immunodeficiency virus type 1. Immunology 1991;72:361–367.

22 Knight SC, Macatonia SE, Patterson S, Gompels M, Pinching AJ: Effect of HIV on antigen presentation by dendritic cells and monocytes (abstract). VII Int Conf on AIDS, Florence, June, 1991, vol 1, p 63.

23 Patterson S, Gross J, Bedford P, Knight SC: HIV infection of bone marrow-derived dendritic cell (DC) (abstract). VII Int Conf on AIDS, Florence, June, 1991, vol 1, p 95.

24 Langhoff E, Terwiliger EF, Poznansky MC, Bos H, Kalland KH, Haseltine WA:

Prolific HIV-1 growth in human dendritic cells (abstract). VII Int Conf on AIDS, Florence, June, 1991, vol 2, p 40.

25 Macatonia SE, Patterson S, Knight SC: Suppression of immune responses by dendritic cells infected with HIV. Immunology 1989;67:285–291.

26 Köhler H, Nara P, Chamat S, Caralli V, Ryskamp T, Haigwood N, Newman R, Kang CY: Analysis of the anti-gp120 antibody response in HIV-1 infected patients (abstract). VII Int Conf on AIDS, Florence, June, 1991, vol 1, p 61.

27 Beretta A, Lopalco L, Longhi R, Ciccomascolo F, Manera E, De Rossi A, Siccardi AG: Neutralizing peptides and neutralizable epitopes of HIV-1 env glycoproteins (abstract). VII Int Conf on AIDS, Florence, June, 1991, vol 2, p 58.

28 Goudsmit J, Wolfs T, Kuiken C, Zwart G, Tersmette T: Genomic diversity and antigenic variation of the V3 domain as determining factors of HIV-1 pathogenesis (abstract). VII Int Conf on AIDS, Florence, June, 1991, vol 1, p 21.

29 Nara P, Garrity RR, Dunlop N, Smit L, de Ronde A, Swart G, Back N, Goudsmit J: Clonal dominance and antigenic variation of the HIV-1 envelope: Evolutionary tools for survival in the humoral immune system (abstract). VII Int Conf on AIDS, Florence, June, 1991, vol 2, p 74.

30 Schulz TF, Stephens PE, Tailor C, Clements G, Thomson S, Hoad JG, Moore J, Kay J, Weiss RA: Is proteolytic cleavage in the V3 loop of HIV-1 gp120 necessary for viral infectivity? (abstract). VII Int Conf on AIDS, Florence, June, 1991, vol 2, p 54.

31 Emini EA, Schleif WA, Murthy K, Eda Y, Tokiyoshi S, Putney SD, Matsushita S, Nunberg JH, Eichberg JW: Passive immunization with a monoclonal antibody directed to the HIV-1 gp120 principal neutralization determinant confers protection against HIV-1 challenge in chimpanzees (abstract). VII Int Conf on AIDS, Florence, June, 1991, vol 2, p 72.

32 Goudsmit J, Back N, Zwart G, Smit L, Dunlop N, Nara P: Characteristics of the HIV-1 neutralizing antibody response during infection and immunization; in Girard M, Valette L (eds): Retroviruses of Human AIDS and Related Animal Diseases. Lyon, Fondation Mérieux, 1991, vol 5, pp 73–79.

33 Herlihy WC, Javaherian K, LaRosa GJ, Bolognesi DP, Matthews TJ, Putney SD: The principal neutralization determinant of HIV-1 elicits broadly neutralizing antibodies (abstract). VII Int Conf on AIDS, Florence, June, 1991, vol 2, p 88.

34 Thali M, Olshevsky U, Furman C, Gabudza D, Posner M, Sodroski J: Structure and function of discontinuous gp120 epitope recognized by a broadly-reactive neutralizing human monoclonal antibody (abstract). VII Int Conf on AIDS, Florence, June, 1991, vol 2, p 54.

35 Tilley SA, Honnen WJ, Racho ME, Hilgartner M, Pinter A: Human monoclonal antibodies against the putative CD4 binding site and the V3 loop act in concert to neutralize virus (abstract). VII Int Conf on AIDS, Florence, June, 1991, vol 1, p 39.

36 Davis MM, Bjorkman PJ: T-cell antigen receptor genes and T-cell recognition. Nature 1988;334:395–402.

37 Berzofsky JA, Shirai M: Promiscuity of cytotoxic T cell epitopes from the HIV-1 envelope (abstract). VII Int Conf on AIDS, Florence, June, 1991, vol 1, p 38.

38 Koenig S, Jones G, Boone E, Brewah A, Newell A, Torbett B, Mosier D, Fauci AS:

Inhibition of HIV replication by HIV-specific CTL (abstract). VII Int Conf on AIDS, Florence, June, 1991;1:77.

39 Mackewicz C, Legg H, Ortega H, Levy JA: CD8+ lymphocytes from HIV-1 seropositive individuals block HIV replication in CD4+ lymphocytes by inhibiting viral RNA synthesis (abstract). VII Int Conf on AIDS, Florence, June, vol 2, p 74.

40 Rivière Y, McChesney M, Porrot F, Tanneau F, Marie C, Sansonetti P, Lopez O, Molereau G, Pialoux G, Kieny MP, Takaia F, Montagnier L: Relation between HIV disease and the activities of HIV specific cytotoxic effector cells (abstract). VII Int Conf on AIDS, Florence, June, 1991, vol 1, p 61.

41 Rodney P, Rowland-Jones S, Gotch F, Nixon DF, Ogunlesi A, Edwards J, McMichael AJ: Mutation in T cell epitopes of HIV gag provides a mechanism for escape from immune control (abstract). VII Int Conf on AIDS, Florence, June, 1991, vol 1, p 21.

42 Tomasselli A, Adams L, Greenberg B, Deibel M, Sawyer T, Thaisrivongs S, Heinrikson R: Toward an understanding of the specificity of the HIV protease through a study of nonviral protein substrates: Application to inhibitor design (abstract). VII Int Conf on AIDS, Florence, June, 1991, vol 2, p 24.

43 Poznansky MC, Langhoff E, Lo S, Haseltine WA, Sodroski J: The use of an HIV-1 based retroviral vector to transfer antiviral constructs to human lymphocytes in vitro (abstract). VII Int Conf on AIDS, Florence, June, 1991, vol 2, p 25.

44 Caruso M, Klatzman D: Inhibition of HIV spreading in CD4+ cells harbouring an HIV-inducible suicide (abstract). VII Int Conf on AIDS, Florence, June, 1991, vol 2, p 102.

45 Zagury D, Picard O, Halbreich A, Bernard J, Carelli C, Bizinni B, Salaun JJ, Lurhuma Z, Mbayo K, Burny A, Frottier J, Imbert JC: Progress report IV on AIDS vaccine in humans: Phase I clinical trial in HIV infected patients (abstract). VII Int Conf on AIDS, Florence, June, 1991, vol 1, p 39.

46 Dorozynsky A: French ban immunotherapy treatment. Science 1991;252:1608.

47 Girard MP, Kieny MP, Gluckman JC, Barré-Sinoussi F, Montagnier L, Fultz PN: Candidate vaccines for HIV; in Mehem A, Spiers RE (eds): Vaccines for Sexually Transmitted Diseases. London, Butterworths, 1989, pp 227–237.

48 Fultz PN, Siegel RL, Brodie A, Mawle AC, Stricker RB, Swenson RB, Anderson DC, McClure HM: Prolonged CD4+ lymphocytopenia and thrombocytopenia in a chimpanzee persistently infected with human immunodeficiency virus type 1. J Infect Dis 1991;163:441–447.

49 Girard MP, Eichberg JW: Progress in the development of HIV vaccines. AIDS 1990; 4(suppl 1):S143–S150.

50 Feichtinger H, Li SL, Grünewald K, Putkonen P, Weyrer K, Biberfeld G, Biberfeld P: Malignant lymphomas in SIV-infected *Cynomolgus* monkeys: A model for EBV-associated lymphomagenesis (abstract). VII Int Conf on AIDS, Florence, June, 1991, vol 1, p 55.

51 Li SL, Kaaya E, Putkonen P, Feichtinger H, Parravicini C, Ekman M, Biberfeld P: Generalized giant cell disease in SIV-infected *Cynomolgus* monkeys (abstract). VII Int Conf on AIDS, Florence, June, 1991, vol 1, p 155.

52 Collignon C, Thiriart C, Langedijk H, Meloen R, Desrosiers R, de Wilde M, Bruck C: Mapping of neutralizing antibodies of the SIV env glycoprotein (abstract). VII Int Conf on AIDS, Florence, June, 1991, vol 1, p 28.

53 Putney SD, Langlois A, LaRosa G, Desrosiers R, Meloen RH, Javaherian K: The principal neutralization determinant of SIV is different than HIV-1 (abstract). VII Int Conf on AIDS, Florence, June, 1991, vol 2, p 59.

54 Corcoran T, Kent KA, Stott EJ, Taffs LF: SIV neutralising antibody: Its assay and role in vaccine-induced protection (abstract). VII Int Conf on AIDS, Florence, June, 1991, vol 2, p 72.

55 Berman PW, Gregory TJ, Riddle L, Nakamura GR, Champe MA, Porter JP, Wurm FM, Hershberg RD, Cobb EK, Eichberg JW: Protection of chimpanzees from infection by HIV-1 after vaccination with recombinant glycoprotein gp120 but not gp160. Nature 1990;345:622–625.

56 Girard M, Kieny MP, Pinter A, Barré-Sinoussi F, Nara P, Kolbe H, Kusumi K, Chaput A, Mullins JI, Muchmore E, Ronco J, Kaczorek M, Gomard E, Gluckman JC, Fultz PN: Immunization of chimpanzees confers protection against challenge with HIV. Proc Natl Acad Sci USA 1991;88:542–546.

57 Mills KHG, Stott EJ, Chan WL, Taffs LF, Cranage M, Greenaway P, Thiriart C, De Wilde M, Bruck C, Page M, Kitchin PA: Vaccination against SIV in macaques (abstract). VII Int Conf on AIDS, Florence, June, 1991, vol 1, p 27.

58 Gardner M, Carlson J, Jennings M, Luciw P, Yilma T, Planelles V, Marhas M, Sutjipto S, Miller C, Yamamoto J, Pedersen N, Steimer K, Haigwood N: Vaccine protection against simian and feline immunodeficiency viruses (abstract). VII Int Conf on AIDS, Florence, June, 1991, vol 2, p 122.

59 Hu SL, Abrams K, Barber G, Moran P, Zarling JM, Kuller L, Morton WR, Benveniste RE: Proection of macaques against homologous SIV infection by immunization with live recombinant vaccinia virus followed by recombinant-made SIV *mne* gp160 (abstract). VII Int Conf on AIDS, Florence, June, 1991, vol 2, p 58.

60 Stover CK, de la Cruz VF, Fuerst TR, Burlein JE, Benson LA, Bennett LT, Bansal GP, Young JF, Lee MH, Hatfull GF, Snapper SB, Barletta RG, Jacobs WR, Bloom BR: New use of BCG for recombinant vaccines. Nature 1991;351:456–460.

61 Aldovini A, Young RA: Humoral and cell-mediated immune responses to live recombinant BCG-HIV vaccines. Nature 1991;351:479–481.

62 Aldovini A, Young RA: Development of an AIDS vaccine candidate: Recombinant BCG vaccine vehicles induce humoral and cell-mediated immune responses to HIV antigens (abstract). VII Int Conf on AIDS, Florence, June, 1991, vol 1, p 38.

63 Schlienger K, Michel ML, Mancini M, Dormont D, Rivière Y, Tiollais P: HIV-1 major neutralizing determinant exposed in HBsAG particles is immunogenic in primates (abstract). VII Int Conf on AIDS, Florence, June, 1991, vol 2, p 88.

64 Putkonen P, Thorstensson R, Albert J, Hild K, Biberfeld G, Norrby E: Prevention of HIV-2 and SIVsm infection by passive immunization in *Cynomolgus* monkeys (abstract). VII Int Conf on AIDS, Florence, June, 1991, vol 1, p 27.

Prof. J.C. Gluckman, CNRS URA1463, CERVI, Hôpital de la Pitié-Salpêtrière, 47, Bld de l'Hôpital, F-75651 Paris Cedex 13 (France)

Rossi GB, Beth-Giraldo E, Chieco-Bianchi L, Dianzani F, Giraldo G, Verani P (eds): Science Challenging AIDS. Basel, Karger, 1992, pp 238–249

Clinical Science and Trials

Jean Dormont

Service de Médecine interne et de Réanimation, Hôpital Antoine-Béclère, Clamart, France

I will focus my paper on a few clinical manifestations of HIV infection and next on clinical trials and related topics. Weller already clarified several difficult points of trial design and interpretation in his keynote lecture [1], thus simplifying my task.

Clinical Manifestations of HIV Infection

Non-Hodgkin Lymphoma

Non-Hodgkin lymphoma (NHL) is an increasing problem in advanced patients. Moore et al. [2], in a large cohort, indicated a probability of NHL of 3.2% after 2 years of therapy. Lietzau [3], in 55 patients followed on a long term and very extensively, gave a probability of tumor of 12% after 2 years and of 29% after 3 years on antiretrovial therapy. The role of zidovudine (azidothymidine; AZT) or other drugs, as tumor inductors, is not demonstrated as the development of NHL is not well correlated with the duration of treatment. The prolongation of survival in a profoundly immunosuppressed state is an alternative and likely hypothesis. Hence, the finding of new methods including more potent antiretroviral regimens is probably crucial in the prevention of this growing manifestation of HIV disease.

Further research has to better define several factors, including Epstein-Barr virus (EBV), which are important in the pathogenesis and clinical course of NHL occurring earlier in HIV infection. Kaplan [4], looking at molecular predictors of survival, demonstrated the longest survival to occur in the EBV-negative polyclonal group.

Kaposi Sarcoma

As NHL, Kaposi sarcoma (KS) is an increasingly common problem in advanced, severely immunosuppressed patients, and most available treatments work poorly at this stage. Gallo, in his keynote lecture [5], remarkably presented new avenues for research and I can only repeat one of his messages: the major role of several cytokines in endothelial cell proliferation. This may lead to new therapeutic interventions, such as blocking of cytokine activity. As an illustration of the role of cytokines, Martinez-Maza [6] showed that the proliferative effect of HIV *tat* protein on KS-derived cells is partially mediated by interleukin-6.

Renal Disease

The role of HIV itself in renal disease is controversial. Kimmel et al. [7], in 2 white patients with IgA nephropathy, found HIV DNA in renal biopsies by the polymerase chain reaction (PCR) and suggested an in situ mechanism of immune-complex formation with IgA, IgM and p24. Another evidence was offered by Kopp et al. [8] in a transgenic mouse model using a deleted provirus. HIV genes were expressed in the kidney as well as HIV proteins such as gp41 and *rev*. These data suggest a direct role of the virus as one of the mechanisms operating in renal manifestations in HIV patients.

Thrombocytopenia

Immune thrombocytopenia can be seen at any stage of HIV infection and is not related to disease progression. Two series reported by Oksenhendler et al. [9] and by Cinque et al. [10] with a total of more than 250 patients with severe thrombocytopenia ($< 50.10^6$ platelets/mm^3) demonstrated sustained response to AZT in 50–60% of cases. The study by Oksenhendler et al. [9] showed, in addition the good response to splenectomy (85%) and the lack of difference in AIDS-free survival between splenectomized and not splenectomized patients. The possible infection by HIV-1 of a human megakaryocytic cell line reported by Raharinivo et al. [11] and the favorable effect of AZT suggest that the virus may be, at least in part, directly responsible for the thrombocytopenia.

Pathology/Brain Alterations

A large autopsy series from d'Arminio-Monforte [12] confirmed that cytomegalovirus (CMV) infections, lymphomas and mycotic diseases are clinically underestimated. In another large autopsy series with brain examination, Vago et al. [13] found multinucleated giant cells (which are supposed to be a cytopathic consequence of HIV infection) in 76% of untreated cases and only in 23% of patients treated with AZT. This confirms the frequency of alterations of the central nervous system (CNS) and is compatible with the conclusions of Chiesi et al. [14] showing a decrease of AIDS-dementia complex after large-scale AZT treatment.

Looking at the virus itself, the data of Sonnerborg et al. [15] suggest a still higher frequency of CNS infection. They found indeed, by culture and/or PCR, HIV DNA in the cerebrospinal fluid of 25 out of 28 patients. These patients belonged to all stages of the Centers of Disease Control (CDC) classification and only 4 of them had overt neurological symptoms.

Perinatal Diagnosis

Early diagnosis of perinatal transmission remains a critical problem. Quinn et al. [16] have confirmed their previous findings on the value of specific IgA antibodies found in infants 4–12 months of age, with excellent sensitivity and specificity. Burgard et al. [17] used PCR–DNA and viral cultures in newborns and found an excellent specificity but a sensitivity of only about 50%. This limited sensitivity is probably due to the low viral load found early in life: as shown by the same group [18] the sensitivity can be enhanced to 70% if cultures are repeated up to the 3rd month.

Clinical Trials: Opportunistic Infections

Pneumocystis carinii *Pneumonia*

Two studies have been reported regarding the primary prophylaxis of *Pneumocystis carinii* pneumonia (PCP). Hirschel et al. [19] conducted a placebo-controlled prophylaxis trial with aerosolized pentamidine in 223 patients and showed a 70% efficacy in noncompliant and a 90% efficacy in compliant patients. Payen et al. [20] compared dapsone (50 mg/day) and pyrimethamine-sulfadoxine (500 mg weekly) in 121 patients and found no difference after a mean follow-up of 6 months.

Concerning the treatment of PCP, Black et al. [21], after demonstrating the efficacy of intravenous clindamycin plus primaquine, have now tried successfully the same treatment but entirely by oral route.

Toxoplasma Encephalitis

The study of Katlama et al. [22], including 336 patients, compared pyrimethamine (50 mg/day) plus sulfadiazine (4 g/day) with pyrimethamine plus clindamycin (2.4 g/day). A preliminary analysis of 225 patients showed no difference in efficacy and toxicity between the two groups. Toxic reactions were, however, dissimilar with each treatment; thus, a switch from one arm to the other occurred in 25% of the cases.

The new drug 566 C80 from Burroughs Wellcome, a hydroxynaphthoquinone, was used by Masur et al. [23] as a salvage treatment in patients intolerant on progressing under conventional therapy. Although the experience is still limited (8 patients), the efficacy and the tolerability seem good, even in conjunction with antiretroviral therapy. This drug is also being tested in PCP.

Herpes simplex Virus and Cytomegalovirus

In cases with acyclovir-resistant herpes simplex virus (HSV), Safrin et al. [24] showed foscarnet to be definitely more active and less toxic than vidarabine. Recurrence after foscarnet discontinuation occurred at a median time of 42 days.

Spector et al. [25] compared immediate versus deferred ganciclovir therapy in CMV peripheral retinitis, and showed that ganciclovir significantly delays progression. This work is important as it is the first well-controlled trial in CMV retinitis and will constitute a good basis for the assessment of oral ganciclovir therapy.

Biological Markers

Many biological markers, often called 'surrogate markers', have been studied in HIV infection for their prognostic significance and their use in clinical trials. Although they will never completely replace clinical end points, they are very useful and justify further studies.

p24 Antigen

Bollinger et al. [26] and Winger et al. [27] showed that acidification of plasma increases the sensitivity of the p24 antigen assay, by dissociation of antigen-antibody complexes. With or without a more sensitive antigen capture assay, they suggest that this may improve the utility of the test as a marker for therapeutic efficacy during a clinical trial. A comparison with viremia would be of primary importance. Indeed, if a sensitized method were a good screening procedure for viremic asymptomatic subjects it would certainly be very helpful. Clearance of antigen under antiretroviral treatment is probably a useful index of treatment efficacy, but the variability of observed data results in a number of difficulties for quantitative conclusions. Conversely, as shown by Bodsworth [28] the persistence of antigen at all times during treatment is an index of treatment failure and poor survival.

CD4 Count

Cohort studies have shown the CD4 count to have a prognostic value, by itself or combined with p24 antigen or other biological markers. Yarchoan et al. [29] showed a CD4 count of less than $50/mm^3$ to be an indicator of mortality. In clinical trials, the magnitude of the increase of the CD4 count over baseline and the number of months necessary for this count to drop and reach again baseline are interesting indices of drug activity, at least for reverse-transcriptase inhibitors. The fact that AZT rarely produces an increase of more than 50 CD4 cells for more than 6–12 months probably means that this drug given in vivo at reasonably tolerated doses is not able to block completely and for a long time HIV replication. It can be also a prognostic index: indeed, Swanson and Cooper [30] demonstrated in 279 non-AIDS patients under AZT that an absence of increase of CD4 count is associated with faster progression to AIDS and shorter survival.

Clinical Virology

During the last 2 years, a tremendous development of clinical virology has been observed and has permitted a new approach to natural history, prognosis and clinical trials. In several countries, including the USA, UK, Netherlands and France, networks of virologists have been encouraged to develop consensus techniques and quality controls. I will briefly summarize the essence of this effort. One type of approach to the virus cycle is qualitative: slow and high replication rate; syncitium formation; macrophage or lymphocyte tropism, etc. The other type, difficult but essential, is quantitative with 3 groups of techniques: (1) blood culture in plasma and peripheral-blood mononuclear cells (PBMC); (2) PCR, from DNA or from RNA, and (3) resistance to drugs, phenotypic or genotypic.

In clinical studies, virus is routinely obtained from blood samples and this is the only way to proceed in most studies. We are not entirely sure, however, that this approach is a good reflection of the site and magnitude of viral replication, especially in early CDC stages. Indeed, Graziosi et al. [31] in CDC group III patients, showed the frequency of HIV-1-infected CD4 cells to be 1/100–1/1,000 in lymphoid tissues, thus 50–100 times greater than in CD4 cells of peripheral blood. This study is essential as it may explain why immune deficiency may appear even with very few infected cells in peripheral blood. We should also remember this finding to avoid any oversimplified interpretation of virological results.

I turn back to blood cultures to underscore the major consequences of quantitative methods. They have already modified our views of the natural history of the disease showing a breakthrough of viral replication and modification of virus properties anticipating clinical progression. They are also currently changing the strategy of drug evaluation. Virology-based trials – although they will never replace longterm clinical evaluation – are indeed the only possibility we have to develop an accelerated screening of new drugs and combinations. We can anticipate many difficulties in the interpretation of the data. However, we have to move ahead and make the necessary adjustments.

Trials with Antiretroviral Drugs

The information presented at the conference is of moderate size, probably because many new drugs trials or combinations are in an early phase. Conversely, large trials [on dideoxyinosine (ddI) and dideoxycytidine (ddC)] are not yet fully analyzed and virological data, although actively rescued, are only available in part. We know, however, from the epidemiological studies of Vella et al. [32] that AZT efficacy, with a median follow-up of 26 months, is similar in intravenous drug users and in homosexuals. We have also a short update of the Concorde and Alpha trials by Aboulker et al. [33, 34] and preliminary information by Youle et al. [35] on a TIBO compound. Concorde, as already mentioned by Weller [1], has the aim to show if early AZT is preferable to deferred treatment and may prolong life. An interim analysis is planned before the end of 1991.

Combined or alternating treatments have been actively developed in the last year but a more important effort is warranted. We know indeed that a single drug has a limited efficacy. With a combination we also hope to avoid enhanced toxicity and prevent the emergence of resistance. The prerequisites for a combined trial are: the finding of in vitro synergism, or at least of an additive effect; the absence of overlapping toxicities, and the likelihood to achieve in vivo concentrations which are synergistic in vitro.

In two-drug combination trials, Collier [36] showed that a combination of AZT and ddI is well tolerated at a moderate dose of both drugs; they observed an increase of CD4 cell count, but could not demonstrate superiority over a single-drug therapy. A phase I-II trial of AZT + ddC was presented prior to this conference by Richmann and Fischel [37]. A sustained increase of CD4 cell count and a decrease of p24 antigen have been observed. This is encouraging and will be controlled in a large efficacy trial.

In the above-mentioned combination trials of dideoxynucleosides, two important findings emerge. First, the pharmacokinetics of one drug is not modified by the other one. Second, the dose of each drug can be similar to that used in monotherapy without marked toxicity, which is a clear justification for the concurrent use in these cases. If combined toxicity was anticipated, sequential use should be preferred.

Frissen et al. [38] compared AZT (1,000 mg) with AZT (500 mg) plus interferon-α (3 *M* t.i.w.), with no demonstrable difference but with somewhat more toxicity in the latter.

A three-drug combination was presented by Yarchoan et al. [39]. They evaluated in 38 patients the alternance every 3 weeks of AZT, ddI and ddC; the tolerability was good and an increase of CD4 cell count was observed up to 9 months. Further data, however, are necessary to conclude on activity and prevention of resistance. The effect of recombinant granulocyte-colony-stimulating factor on drug-induced neutropenia seems consistent in 11 patients, as shown by de Wit et al. [40].

New attempts at bone marrow transplantationhave been presented, one in an AIDS patient [41] and another [42] in a much less advanced patient. The donor was an identical twin in both cases and the preparation included intensive antiretroviral therapy (AZT, rCD4-IgG, interferon-α) and the infusion of immunized donor cells. One patient died after 10 months with negative viral cultures but a poor immunologic restoration. The other patient [42] is alive and well with an increase in the CD4 count. Further transplantations are planned and will bring more information on this important experimental approach.

Drug Resistance

Since the pioneer work of Larder et al. [43] and Larder and Kemp [44] who demonstrated resistance to AZT and showed point mutations of reverse transcriptase at codons 67, 70, 215 and 219 in patients progressing under AZT therapy, a tremendous effort has been devoted to the study of resistance.

Several methods are still used in exploring phenotypic resistance and there is yet no general agreement. With the HeLa plaque assay, for instance, only a minority of virus will grow after adaptation to this system. Other scientists such as Bach et al. [45] and Brun Vézinet et al. [46] will thus prefer PBMC cultures with evaluation of reverse-transcriptase activity or p24 level. The reality may also be more sophisticated than we think: for instance, d'Aquila et al. [47] have found differences in sensitivity in viruses obtained from plasma and from PBMC in the same individual.

The sequence of mutations in asymptomatic patients on long-term AZT therapy has been extensively studied by Boucher et al. [48]. They showed mutations at codons 70 and 215 to appear in that order, often with temporary disappearance of mutation 70 when mutation 215 appears. Mutations 67 and 219 have not been found. These patients are thus only partially resistant after 2 years.

What happens when AZT is stopped or replaced by ddI? Boucher et al. [48] observed a persistence of partial resistance and Bach et al. [45] a more or less complete reversion of resistance but with the persistence of mutation at codon 215. In fact, a broad spectrum of situations is to be expected as each individual has several virus strains with different AZT sensitivities. Further studies will be important to decide if a new challenge with AZT, after reversion of resistance, is warranted and if it is useful in vivo (as it is the case in vitro) to maintain a two-drug combination when the patient looks partially or completely resistant to one of them.

As expected, resistance to ddI has been observed after a few months in symptomatic patients by Bach et al. [45] and by Reichman et al. [49] and is genotypically different from AZT resistance as a point mutation is found at reverse-transcriptase codon 74. Shirasaka et al. [50] in patients receiving an alternating regimen of AZT and ddC; found no resistance to ddC whereas resistance to AZT developed quickly. But only a few patients were studied.

As shown by Richman at the San Francisco Conference [51], resistance to AZT appears later and only in a minority of asymptomatic patients with more than 400 CD4 cells. This was confirmed by Broadhurst et al. [52]. However, no correlation has yet been demonstrated between progression and resistance. Boucher et al. [48] showed that progression to AIDS can occur before the emergence of fully resistant isolates. Many questions are thus raised including that of other mechanisms able to explain treatment failure. Hill et al. [53] showed a decreased intracellular activation of AZT occurring after 3 months of therapy and before virus strains become truly resistant. Further research is certainly warranted in this new direction.

I will conclude only by expressing my wish to observe in the next year major advances not only due to new drugs but also to novel designs and strategies, enhanced international cooperation and the best combined use of already known medications.

References[1]

1 Weller I: Present outlook of clinical trials. Keynote lecture. June 18, 1991.
2 Moore R, Kessler H, Richman DD, Chaisson RE: Non Hodgkin lymphoma in patients with advanced HIV infection treated with zidovudine (abstract T.U. 81).
3 Lietzau J: Markers predictive for the development of opportunistic non-Hodgkin's lymphoma (NHL) in patients with severe HIV infection on long-term antiretroviral therapy (abstract TU.B. 4).
4 Kaplan LD: Molecular predictors of survival in HIV associated non Hodgkin's lymphoma (abstract RU.A.7).
5 Gallo R: Aids-related malignancies. Keynote lecture. June 18, 1991.
6 Martinez-Maza O: HIV-*TAT* increases IL–6 production by and proliferation of AIDS-KS derived cells (abstract TU.A.5).
7 Kimmel PL, Phillips TM, Abraham AA, Szallasi T, Ferreira-Centeno A, Garrett CT: Molecular biological demonstration of viral genome in renal biopsies of HIV infected patients with IgA nephropathy (abstract TU.B.32).
8 Kopp JB, Weeks BS, Marinos NJ, Bryant JL, Dickie P, Notkins AL; Klotman PE: HIV associated nephropathy in transgenic mice (abstract TU.B.31).
9 Oksenhendler E, Bierling P, Delfraissy JF, Peynet J, Chevret S, Seligmann M: HIV-related immune thrombocytopenia cohort study: 174 patients (abstract W.B.82).
10 Cinque P, Landonio G, Argenteri B, Cargnel A, Coen M, Gringeri A, Lazzarin A, Meraviglia P, Nosari AM; Ruggieri A, Santagostino E: Long-term efficacy of zidovudine on HIV related thrombocytopenia (abstract W.B. 83).
11 Raharinivo B, Monte D, Auriault C, Capron A, Ameisen JC: Productive HIV-1 infection of the CD4-negative megakaryocytic cell line DAMI (abstract W.B. 81).

[1] References are quoted as they stand in the program and abstracts books of the VII International Conference on AIDS, Florence, June 17–21, 1991, except for references 37, 43, 44 and 51.

12 D'Arminio Monforte A, Vago L, Lazzarin A, Formenti T, Rizzardi GP, Guzzeti S, Musicco M, Costanzi G, Moroni M: Clinical and autoptical diagnosis in AIDS: Autopsy warns the clinical approach (abstract M.B. 81).
13 Vago L, Castagna A, Boldorini R, Christina S, Cinque P, Trabattoni GR, Lazzarin A, Moroni M, Costanzi G: Effect of zidovudine on brain HIV-associated pathological lesions (abstract W.B. 28).
14 Chiesi A, Agresti MG, Dally LG, Zacarelli M, Tomino C, Floridia M, Vella S: Decline of Aids-dementia complex notifications in the period 1989–1990 in Italy: possible role of early zidovudine treatment (abstract M.B. 2036).
15 Sonnerborg A, Johansson B, Strannergard O: Detection of HIV-1 DNA and infectious virus in cerebrospinal fluid (abstract W.B. 29).
16 Quinn T, Kline R, Halsey N, Ruff A, Boulos R, Butz A, Hutton N, Modlin J: Diagnosis of perinatal HIV infection by detection of viral-specific IgA antibodies in US and Haitian pediatric populations (abstract TU.B. 33).
17 Burgard M, Rouzioux C, Blanche S, Mayaux MJ, Kouznetzoff A, Griscelli C: Early diagnosis of HIV-1 infection in newborns: Comparison of PCR-DNA and viral culture (abstract TU.B. 35).
18 Rouzioux C, Burgard M, Mayaux MJ, Blanche S, Ferroni A, Griscelli C: Viral culture: An efficient method for early diagnosis of infection in newborns (abstract W.B. 2063).
19 Hirschel B, Chopard P, Heald A, Lazzarin A, Opravil M: A controlled study of inhaled pentamidine for primary prevention of Pneumocystis carinii pneumonia (abstract TH.B. 41).
20 Payen MC, Franchioly P, Gerard M, Sommereijns B, de Wit S, Clumeck N: Daily dapsone versus pyrimethamine-sulfadoxine (Fansidar) as primary prophylaxis against pneumocystis carinii pneumonia (PCP) (abstract TH.B. 43).
21 Black JR, Akil B, Murphy R, Fass R, Feinberg J, Mills J, Sattler F: Oral clindamycin plus primaquine therapy for Pneumocystis carinii pneumonia in AIDS (abstract TH.B. 42).
22 Katlama C, de Wit S, Guichard A, Mignolet F, van Glabeke M, Gentilini M, Clumeck N: Pyrimethamine-clindamycin versus pyrimethamine-sulfadiazine in toxoplasma encephalitis in Aids: A randomized prospective multicentric European study (abstract W.B. 30).
23 Masur H, O'Neill D, Feuerstein I, Baird B, Walker R, Davey RT, Polis MA, Falloon J, Galetto G, Seidel E, Milwee S, Lane HC, Kovacs JA: 566C80 is effective as salvage treatment for toxoplasma encephalitis (abstract W.B. 31).
24 Safrin S, Crumpaker C, Chatis P, Davis R, Hafner R, Rush J, Kessler H, Landry B, Mills J: ACTG 095: A randomized comparison of foscarnet vs vidarabine for treatment of acyclovir-resistant mucocutaneous herpes simplex virus infection in patients with Aids (abstract M.B. 85).
25 Spector SA, Barker C, Buhles W, Feinberg J, Montague P, Weingeist T, de Armond B: A randomized, controlled study of immediate vs deferred ganciclovir therapy in Aids patients with cytomegalovirus peripheral retinitis (abstract M.B. 86).
26 Bollinger RC, Kline R, Francis H, Moss M, Bartlett JG, Quinn T: Acid hydrolysis increases the sensitivity of p24, antigen detection for the evaluation of antiviral therapy in asymptomatic HIV-1 infected individuals (abstract M.B. 34).

27 Winger EE, Reddy MM; Hargrove D, Hugh T, McKinley GF, Grieco MH: A sensitive method for monitoring efficacy of antiretroviral therapy in HIV-infected individuals. A highly sensitive p24 antigen assay (abstract M.B. 38).

28 Bodsworth NJ: Influence of pre-treatment HIV-1 p24 antigen level and subsequent clearance on the response to zidovudine therapy in AIDS patients (abstract TU.B. 3).

29 Yarchoan R, Venzon D, Lietzau J, Pluda JM, Wyvill KM, Steinberg S, Broder S: Nearly all deaths in patients on anti-HIV therapy occurred in individuals with less than 50 CD4 cells/mm^3 in the adult NCl experience: Implications for the development of CD4 as a mortality risk indicator (abstract M.B. 37).

30 Swanson CE, Cooper D: Changes in CD4+ lymphocyte counts as a surrogate endpoint in the evaluation of long-term zidovudine therapy: A national study of non-Aids HIV seropositive patients in Australia (abstract TH.B. 33).

31 Graziosi C, Pantaleo G, Schnittman S, Grennhouse J, Butini L, Kotler D, Fauci AS: Lymphoid tissues function as 'reservoirs' for HIV infection (abstract TU.B. 28).

32 Vella S, Floridia M, Giuliano M, Pezzotti P, Agresti MG, Chiesi A: Progression to Aids and predictors of disease progression in a cohort of 859 zidovudine-treated ARC patients: A study comparing homosexuals and IV drug users (abstract W.C. 44).

33 Aboulker JP, Swart AM: Update on the MRC/INSERM Concorde trial (a multicentre double blind placebo controlled trial of zidovudine in asymptomatic individuals) (abstract W.B. 2117).

34 Aboulker JP, Valentine CB, Tournerie C, Darbyshire JH: Update of the MRC/INSERM Alpha trial (abstract W.B. 2096).

35 Youle M, Pialoux G, Dupont B, Gazzard B, Cauwenbergh C, Janssen PAJ: Early clinical evaluation of R 82913 TIBO compound in patients with severe HIV infection (abstract TH.B. 86).

36 Collier AC: Effect of combination therapy with zidovudine and didanosine on surrogate markers (abstract TU.B. 2).

37 Richman D, Fischel M: Communications at the Satellite Symposia International AIDS Therapy Group, Florence, June 15, 1991, and Advances in HIV Treatment, Florence, June 16, 1991.

38 Frissen J, van der Ende ME, Kaufmann RM, Weigel H, Schreij GS, ten Napel CHH, Hoepelman IM, Koopmans PP, Haverkamp G, Kuiken C, de Wolf F, Lange JMA: Zidovudine and interferon-alpha versus zidovudine in symptomatic HIV-infection (CD4+ count 150/mm^3):24 weeks follow up (abstract TH.B. 35).

39 Yarchoan R, Pluda JM, Shay LE, Wyvill KM, Mitsuya H, Brawley O, Broder S: Treatment of Aids or ARC with an alternating combination regimen of three dideoxynucleosides: A pilot study (abstract TH.B. 84).

40 de Wit S, Rahir F, Franchioly P, Sommereijns B, Clumeck N: Recombinant granulocyte colony stimulating factor (rG-CSF) in HIV patients (P) with zidovudine (Z) related neutropenia (abstract W.B. 84).

41 LaNasa G, Carcassi C, Pizzati A, Arras M, Vacca A, Ledda A, Pintus A, Atzeni S: Allogeneic bone marrow transplantation combined with zidovudine, alfa 2 interferon and donor HIV-1 specific T-cell clones in an Aids patient (abstract TH.B. 81).

42 Zunich KM, Easter M, Owen C, Walker R, Davey R, Wilson W, Leittman S, Fauci AS, Lane HC: Bone marrow transplantation and adoptive transfer of peripheral blood lymphocytes from gp160-immunized syngenic donors combined with zidovudine, interferon-α, and recombinant CD4-IgG in the treatment of HIV infection (abstract TH.B. 82).
43 Larder BA, Darby G, Richman DD: HIV with reduced sensitivity to zidovudine (AZT) isolated during prolonged therapy. Science 1989;243:1731–1734.
44 Larder BA, Kemp SD: Multiple mutations in HIV-1 reverse transcriptase confer high level resistance to zidovudine (AZT). Science 1989;246:1155–1158.
45 Bach MC, St Clair M, King D, Vavro C: Long-term follow-up of patients with HIV isolates resistant to zidovudine (AZT) treated with dideoxyinosine (ddI). (abstract TU.B. 91).
46 Brun-Vézinet F, Fleury HJ, Buffet-Janvresse C, Deforges L, Ferchal F, Ingrand D, Schnmitt MP, et al: Sensitivity of HIV-1 isolated to zidovudine: Comparison of two culture methods (abstract W.B. 2086).
47 d'Aquila RT, Videler JA, Eron J, Johnson VA, Bechtel LB, Merrill DP, Gorczyca P, Kaplan JC, Hirsch MS: Plasma and PBMC derived HIV-1 isolates differ in RT genotype and AZT susceptibility (abstract W.B. 2084).
48 Boucher C, Goudsmit J, O'Sullivan E, Ramautarsing C, Lange J, Larder B: Zidovudine sensitivity of HIV during the development from the asymptomatic stage to AIDS (abstract TU.B. 89).
49 Reichman R, Lamber J, Strussenberg J, Dolin R: Decreased dideoxyinosine (ddI) sensitivity of HIV isolated obtained from long term recipients of ddI (abstract TU.B. 92).
50 Shirasaka T, Yarchoan R, Husson R, Shimada T, Wyvill DM, Broder S, Mitsuya H: HIV may develop resistance preferentially to azidothymidine as compared to dideoxycytidine and dideoxyinosine in patients receiving antiviral therapy (abstract W.A. 9).
51 Richman D: Introductory remarks at the abstract session B11: Resistance to antiretrovirals. Sixth International Conference on AIDS, San Francisco, June 20–24, 1990.
52 Broadhurst K, Ball A, Stein CA, Levantis P, Forster GE, Goh BT, Colvin B; Jackson GG, Oxford JS: Quantitation of AZT resistant viruses in the HIV cohort at the Royal London Hospital (abstract TU.B. 87).
53 Hill MD, Miles S, Gomperts E, Holcenberg JS, Woods L, Avramis VI; Decreased activtion of zidovudine (AZT) in peripheral blood lymphocytes from patients with Aids chronically treated with AZT (abstract TH-B. 36).

Prof. J. Dormont, Hôpital Antoine-Béclère, F–92141 Clamart Cedex (France)

Rossi GB, Beth-Giraldo E, Chieco-Bianchi L, Dianzani F, Giraldo G, Verani P (eds): Science Challenging AIDS. Basel, Karger, 1992, pp 250–264

Epidemiology and Prevention Commentary

J.M. Mann[a], *S.J. Heymann*[a], *R.A. Royce*[a], *K. Kay*[a], *C. Mantel*[a], *J.F. McGuire*[a], *E. Muhondwa*[b]

[a] Harvard School of Public Health, Boston, Mass., USA;
[b] Muhilimbili Medical Center, Dar es Salaam, Tanzania

Each year, at the International AIDS Conference, we come together and help to shape the pandemic by pooling our collective research and understanding. In that sense, our views about the pandemic and our progress against it help give meaning and direction to the future. This year, in reviewing the more than 100 oral presentations and over 700 posters in the epidemiology and prevention track, it is the pandemic's dynamic quality and unpredictability which is most impressive, and which should be emphasized.

The unstable and volatile quality of the pandemic is manifested in several ways. First, HIV continues to spread in all already affected communities and areas of the world. In the industrialized world, Crosier et al. [1] documented an increase in HIV prevalence among injecting drug users in London from 6% or less to 13% in 1990. In the United States, where 40,000–80,000 new HIV infections are anticipated in 1991, Berkelman et al. [2] reported that AIDS in women is increasing rapidly, with 55% of all US AIDS cases in women during the last decade occurring the last 2 years. Among United States army personnel who are repeatedly tested for HIV infection, overall HIV seroincidence declined since 1985 and then stabilized, yet among soldiers who were black women or teenagers, the incidence of new infections rose [3]. In France, Rude et al. [4] found no deficit of new AIDS cases among heterosexuals and injecting drug users, when reported cases are compared with projections based on earlier trends. Methodologic issues can also strongly influence what parts of the epidemic are visible. For example, in the United States, analysis by Pagano et al. [5] of CDC surveillance data suggested that the apparent 'slowdown' in the increase in numbers of new AIDS cases has been strongly influenced by the increasing delay in reporting; when this factor was accounted for, the number of cases was still growing. McDougall et al. [6], working in Vancouver, British Colombia, found that in a prospective follow-up study those with higher levels of risk behavior were more likely to miss follow-up appointments, which could potentially lead to an underestimate of seroincidence rates. In Amsterdam, Fennema et al. [7] found that when projections of infection rates for an entire community of injecting drug users were based on results from a convenience sample, the projected rate was substantially lower than results from a more representative sample – the predicted seroprevalence was 21%, while observed seroprevalence was 37%. The net effect is that projections based on these studies may be overly optimistic, and consequently misleading, regarding the actual evolution of community experience with HIV.

Another manifestation of epidemic volatility is its spread, sometimes quite rapid, into communities and countries previously unaffected or little affected by the pandemic. In Paraguay, Cabello et al. [8] reported that 9% of homosexuals in the capital city were HIV infected; Melbye et al. [9] reported on the arrival of HIV infection and AIDS among Eskimos in Greenland; and in Lagos, Nigeria, over 10% of women sex workers were found to be HIV infected [10]. Chin [11] noted that an estimated 400,000 Thais are now HIV infected, including a nationwide average of over 15% of women sex workers and 6% of military recruits. Also, recent reports from India (unfortunately not presented in detail at this Conference) describe a rapid increase in HIV infections among women sex workers, men attending sexually transmitted disease clinics, and injecting drug users.

The extent of global vulnerability to further spread of HIV infection provides another reason to avoid complacency about the pandemic. Only 5–10% of the world's injecting drug users have thus far been HIV infected [11], yet needle-sharing remains a common practice and the potential therefore remains for new, explosive epidemics in communities of injecting drug users. In addition, global vulnerability to sexual transmission of HIV is reflected by the World Health Organization's estimate that at least 250 million new sexually transmitted infections occur each year [11]. Among women sex workers in communities with low HIV seroprevalence like cities of central Mexico, a high rate of genital ulcers and other sexually transmitted diseases provides conditions favorable to HIV introduction and spread [12]. Furthermore, behaviors creating opportunities for HIV transmission can resurge as a result of recidivism or entry of new susceptibles into a population. For example, the report by Hays et al. [13] from the young men's survey group in California found that 36% of 18- to 27-year-old homosexual and bisexual men reported unprotected sexual intercourse in the past 2 months; in San Francisco, Stall et al. [14] found that young gay men reported nearly twice as much unprotected anal intercourse as older gay men.

Yet regardless of which persons, with whatever behaviors, may be first affected in a given community, as the epidemic matures there will be a tendency towards further spread within and throughout society. The extraordinary differentiation and heterogeneity of both HIV infection dynamics and transmission-associated behaviors within and among populations should be better appreciated. For example, substantial inter-village variations in HIV seroprevalence are found in rural Uganda [15], increasing numbers of women with AIDS are reported in Brazil [16, 17] and southern Spain [18], incidence rate and risk factors for HIV infections are found to differ markedly between men and women injecting drug users in Baltimore [19], HIV prevalence differs significantly among white, African-American and Latino homosexual and bisexual men in a single large city [20], and foci of HIV infection are reported among the homeless of New York City [21] or San Francisco [22], and among aboriginal street people in Vancouver, Canada [23]. It is reasonable to assume that HIV will eventually reach most if not all human communities and within each community and country, the epidemic intensifies, evolves, and remains a threat.

It is also essential to avoid static thinking about the epidemic in temporal terms. Just as a period of increase in HIV infection among men donating blood in Port-au-Prince, Haiti, from 3.7 to 6.9% occurred from 1986 to 1988 [24], so the prevalence can stabilize, at least temporarily, or even decline, as it apparently has in Port-au-Prince both among blood donors [24] and women attending antenatal clinics [25]. It is important to avoid misinterpreting a levelling of new reported AIDS cases as a direct result of current HIV transmission patterns. Not only do AIDS cases reflect infections acquired in the past, but in some parts of the world the effects of antiretroviral treatment and prophylaxis and improved treatment for opportunistic infections will alter and obscure the relationship between current disease and the dynamics of past HIV infection.

For all these reasons, it is neither prudent nor appropriate to describe the HIV pandemic as a stable or fully predictable phenomenon. Based on increasing experience, we can move beyond fears of explosive epidemics in certain populations without going to the opposite extreme of declaring premature victory in any population. The pandemic is still too new and its interaction with personal and societal life too complex to allow us the mental relief of predictability or complacency.

Advances in knowledge in several specific areas should be highlighted. Compared with previous years and conferences, the impact of HIV/AIDS on women has been more extensively addressed. The data presented during this conference on survival suggests that the experience of women with AIDS in certain industrialized country settings is likely to be similar to men. Reports from Italy, Spain and the United States found no significant difference in relative survival after controlling for confounders [26–30]. Considerable work was also reported on manifestations of HIV infection in women. Carpenter [31] found that among 200 North American women, esophageal candidiasis (33%) and mucocutaneous herpes accounted for more than 50% of AIDS-defining illnesses. Thompson et al. [32] reported that among HIV-infected women who met the CDC criteria for AIDS, candida esophagitis, *Mycobacterium avium* complex infections, and wasting syndrome were more commonly observed than among men. However, others reported that the presenting diagnosis was similar among men and women [33] and United States national surveillance data showed that less than 20% of both men and women have esophageal candidiasis and fewer than 5% have mucotaneous herpes as the first AIDS-defining event [34].

In addition, there appears to be significant morbidity specific to women that is not included in the CDC/WHO AIDS definition, such as cervical squamous intraepithelial lesions, as reported both by Chiphangwi et al. [35] in Malawi and La Guardia et al. [36] in New York.

Regarding vertical transmission, substantial differences continue to be observed between rates observed in industrialized and developing countries; compared with the 13% rate in the European Collaborative Study [37], Lepage et al. [38] reported a 30% rate in Rwanda, while Datta et al. [39] reported a 45% rate of vertical transmission in Nairobi, Kenya. During this conference, evidence was presented that greatest risk for vertical transmission occurs when the mother is in the first year after HIV seroconversion, has advanced disease or low percentage of T4 lymphocytes, or has an inflammation of the placenta and umbilical cord [40–43].

On Monday, Datta et al. [39] also showed that non-infected children of HIV-infected mothers are at risk for substantial excess morbidity and mortality compared with children of uninfected mothers. For example, in her series, the mortality rate at 5 years was estimated to be 24% for HIV-infected children, 12% for noninfected children of infected mothers, and 5% among control children born to HIV-negative mothers. In Rwanda, Bulterys et al. [44] reported increased mortality among noninfected children of infected mothers during the first 3 months of life as well as low birth weight, a finding also noted by Kamenga et al. [45] in Zaire and Johnstone et al. [46] in the United Kingdom.

In addition, data presented during this conference suggest that breast-feeding must receive additional attention. Ruff et al. [47] detected evidence of HIV by PCR technique in breast milk from 71% of HIV-infected women tested, with a higher rate of PCR positivity in colostrum than in later milk; similar findings were also noted by Vonesch et al. [48]. In Italy, Gabiano et al. [49] found a significantly higher vertical transmission rate in breast-fed, compared with bottle-fed infants of HIV-infected women. Two other studies also suggested that HIV transmission to newborns may occur during or after birth. In Rwanda, Van de Perre et al. [50] reported that among mothers who seroconverted to HIV after delivery, 53% of their infants were infected, although the meaning of this finding is obscured by lack of information about the actual time of maternal infection. Similarly, Kreiss et al. [41] suggested that intrapartum or postpartum transmission may be important, finding that while 10% of infants were infected at birth (using PCR technique), 38% were infected by 9 months of age. These studies do not warrant any changes in policy at present, but the issue of breast-feeding and HIV infection merits further, and definitive research effort [51].

As pointed out by Johnson [52], in the industrialized world, so-called 'natural history' studies have become studies of the 'unnatural history of disease' as both progression from infection to clinical disease and survival are influenced by therapeutic interventions. At this conference, several observational studies examined the effect of prophylaxis and treatment. For example, in separate reports using data from the Multicenter AIDS Cohort study (MACS), Munoz et al. [53] and Graham et al. [54] found a positive effect of zidovudine and/or aerosolized pentamidine in prolonging the length of AIDS-free time in HIV-infected people. In addition, three studies reported a positive effect of treatment on survival after an AIDS diagnosis. Lemp et al. [55], using City of San Francisco surveillance data, found a 'gain' of 9.6 months in median survival among people receiving zidovudine compared with those receiving no therapy. Controlling for age, initial diagnosis and year of diagnosis, the relative hazard for those receiving zidovudine, aerosolized pentamidine or dideoxyinosine (ddI) was reduced to about 0.75. Saah et al. [56], using MACS data, and controlling for initial diagnosis, CD4 count and hemoglobin, found a relative mortality hazard of 0.38 among people receiving zidovudine after diagnosis of AIDS. Osmond et al. [57] combined data from two cohorts in San Francisco and found that the median duration of survival postdiagnosis increased from 15 months in 1983–1986 to 19 months in 1987–1988, yet declined subsequently in 1989–1991 to 16 months (although this last figure is an estimate). Osmond and co-workers provide some evidence that the level of immune competence at diagnosis may be changing over time; men may be more immunocompromised at diagnosis in the more recent, as compared with earlier, periods. While each of these studies have important methodological caveats, especially involving loss to follow-up, they offer support in population-based studies to results of clinical trials using antiretroviral agents and *Pneumocystis carinii* prophylaxis.

Unfortunately, in the developing world, the history of HIV infection is still 'natural', that is, unaffected by availability of antiretroviral agents or specific prophylaxis for opportunistic infections. Yet in the developing world, even the natural history of HIV infection, so critical for projecting health and social service needs and assessing the impact of prophylactic and therapeutic interventions, remains unclear. Anzala et al. [58] followed 144 women sex workers with known time of seroconversion: the median time from HIV infection to occurrence of AIDS was only 44.6 months, although this estimate only applies to women developing AIDS during 60 months or less of follow-up. Several other studies of natural history among HIV-infected adults in the developing world were also reported, but each suffers from uncertainty about the actual time of HIV infection. For example, Pape et al. [59] followed 242 HIV-infected yet asymptomatic spouses; during a mean follow-up period of 2 years, 48% developed AIDS and 30% died. In Rio de Janiero, investigators followed 132 asymptomatic infected people for an average of about 3 years, during which time 36% developed AIDS [60]. It is imperative that well-designed natural history studies be carried out in the developing world, including Asia and Latin America, in addition to Africa.

One barrier to conducting these studies on a global basis is the lack of an internationally useful and agreed upon definition of AIDS. In 1985,

WHO developed a clinical definition of AIDS for surveillance purposes in Africa; this definition (called the 'Bangui' definition after the capital of the Central African Republic where the relevant meeting was held) relied exclusively upon clinical findings, and did not require laboratory support. In early 1989, the 'Caracas' case definition was developed for use in Latin America and the Caribbean, to take advantage of available laboratory capacity in the region yet also take account of regional differences in clinical expression of HIV disease [61]. Therefore, at present, AIDS surveillance is based on at least three definitions: the CDC/WHO, Caracas and Bangui definitions. Finally, while not reported in a formal communication at this conference, the US Centers for Disease Control is apparently considering a new AIDS case definition based on CD4 counts. It is clear that the basic issues of disease definition and surveillance urgently require global discussion to strengthen collective research, prevention and treatment capabilities.

A benefit of international discussion has been the emerging capacity to make comparisons among populations, as well as over time. Des Jarlais et al. [62] presented a beautiful example of this approach in his comparison of the HIV epidemic in New York City and Bangkok. Among injecting drug users in both cities, a period of HIV introduction was followed by rapid spread, followed by an early and then a more advanced stabilization phase. Since 1984, in New York City, HIV seroprevalence has been maintained at about 50%; in Bangkok, a 40% seroprevalence has been sustained for the past 2 years. Such comparisons create an unparalleled opportunity for discovery of key elements of policy, program and practice associated with more effective prevention efforts. Other comparisons with potential for a rich harvest of preventive information include comparisons of differential dynamics of infection in different vertical transmission settings and in open populations. For example, in Kinshasa, Zaire, studies in several adult populations have documented a stable or nearly stable seroprevalence, associated, as in Kaseka et al.'s [63] report at this conference, with a low seroincidence rate of about 0.5 per 100 person-years of observation. Similarly, Boulos et al. [25] and Metelus-Chalumeau et al. [24] reported on two Haitian populations in whom HIV seroprevalence has apparently stabilized after a phase of substantial increase in women attending prenatal clinics and blood donors in Port-au-Prince. Yet in other populations, including antenatal clinics in Kampala, Uganda [64], Nairobi, Kenya [65], or Abidjan, Cote d'Ivoire [66], HIV seroprevalence continues to increase. Second-generation epidemiological studies are now needed to take full advantage of the longitudinal and community-comparative data which are available.

Finally, it is important to highlight some of the achievements in preventing HIV infection which were reported in the Epidemiology and Prevention track of this conference. From Zaire, Mulivanda et al. [67] showed how a program which combined treatment of sexually transmitted diseases and condom promotion could reduce the annual incidence of HIV among women sex workers from 18 to 2%, as well as substantially reduce STD incidence. Other resounding successes in preventing transmission among discordant couples were reported by Musicco et al. [68], DeVincenzi and Ancelle-Park [69], and in the study of Moore et al. [70] among 86 such couples, among whom support and counselling led to zero seroconversions in over 100 person-years of observation and to equal levels of condom use when the infected partner was either a man or a woman.

Concerned with prevention, a global perspective is needed, as we look beyond numbers to meaning, in this dynamic, unstable and expanding epidemic. In 1991, as in each previous year, based on new data and ever greater experience, this conference has surveyed both the envolving pandemic and our collective response.

We are entering into a new era in our global confrontation with AIDS. This conference was held at a time of crisis, as always with both its clear dangers and latent opportunities. We are in crisis for four reasons.

First, the pandemic not only remains volatile, dynamic and unstable, but is gaining momentum and its major impacts, in all countries, are yet to come.

Second, public complacency is rising and societal commitment to HIV/AIDS is declining. A new cycle of public blaming has arisen, discrimination is resurgent, and frontiers threaten to close. Denial, which has been an integral part of the pandemic since the beginning, remains as powerful as ever, all over the world.

Third, despite the extraordinary, creative, courageous and innovative efforts undertaken against AIDS, particularly at the community level, we can see major limitations in the first decade of AIDS policies and programmes.

In responding to AIDS we have been slow, reactive and late, and have therefore been behind rather than ahead of the pandemic in our thinking and action. We have not derived full advantage from available worldwide knowledge and experience in HIV/AIDS prevention and care. We have been profligately wasteful of what has been learned, for we have not developed the sustained and growing capacity to improve steadily our prevention and control efforts through a process of global learning.

Fourth, we now understand that the basic, pre-existing, underlying health and social issues which AIDS has unveiled must be addressed, or we will never succeed in controlling the pandemic. Advancing the areas of human rights, the role and status of women, access to drugs and an eventual vaccine, availability of health, education and social services is essential to our ability to confront AIDS, as well as for progress in health.

Therefore, at the start of the second decade, and in the face of competing health and social priorities, HIV/AIDS programmes are increasingly uncertain about the best ways to proceed. Within our midst, there is growing and appropriate concern, despite the unprecedented and truly remarkable efforts already undertaken, about whether what we are doing and how we are thinking and acting will be sufficient for the task ahead. Our own 'status quo' frightens us.

A revitalization, a creative renewal of our efforts, is needed to build upon success and to become capable of facing the evolving circumstances and even greater challenges of the 1990s.

We realize also that we face, here and in the days ahead, a threat to our fundamental solidarity and sense of common purpose. The pandemic has become highly differentiated and locally specific; yet if we lose our unity, our capacity to support and learn from each other, then we are lost in a world which needs our help in order to care about AIDS, about human rights and dignity, and about health.

AIDS is the world's first truly global epidemic, it has called forth the world's first truly global strategy and response, and the connection between individuals, the community and the world at large is being newly forged. To express our hope, our grief and our commitment, to link individuals and communities around the world, we may need a new international movement. Organizations like the International Physicians for the Prevention of Nuclear War (IPPNW) have shown how powerful the joined skill and purpose of health workers worldwide can be; organizations like Doctors of the World and Doctors without Borders have articulated a new principle, the 'right to interfere', to be involved and to respond to the suffering of others regardless of, beyond and across national borders. This new, global humanitarianism will require innovation to develop new bridges between individuals, their community, and the peoples of the world. Unofficial, nongovernmental bridges are now needed, to respond to the visceral need to be together, to express our human solidarity in the face of all challenges; a grand alliance of people, a global Charter 77 of health.

Science against AIDS – yes. But we should not yield to the temptation to wait for technology to save us from ourselves. AIDS has forced us, and will continue to force us, to be pioneers, and to move onwards into the unforeseeable future. We carry with us collective knowledge, extraordinary experience, and the courage we can gain only from each other.

A creative renewal may take time, but it will arise, as strength arose during the first decade against AIDS, from individuals and communities: those living with HIV/AIDS and those working to improve the conditions of our community, national and international life and against the apartheids of race, sex, sexual preference, national and religious origin. We can cultivate our diversity, but without losing our global view, our demands for equity, our solidarity.

For today, let alone tomorrow, the future is uncertain, except that we know it will be hard. Yet we are a decade down a path we will not abandon – a path which gives us strength and love – facing discrimination, disease and death. A path we only dare to walk – together.

References

1 Croisier AN, Stimson GV, Ettorre Em, Stephens SF: Prevalence of HIV infection among injecting drug users in London. VII Conf on AIDS, Florence, June 1991. Abstract Book, vol 1, p 34, abstr MC 48.

2 Berkelman R, Fleming P, Chu S, Hanson D: Women and AIDS: The increasing role of heterosexual transmission in the United States. VIIth Int Conf on AIDS, Florence, 1991. Abstract Book, vol 2, p 49 abstr WC 102.

3 McNeil J, Brundage J, Gardner L, Wann F, Renzullo P, Burke D: United States army: 1985–90. VIIth Int Conf on AIDS, Florence, 1991. Abstract Book, vol 1, p 33, abstr MC 44.

4 Rude N, Costagliola D, Valleron AJ: Deficit in AIDS incidence: The case of France. VIIth Int Conf on AIDS, Florence, 1991. Abstract Book, vol 2, p 32, abstr WC 39.

5 Pagano M, Tu XM, DeGruttola V, MaWhinney S: The 'peak' in incidence: A slowdown in progression to AIDS or in reporting of AIDS. VIIth Int Conf on AIDS, Florence, 1991. Abstract Book, vol 2, p 307, abstr WC 3046.

6 McDougall L, Schechter MT, Sheps S, Craib KJP, Le TN, Fay S, Montaner JSG: Bias in AIDS cohort studies and an empiric demonstration of volunteer bias in an investigation of homosexual men. VIIth Int Conf on AIDS, Florence, 1991. Abstract Book, vol 2, p 47, abstr WC 95.

7 Fennema JSA, van Ameijden EJC, van den Hoek JAR, Coutinho RA: Incorrect prediction on HIV seroprevalence among intravenous drug users (IDU) contacted on the street in Amsterdam. VIIth Int Conf on AIDS, Florence, 1991. Abstract Book, vol 2, p 306, abstr WC 3041.
8 Cabello A, Cabral M, Vera E, Arrom C, Kiefer R, Zoulek G: The risk of sexually acquired HIV infection in Paraguay. VIIth Int Conf on AIDS, Florence, 1991. Abstract Book, vol 1, 354, abstr MC 3225.
9 Melbye M, Misfeldt J, Olsen J, Schou G: AIDS and HIV epidemiology in Eskimos. A national person identifiable HIV registry and active contact tracing in Greenland. VIIth Int Conf on AIDS, Florence, 1991. Abstract Book, vol 1, p 367, abstr MC 3276.
10 Dada AJ, Ajayi J, Ransome-Kuti O, Williams G, Abebe E, Levin A, Quinn T, Blattner W: Prevalence survey of HIV-1 and HIV-2 infections in female prostitutes in Lagos State, Nigeria. VIIth Int Conf on AIDS, Florence, 1991. Abstract Book, vol 1, p 371, abstr MC 3295.
11 Chin J: Present and future dimension of the HIV/AIDS pandemic. Science challenging AIDS. Proc 7th Int Conf on AIDS, Florence, 1991. Basel, Karger, 1992, pp 33–50.
12 Valdespino JL, Loo E, Cruz C, Garcia ML, Magis C, Herrera C, Sepulveda J: Risk factors interrelated between AIDS and STD among female prostitutes in Mexico. VIIth Int Conf on AIDS, Florence, 1991. Abstract Book, vol 1, p 354, abstr MC 3226.
13 Hays RB, Kageles S, Coates T: Understanding the high rates of HIV risk-taking among young gay and bisexual men: The young men's survey. VIIth Int Conf on AIDS, Florence, 1991. Abstract Book, vol 1, p 48, abstr MC 101.
14 Stall R, Bye L, Hays R, Kegeles S, Catania J: Are young gay men being reached by health education messages? A comparison of sexual risk-taking between young and older men. VIIth Int Conf AIDS, Florence, 1991. Abstract Book, vol 2, p 297, abstr WC 3004.
15 Kengeya-Kayondo JF, Kamali A, Nunn AJ, Mulder DW: Intervillage variations in HIV seroprevalence in a rural Ugandan community. VIIth Int Conf on AIDS, Florence, 1991. Abstract Book, vol 1, p 32, abstr MC 40.
16 Castilho E, Guimaraes MCD, Chequer P, Rodrigues L: Features of heterosexual exposure category in Brazil, 1980–1990. VIIth Int Conf on AIDS, Florence, 1991. Abstract Book, vol 1, p 361, abstr MC 3252.
17 Matida A, Sanches K, Valente K: Changes in the profile of AIDS epidemic in Rio de Janeiro, Brazil. VIIth Int Conf on AIDS, Florence, 1991. Abstract Book, vol 1, p 375, abstr MC 3308.
18 Rivero A, Vergara A, Aguando I, Santos J, Key C, Leal M: Current prevalence of anti-HIV-1 in female prostitutes from southern Spain. VIIth Int Conf on AIDS, Florence, 1991. Abstract Book, vol 2, p 328, abstr WC 3129.
19 Solomon L, Astemborski J, Warren D, Vlahov D, Cohn S, Munoz A: Differences in risk factors for seroconversation among female and male IVDUs. VIIth Int Conf on AIDS, Florence, 1991. Abstract Book, vol 1, p 310, abstr MC 3050.
20 Kellogg TA, Wilson MJ, Lemp GF, Hernandez S, Reardon J, Ruiz J, Elcock M, Bolan G, Obata B: Prevalence of HIV-1 among homosexual and bisexual men in the

San Francisco bay area: Evidence of infection among young gay men. VIIth Int Conf on AIDS, Florence, 1991. Abstract Book, vol 2, p 298, abstr WC 3010.

21 Cournos F, Empfield M, McKinnon K, Weinstock A, Horwath E, Meyer I, Currie C: HIV seroprevalence and risk behaviors among homeless and hospitalized mentally ill. VIIth Int Conf on AIDS, Florence, 1991. Abstract Book, vol 1, p 351, abstr MC 3212.

22 Zolopa A, Gorter R, Meakin R, Keffelew A, Wolfe H, Moss AR: Homelessness and HIV infection: A population based study. VIIth Int Conf on AIDS, Florence, 1991. Abstract Book, vol 1, p 359, abstr MC 3244.

23 Rekert ML, Chan S, Barnett J, Lawrence C, Manzon L: HIV and north American aboriginal peoples. VIIth Int Conf on AIDS, Florence, 1991. Abstract Book, vol 1, p 357, abstr MC 3237.

24 Metelus-Chalumeau E, Larco P, Poumerol G: Four years (April 1986-June 1990) of epidemiological trends of HIV infection among blood donors in Port-au-Prince, Haiti. VIIth Int Conf on AIDS, Florence, 1991. Abstract Book, vol 1, p 302, abstr MC 3018.

25 Boulos R, Holt E, Kissinger P, Jaar, S, Metteyer U, Desormeaux J, Coberly J, Quinn T, Halsey N: Stable HIV-1 seroprevalence rates in pregnant women residing in a Haitian urban slum. VIIth Int Conf on AIDS, Florence, 1991. Abstract Book, vol 1, p 301, abstr MC 3015.

26 Araneta MR, Lemp GF, Cohen JB, Derish PA, Carmona I, Clevenger AC: Survival trends among women with AIDS in San Francisco. VIIth Int Conf on AIDS, Florence, 1991. Abstract Book, vol 1, p 328, abstr MC 3122.

27 Royce RA, Tu X, Pagano M: Gender differences in survival after AIDS diagnosis: US surveillance data. VIIth Int Conf on AIDS, Florence, 1991. Abstract Book, vol 1, p 331, abstr MC 3135.

28 Horsburgh CR, Hanson D, Fann SA, Havlik JA, Thompson SE: Predictors of survival in HIV infection include CD4+ cell count, AIDS defining condition and therapy but not sex, age, race or risk activity. VIIth Int Conf on AIDS, Florence, 1991. Abstract Book, vol 1, p 341, abstr MC 3175.

29 Vergara-Campos A, Perez-Guzman E, Torres-Tortosa M, Lucero M, Gonzales-Serrano M, Perez-Cortes S, Perez-Moreno JM: Natural evolution of AIDS in women and its relationship with heterosexual transmission in the province of Cadiz – a multicenter study. VIIth Int Conf on AIDS, Florence, 1991. Abstract Book, vol 1, p 347, abstr MC 3196.

30 Pezzotti P, Rezza G, Zerboni R, Lazzarin A, Angarano G, Sinisso A, Aiuti F, Ricchi E, Canessa A, Barbanera M, Carbonari P, LoCaputo A, Tirelli U: Influence of gender, age and transmission category on the progression from HIV seroconversion to AIDS. VIIth Int Conf on AIDS, Florence, 1991. Abstract Book, vol 1, p 67, abstr TuC 43.

31 Carpenter C, Liebman BD, Stein M, Fisher A, Fiore TC, Mayer RH: Distinctive features of HIV infection in 200 north American Women. VIIth Int Conf on AIDS, Florence, 1991. Abstract Book, vol 1, p 298, abstr MC 3000.

32 Thompson M, Whyte B, Morris A, Rimland D, Thompson S: Gender differences in the spectrum of HIV disease in Atlanta. VIIth Int Conf on AIDS, Florence, 1991. Abstract Book, vol 1, p 326, abstr MC 3115.

33 Grant IH, Anastos K, Ernst J: Gender differences in AIDS-defining illness. VIIth Int Conf on AIDS, Florence, 1991. Abstract Book, vol 1, p 346, abstr MC 3192.

34 Fleming PL, Clesielski CA, Berkelman RL: Sex-specific differences in the prevalence of reported AIDS-indicative diagnoses, United States, 1988–1989. VIIth Int Conf on AIDS, Florence, 1991. Abstract Book, vol 1, p 350, abstr MC 3210.

35 Chiphangwi J, Dallabetta G, Miotti P, Liomba G, Wangel AM, Saah A: Cervical squamous intraepithelial lesions (CSIL) and HIV-1 infection in Malawian women. VIIth Int Conf on AIDS, Florence, 1991. Abstract Book, vol 1, p 47, abstr MC 98.

36 LaGuardia K, McGuinness K, Hunter D: Cervical disease among HIV-infected women by immune status. VIIth Int Conf on AIDS, Florence, 1991. Abstract Book, vol 1, p 47, abstr MC 97.

37 European Collaborative Study: Children born to women with HIV-1 infection: Natural history and risk of transmission. Lancet 1991;ii:253–260.

38 Lepage P, Van de Perre P, Msellati P, Hitimana DG, Van Goethem C, Bazubagira A, Stevens AM, Mukamabano B, Simonon A, Dabis F: Natural history of HIV-1 infection in children in Rwanda: A prospective cohort study. VIIth Int Conf on AIDS, Florence, 1991. Abstract Book, vol 2, p 34, abstr WC 46.

39 Datta P, Embree J, Ndinya-Achola JO, Kreiss J, Plummer FA: Perinatal HIV-1 transmission in Nairobi, Kenya: 5 year follow-up. VIIth Int Conf on AIDS, Florence, 1991. Abstract Book, vol 1, p 20, abstr MC 3.

40 St. Louis ME, Kabagabo U, Brown C, Kamenga M, Davachi F, Behets F, Batter V, Nelson AM, Manzila T, Quinn TC, Oxtoby M, Heyward WL: Maternal Factors associated with perinatal HIV transmission. VIIth Int Conf on AIDS, Florence, 1991. Abstract Book, vol 1, p 304, abstr MC 3027.

41 Kreiss J, Datta P, Willerford D, Embree J, Bwayo J, Hensel M, Gallatin M, Roberts P, Ndinya-Achola JO, Holmes K, Plummer F: Vertical transmission of HIV in Nairobi: Correlation with maternal viral burden. VIIth Int Conf on AIDS, Florence, 1991. Abstract Book, vol 1, p 313, abstr MC 3062.

42 Lallemant M, Lallemant Le Coeur S, Samba L, Cheynier D, Nzingoula S, Larouze B, M'Pele P: Assessing the risk for mother-infant HIV-1 transmission: A challenge in developing countries. VIIth Int Conf on AIDS, Florence, 1991. Abstract Book, vol 1, p 317, abstr MC 3078.

43 Hague RA, Mok JYQ, MacCallum L, Burns S, Yap PL: Do maternal factors influence the risk of vertical transmission of HIV? VIIth Int Conf on AIDS, Florence, 1991. Abstract Book, vol 2, p 355, abstr WC 3237.

44 Bulterys M, Chao A, Kurawige JB, Munyemana S, Mbarutso E, Mukantabana S, Kageruka, M, Nawrocki P, Mukafaranswa B, Dushimimana A, Saah A: Maternal HIV infection and intrauterine growth: A prospective cohort study in Butare, Rwanda. Abstract Book, vol 2, p 354, Abstr WC 3234.

45 Kamenga M, Manzila T, Behets F, Oxtoby M, Brown C, Sulu M, Nelson AM, Editi B, Batter V, Davachi F, Quinn TC, St Louis ME: Maternal HIV infection and other sexually transmitted diseases and low birth weight in Zairian children. VIIth Int Conf on AIDS, Florence, 1991. Abstract Book, vol 2, p 357, abstr WC 3244.

46 Johnstone FD, MacCallum LR, Brettle RP, Hamilton BA, Peutherer J, Burns SM, Gore SM: Population based, controlled study; effects of HIV infection on pregnancy. VIIth Int Conf on AIDS, Florence, 1991. Abstract Book, vol 2, p 355, abstr WC 3239.

47 Ruff A. Coberly J, Farzadegan H, Desormeaux J, Losikoff P, Bureau J, Jules D, Quinn T, Boulos R, Halsey NA: Detection of HIV-1 by PCR in breast milk. VIIth Int Conf on AIDS, Florence, 1991. Abstract Book, vol 1, p 300, abstr MC 3009.

48 Vonesch N, Caprilli F, Castiglione HA, Cordiali P, Sturchio E, Pezella M: The human colostrum as a route of HIV-1 transmission. VIIth Int Conf on AIDS, Florence, 1991. Abstract Book, vol 1, p 304, abstr MC 3024.

49 Gabiano C, Tovo PA, de Martino M, Galli L: HIV-1 transmission rate in first born children to seropositive mothers and interfering factors. VIIth Int Conf on AIDS, Florence, 1991. Abstract Book, vol 2, p 356, abstr WC 3241.

50 Van de Perre P, Simonon A, Msellati P, Hitimana DG, Vaira D, Bazubegin A, Van Goethem C, Mukamabano B, Karita E, Sonday-Thull D, Dabis F, Lepage P: Mother-to-infant postnatal transmission of HIV-1: A cohort study. VIIth Int Conf on AIDS, Florence, 1991. Abstract Book, vol 2, p 31, abstr WC 33.

51 Heymann SJ: Modeling the impact of breast-feeding by HIV-infected women on child survival. Am J Publ Health 1990;80:1305–1309.

52 Johnson MM, Schechter MT, Craib KJI, Le TN, Montaner JSG: Zidovudine and slowing of progression to AIDS: The unnatural history of disease. VIIth Int Conf on AIDS, Florence, 1991. Abstract Book, vol 1, p 325, abstr MC 3109.

53 Munoz A, Hoover D, He Y, Detels R, Kingsley L, Graham NMH, Vermund SH, Phair J: Effect on time to AIDS of symptoms, immune activation markers and prophylaxis at different levels of CD4. VIIth Int Conf on AIDS, Florence, 1991. Abstract Book, vol 1, p 83, abstr TuC 99.

54 Graham N, Zeger SL, Park LP, Phair JPP, Detels R, Vermund SH, Ho M, Saah A: Effect of therapy on risk of HIV-1 progression to AIDS and PCP. VIIth Int Conf on AIDS, Florence, 1991. Abstract Book, vol 2, p 32, abstr WC 41.

55 Lemp GF, Hirozawa AM, Ananeta MR, Young K, Nieri G: Improved survival for persons with AIDS in San Francisco. VIIth Int Conf on AIDS, Florence, 1991. Abstract Book, vol 1, p 66, abstr TuC 41.

56 Saah A, Hoover D, Munoz A, Detels R, Phair J, Rinaldo C, Vermund SH: The correlates of survival after diagnosis of AIDS. VIIth Int Conf on AIDS, Florence, 1991. Abstract Book, vol 1, p 83, abstr TuC 100.

57 Osmond D, Shiboski S, Moss AR: Changes in AIDS survival from 1984 to 1990 in the San Francisco general hospital cohort study. VIIth Int Conf on AIDS, Florence, 1991. Abstract Book, vol 2, p 33, abstr WC 43.

58 Anzala AO, Plummer FA, Wambugu P, Nagelkerke NJD, Ndinya-Achola JO, Bwayo J, Ngugi EN: Incubation time to symptomatic disease (SD) and AIDS in women with a known duration of infection. VIIth Int Conf on AIDS, Florence, 1991. Abstract Book, vol 1, p 84, abstr TuC 103.

59 Pape JW, Deschamps MM, Haffner A, Johson WD: Natural history of HIV infection in HIV-infected asymptomatic spouses of AIDS patients in Haiti. VIIth Int Conf on AIDS, Florence, 1991. Abstract Book, vol 1, p 326, abstr MC 3114.

60 Morais-de-Sa CA, Sion FS, Almeida MCR, Feitosa CC, Eyuer-Silva W, Castilho EA, Guimaraes MCD, Rachid de Lacerda MC, Quinhoes Ep, Signorini D, Rubini N: Progression to AIDS in asymptomatic HIV-infected individuals and persistent generalized lymphadenopathy in Rio de Janeiro, Brazil. VIIth Int Conf on AIDS, Florence, 1991. Abstract Book, vol 1, p 371, abstr MC 3292.

61 Werneck E, Kritski AL, Castilho E, Calvao-Castro: Evaluation of the revised Caracas case-definition of AIDS among patients with active pulmonary tuberculosis in Rio de Janeiro State, Brazil. VIIth Int Conf on AIDS, Florence, 1991. Abstract Book, vol 2, p 315, abstr WC 3077.
62 Des Jarlais D, Choopanya K, Wenston J, Vanichseni S, Sotheran JL, Plangsringarm K: Risk reduction and stabilization of HIV seroprevalence among drug injectors in New York City and Bangkok, Thailand. VIIth Int Conf on AIDS, Florence, 1991. Abstract Book, vol 1, p 20, abstr MC 1.
63 Kaseka N, Batter V, Eleka M, Ryder R: Incidence of extramarital sex, STDs, HIV-1 infection and tuberculosis in 10,325 individuals followed for 2 years in Kinshasa, Zaire. VIIth Int Conf on AIDS, Florence, 1991. Abstract Book, vol 2, p 49, abstr WC 101.
64 Hom D, Guay L, Mmiro F, Ndugwa C, Goldfarb J, Olness K: HIV-1 seroprevalence rates in women attending a prenatal clinic in Kampala, Uganda. VIIth Int Conf on AIDS, Florence, 1991. Abstract Book, vol 2, p 361, abstr WC 3262.
65 Ndinya-Achola JO, Datta P, Embree J, Kreiss JK, Maitha GA, Plummer FA: Increasing seroprevalence of HIV-1 in pregnant women in Nairobi, 1986–1990. VIIth Int Conf on AIDS, Florence, 1991. Abstract Book, vol 2, p 362, abstr WC 3264.
66 Doorly R, Kadio A, Brattegaard K, Gnaore E, Coulibaly IM, De Cock KM: Trends in HIV-1 and HIV-2 infections in Abidjan, Cote d'Ivoire, 1987–1990. VIIth Int Conf on AIDS, Florence, 1991. Abstract Book, vol 1, p 32, abstr MC 42.
67 Mulivanda T, Manoka AT, Nzila N, Way Way, St. Louis ME, Piot P, Laga M: The impact of STD control and condom promotion on the incidence of HIV in Kinshasa prostitutes. VIIth Int Conf on AIDS, Florence, 1991. Abstract Book, vol 1, p 20, abstr MC 2.
68 Musicco M, Saracco A, Nicolosi A, Gervasoni I, Vaccher E, Quirino T, Sinicco A, Turbessi G, Costigliola P, Gafa S: Italian partner study. VIIth Int Conf on AIDS, Florence, 1991. Abstract Book, vol 1, p 20, abstr MC 4.
69 De Vicenzi I, Ancelle-Park R: Heterosexual transmission of HIV: Follow-up of a European cohort of couples. VIIth Int Conf on AIDS, Florence, 1991. Abstract Book, vol 1, p 305, abstr MC 3028.
70 Moore L, Padian N, Shiboski S, O'Brien TR: Behavior change in a cohort of heterosexual couples with one HIV-infected partner. VIIth Int Conf on AIDS, Florence, 1991. Abstract Book, vol 2, p 49, abstr WC 103.

J.M. Mann, MD, Harvard School of Public Health, Boston, MA 02115 (USA)

Rossi GB, Beth-Giraldo E, Chieco-Bianchi L, Dianzani F, Giraldo G, Verani P (eds): Science Challenging AIDS. Basel, Karger, 1992, pp 265–280

Summary Report: Social and Behavioral Science

Elio Guzzanti

Medical Research Institute, Bambino Gesù Children's Hospital, Rome, Italy

Introduction

Social and behavioral science, covering different but closely connected topics, provides an important contribution to the understanding of the social causes of the HIV epidemic and to social response, through the foundation of a sound social policy, well balanced between prevention and care.

During the VIIth International Conference on AIDS, presentations, posters and the voices of people attending the parallel sessions, mainly persons with AIDS/HIV and Non-Governmental Organizations (NGOs), produced a global and updated picture of the situation worldwide.

Information and Education

Since prevention continues to be in the forefront in the struggle against HIV infection, the population must receive information that is clear and correct, from a scientific and ethical point of view. To achieve this, formal relations need to be established between institutions and the mass media, as shown by the experience in Manila [1]. However, this information must take into account the social context in which it will operate, not only regarding the content of the messages, but also to be able to reach the population that is marginalized for various reasons, ranging from widespread illiteracy to the lack of mass media in the communities, as in Ethiopia [2], Ivory Coast [3] and Zaire [4], and the tendency to not use these means even if available, as happens in certain social groups in industrialized countries. Today the general population is more aware of the routes of transmission, but is not as well informed about how the infection is not transmitted and what is particularly evident is the inconsistency between the increased level of knowledge and the relatively unmodified personal behavior leading to risk conditions [5–8].

So, mass information campaigns are not sufficient; what is needed is a widespread educational effort that reaches the population in the community, and especially the persons considered to be at high risk [9, 10]. Physicians should have an important role in educating about prevention. However, in areas of high HIV infection prevalence, as in San Francisco, only 18% of physicians routinely recommend HIV tests to pregnant women [11]. It is therefore necessary to increase the level of knowledge about this problem among physicians.

In any event, educational activities within the schools continue to be essential because, as shown by the experience in high schools in the United States [12], many students continue with behavior that puts them at risk, and, as proven by the experience of 30 women in Zaire [13] responsible for addressing the question of STDs, AIDS and the prevention of other diseases in the schools, because youths show not only interest, but also new attitudes, including the diffusion of the notions received in school to other peers.

Also necessary are community-level education programs to reinforce and integrate the activities in the schools and reach youths who do not attend school and are very vulnerable [14]. Very important for information and education is the work of NGOs, since they operate at a community level and can reach different groups of the population with whom they use the most appropriate means, in cooperation with the activities of the institutional organizations [15, 16], as in the case of the AIDS Prevention and Control Coalition in Ghana [17].

Homosexuals and Bisexuals

Since the early phase of the epidemic, behavioral change has been frequently reported in gay men [18]. However, there is some evidence that risk behaviors can relapse and that social precarity makes long-term compliance to safer sex particularly difficult [19, 20].

In developing countries, only a few studies have targeted the issue of gay men. In spite of information about HIV/AIDS and an accurate perception of the risk, consistent adoption of safer sexual practices has been limited among homosexual and especially bisexual adolescents and men in Brazil that can constitute a bridge to the general population [21–23]. The relative lack of community-based support structures and intervention programs has failed to provide a context of social support for behavior change [24]. Health promotion for homosexual and bisexual adolescents and men remains an urgent priority in Latin America.

Intravenous Drug Users

Behavioral change has been reported also in i.v. drug users in different geographical areas of the world, either as a consequence of the fear of AIDS or as the effect of specific educational programs. However, a gap between knowledge and behavioral change has been observed in studies conducted in Europe and the United States [25–27]. A study conducted in 50 cities in the US showed that there is denial of risk for HIV infection in this population, and that there was no recognition of the relationship between engaging in high-risk behavior and the probability of HIV infection [28].

The type of drug used may be a factor determining differences in the rate of behavioral change. In a study conducted in the US, a marked difference in perceived vulnerability to HIV infection was observed between heroin and amphetamine users [29]. Heroin users were more likely to believe themselves to be at risk through sex than amphetamine users, despite the significantly greater sexual activity of the latter.

Very few amphetamine users were in contact with drug services and they did not feel they were a 'risk group'. It has been confirmed that i.v. drug users are more likely to change drug-related behavior than sexual behavior, and that female i.v. drug users exchanging sex for money constitute another bridge to the general population. More intense intervention, perhaps involving couples, may be needed to sustain sexual behavioral change.

Women

As regards the HIV epidemic, women are vulnerable not only as physical persons but also as members of society, because in many parts of the world their role in subordinate and not protected by rights or organizations, and so they tend to have less knowledge of the problems, the risks and the ways to avoid them, and also less social possibilities of opposing risk conditions [30–33]. The experiences described in Uganda [11, 34–35], Zambia [32] and Mexico [35, 36] are clear examples of this; the same can be said for particular groups of the female population, such as the prostitutes in Thailand [38] and India [39], but in India the situation is similar also for women who are the steady partners of laborers belonging to lower socio-economic status. In fact, what emerges is the difficulty women have in obtaining safer sex through protection, as well as the risks they run with highly promiscuous partners in social contexts in which this is ordinary.

The problem of women as drug users or partners of HIV-positive men poses further issues because many of them want to have children and the latter reflect the very female quality of always being in the forefront when it comes to providing support or help to persons in difficulty, but also in not depriving loved ones of usual affection [40–42].

The result is a dramatic impact on society, first because of the widespread diffusion of vertical transmission, but also because of the profound effect on the community, especially in Africa [30], produced by the loss of women who have a strong economic importance in this context, as well as their unique role in society and the family. The possibility that, by the year 2000, six to ten million children will be orphaned by AIDS in Africa, indicates the extreme gravity of the situation.

Children

The impact of neonatal HIV seroprevalence [43, 44] has consistent, direct effects on the infant mortality rate, but also, and especially, indirect effects in high prevalence areas [45, 46]. In the North American and European experience, children with AIDS (CWAs) and other HIV diseases use large amounts of hospital services and lengths of stay are not always in decline [47–49]. There is a shift from hospital to out of hospital care, but this raises many problems because the family must deal not only with the disease in the child, but also in the mother and often in the father, and in this family noninfected siblings could be orphaned [50–52].

From Eastern European countries there have been many reports of cases of HIV infection associated with nosocomial injections with nonsterile needles and/or microtransfusions with infected blood. Interventions are needed to reduce the spread of nosocomial HIV infection by stopping microtransfusions and increasing the supply of sterile devices. Particularly in Romania, children in orphanages received a similar 'therapy' with the consequence of a high rate of infection among orphans in these institutions [53].

Care Services

The reference model for care is one of the integrated services where NGOs can be included in plans that are appropriate for the needs of the specific community [16, 54, 55], perhaps changing the initial objectives of the groups if, as in the case of San Francisco, the target groups have changed since the start of activity.

In this model, home care is the most significant human and social element, as shown by the experience of a rural community in Zambia [56], provided that there exists close integration with hospital services to assure the best quality of care and avoid rehospitalization [57, 58].

On the other hand, the primary health care system in itself is supported and helped to grow by new programs in the fight against AIDS, as shown by the experience of the Zambian National AIDS Prevention and Control Programme, which is the result of collaboration between the government and NGOs [59, 60].

In Western countries, hospital care shows an increase in number of patients, and despite the reduction in the number of hospital days for each single patient, the cost increases in relation to the introduction of new and expensive diagnostic and therapeutic procedures, as confirmed also by Brazil [61–64].

When home care is not possible, there is no doubt that human, social and economic reasons tend to favor residential solutions for AIDS patients that need long-term care [65–68]. The cost, as in the case of Barbados, is about one-third that of hospital care [69]. It is important to assure correct follow-up, by preparing the patients already during their stay in hospital to live their disease with the greatest possible capacity for adaptation, and, as shown by a North American experience with patients who had had AIDS for 3 years, it is possible to create conditions for a life that, even though limited, is lived with great realism and awareness, challenging the difficulties of everyday life [70]. People with AIDS are valuable educators; health care workers hear the personal stories behind the numbers and are impressed with the courage of the specific experience [71]. Together, health care workers and people with HIV forge a powerful coalition in the struggle against the HIV epidemic.

Especially serious is the condition of persons with HIV/AIDS in penal institutions: first, because AIDS is not compatible with detention and secondly, because these persons need particular support inside and outside prison, deriving from their psychosocial problems, and often also drug abuse [72]. Another problem is represented by immigrants because their particular socio-economic status constitutes a serious obstacle for compliance with therapeutic regimens. Therefore, importance must be given to the community based primary care services and the participation of family physicians, who must receive additional training in problems raised by HIV infection, including counselling [73–76]. But these services must be developed in terms of personnel and financing, because the special needs of AIDS patients, especially those with neurologic symptoms and/or in the terminal stage, can raise the average of the daily care charge.

Therefore, it is important to use a reliable indicator [77, 78], to assess the condition of the patient also in relation to his care needs. In this way, hospital, residential, home and outpatient services can be used according to the conditions of the single patient to assure the continuity of care and an integrated management of these services. All this has evident implications on policy interventions, resource allocation and the evaluation of the functioning, cost and benefit of care services.

Health Care Worker Issues

Several studies have been conducted in different countries to assess level of knowledge, attitude and practice concerning AIDS and HIV infection among health care workers, showing that significant deficiencies in key areas of HIV knowledge still exist, mostly related to rapid changes in scientific knowledge and standards of care [79–81].

Compliance to Universal Precautions

Several studies confirm that compliance with universal precautions is not common in health care settings, also in emergency departments where a high rate of blood-borne transmissible pathogens has been documented [82, 83].

Stress/Burnout

Burnout is a relevant problem among staff in charge of care for AIDS/HIV infected patients. Studies confirm that age (inversely) and time in the unit affected burnout [84]. Intervention programs are recommended to assist staff in dealing with these issues.

Education

New programs to better disseminate the latest clinical information about HIV/AIDS need to be developed, and methods such as teleconferencing, as in the experience of the USA, can provide education on HIV infection to physicians [85, 86].

Risk of Occupational HIV Infection

Two ongoing large prospective studies conducted in the USA [87, 88] and Italy [89] confirmed that risk of seroconversion after a single at-risk exposure is less than 0.5%.

Needlesticks represent the most common way of exposure. Percutaneous injuries frequently occur during surgical procedures and expose surgeons to infection risks [90].

As the patient population of HIV-infected persons continues to grow, the probability of occupational exposure to HIV will also increase. Additional measures, education, precautions and safety recommendations can protect patients and health care workers and minimize the risk of transmission in health care settings.

AZT Post-Exposure Prophylaxis

A total of about 250 cases of AZT post-exposure prophylaxis have been presented in studies conducted in the USA, Canada and Italy [88, 91, 92]. The efficacy and toxicity of prophylaxis require further studies.

HIV-Positive Health Care Workers

Seropositive caregivers represent a troubling matter, either for the risk of HIV nosocomial transmission to the patient, or for the problems posed in health care organization [93].

In the USA, some patients cared for by the same dentist were infected with HIV during their dental treatment [94]. This transmission has important implications for developing additional guidelines for the management of HIV-infected dentists and surgeons [95].

Dentists

As regards dentists in general, studies carried out in the western countries, Brazil and Thailand [96–98], confirm that dentists' perception of HIV infection as a relevant matter in their practice is high, but that most of them have a moderate level of knowledge concerning AIDS and HIV. Adoption of measures to reduce transmission during dental procedures has increased over the years. It is now necessary to prepare dentists to provide services to AIDS patients and HIV-infected persons, giving them more information and specific training.

Psychological Aspects

The importance of the psychological aspects of AIDS has increased as a consequence of widespread HIV testing and longer survival of AIDS patients.

Studies on HIV-seropositive patients have shown that these subjects are likely to be depressed, but the depression resulting from stressful events is also at the origin of risk behavior on the part of many adolescents [99–102]. And it is again the depression that derives from witnessing the illness and death of many members of the same community that leads to multiple loss syndrome, for which it is necessary to apply appropriate coping strategies and promote affective renewal.

However, the need and modalities of psychological support may vary according to the cultural and social background of different geographical areas.

Ethical Issues

An important ethical issue is represented by the attitudes of the general population toward people with AIDS (PWAs) that are not consistent with stated beliefs about casual transmission and reflect fears about contact proximity.

Therefore, HIV/AIDS-related discrimination cannot be addressed only through legislative prohibitions; active steps must be taken to overcome the fear and ignorance that leads to discrimination [103]. This is significant because voluntary testing policy may be efficient if confidentiality and absence of discrimination are guaranteed [104, 105].

Another issue is the unequal access to care and treatment existing between developing and developed countries, but also, within the latter, among social classes [106].

And still another important issue is the ethics of clinical AIDS research, mainly influenced by local disease patterns, values and economy, to improve the access of women and minorities and to influence research priorities and trial design.

Conclusion

In conclusion, during this conference we have heard many desperate, and sometimes angry, voices speaking out about this epidemic, on the part of persons, organizations and nations. Nevertheless, despite the great differences between one part of the world and another, at the beginning of this second decade of the epidemic we are starting to see signs, here and there, that something is beginning to change in the behavior of the general population and also those responsible for social policy.

Better knowledge of international problems, and these conferences are an excellent forum, increases the possibility of collaboration between the so-called developed and the developing countries. Important ethical principles are being stated, such as tolerance, informed consent and solidarity.

All that we are going to set up, the ethical principles, the scientific knowledge, the diffusion of health education, the participation of HIV/AIDS patients and NGOs in decision-making processes, and the care services, will continue to be with us to confront this crisis and any other crisis that humanity might face in the future.

References

1 Hernandez E, MacDonald G: Educating the media to play a responsible role in AIDS prevention. Abstract Book VIIth Int Conf on AIDS, Florence, June 16–21, 1991, WD 4265.

2 Eshete H: Factors affecting the dissemination of information on AIDS. Abstract Book VIIth Int Conf on AIDS, Florence, June 16–21, 1991, WD 4249.

3 Messou E, Carballo M: Information sources about AIDS in the Ivory Coast. Abstract Book VIIth Int Conf on AIDS, Florence, June 16–21, 1991, WD 4249.

4 Convisser J, Kyungu M, Kambamba SA, Eiger R: AIDS communication in the Afri-

can context: Integrating mass media and interpersonal approaches in Zaire. Abstract Book VIIth Int Conf on AIDS, Florence, June 16–21, 1991, WD 4267.

5 Ollivier B, Reboulot B, Cassuto JP: Psychosocial aspects during a campaign to information on the education and prevention of AIDS: July-August 1990. Abstract Book VIIth Int Conf on AIDS, Florence, June 16–21, 1991, WD 4254.

6 Margo G, MacDonald G, Schneider A, Dayrit M, Abad M, Consunji B: The importance of social marketing techniques in creating effective media campaigns on AIDS. Abstract Book VIIth Int Conf on AIDS, Florence, June 16–21, 1991, WD 4273.

7 Bortolotti F, Stivanello A, Noventa F, Forza G, Pavanello N: Sexual behaviors of young people in an urban area of northern Italy. Abstract Book VIIth Int Conf on AIDS, Florence, June 16–21, 1991, MD 4033.

8 Dube L, Reaidi GB: Communication factors influencing message retention and perceived effectiveness of AIDS prevention. Abstract Book VIIth Int Conf on AIDS, Florence, June 16–21, 1991, TH D 114.

9 De Vroome E, Sandfort TH, Tielman R: Sources of information used and the adoption of safer sex among gay men. Abstract Book VIIth Int Conf on AIDS, Florence, June 16–21, 1991, WD 4248.

10 Longo PH, Van Buuren NG, Pereira R, Basilia C, Macedo A: How do males prostitutes get information about AIDS? Abstract Book VIIth Int Conf on AIDS, Florence, June 16–21, 1991, WD 4256.

11 Partridge JC, Milliken N: HIV antibody test in women physicians and informed consent. Abstract Book VIIth Int Conf on AIDS, Florence, June 16–21, 1991, MD 4245.

12 Deborah Holtzman, Lowry R, Kann L, Collins J, Truman B, Kolbe L: Characteristics of HIV-related instruction, knowledge, and behaviors among high school students in the United States, 1989 and 1990. Abstract Book VIIth Int Conf on AIDS, Florence, June 16–21, 1991, WD 108.

13 Bishagara K, Kasali M, Tuliza M, Mulanga K, Grandpierre J, Piri Piri L: Mobilization of women in the struggle against AIDS in urban areas: Action among young people. Abstract Book VIIth Int Conf on AIDS, Florence, June 16–21, 1991, TU D 111.

14 O'Byrne D (moderator): AIDS education in the schools. Int Conf on AIDS, Florence, June 17, 1991, Communities Challenging AIDS.

15 Grose R, O'Malley J, Gonzales del Valle A: NGO networks: Realizing the international potential of community-based organisations. Abstract Book VIIth Int Conf on AIDS, Florence, June 16–21, 199, WD 4010.

16 Crane SF, Filgueiras A, Row Kavi A, Swaminathan S: Reaching marginalized groups through local non-governmental organisation. Abstract Book VIIth Int Conf on AIDS, Florence, June 16–21, 1991, WD 56.

17 Dunnigan KP, Antwi P, Arday-Kotey M, Bentsi C, Dowling T: Grassroots mobilization in Ghana: The formation of Ghana AIDS prevention and control coalition. Abstract Book VIIth Int Conf on AIDS, Florence, June 16–21, 1991, WD 4022.

18 Ruth MT (moderator): Safe sex education and sexual prevention. Int Conf on AIDS, Florence, June 17, 1991, Communities Challenging AIDS.

19 Cohen M: Using theoretical frameworks to evaluate HIV prevention programs. Abstract Book VIIth Int Conf on AIDS, Florence, June 16–21, 1991, WD 51.

20 Pollak M, Schiltz MA: Changing determinants of risk behaviour in the French gay community. Abstract Book VIIth Int Conf on AIDS, Florence, June 16–21, 1991, MD 57.

21 Parker RG, Struchiner CJ, Bechelli AP, Abdalad P, Guimaraes CD: Sexual behavior and AIDS awareness among homosexual and bisexual men in Rio de Janeiro, Brasil. Abstract Book VIIth Int Conf on AIDS, Florence, June 16–21, 1991, WD 4141.

22 Campos R, Raffaelli M, Siqueira E, Paine A, Greco M, Kendall C, Parker R: Sexual practices and risks for HIV infection among Brazilian street youths. Abstract Book VIIth Int Conf on AIDS, Florence, June 16–21, 1991, MD 4051.

23 Fernandes E, Camargo I, Abbud N, Ayres C, Cecchini G, Hughes V, Lima V, Ferreira A: HIV infection among street kids in Sao Paulo, one more form of violence. Abstract Book VIIth Int Conf on AIDS, Florence, June 16–21, 1991, MD 4174.

24 Reisbeck G, Hutner G, Oliveri G, Seidl O, Ermann M: Social support networks of HIV-infected homosexuals. Abstract Book VIIth Int Conf on AIDS, Florence, June 16–21, 1991, WD 4212.

25 Hartgers C, Krijnen P, Van den Hoek JAR, Verster AD, Coutinho RA, Van der Pligt J: Determinants of safe sex among HIV-negative drug users. A test of the protection motivation theory. Abstract Book VIIth Int Conf on AIDS, Florence, June 16–21, 1991, MD 4031.

26 Rugg D, MacGowan R, Stark K: Self-reported changes in sexual and drug-using behaviors in methadone clients following HIV counselling and testing. Abstract Book VIIth Int Conf on AIDS, Florence, June 16–21, 1991, MD 4016.

27 Carballo M (moderator): Substance abuse and HIV. Int Conf on AIDS, Florence, June 20, 1991, Communities Challenging AIDS.

28 McBride D, McCoy C, Inciardi J, Chitwood D: Perception of risk for HIV infection among intravenous drug users in 50 cities in the United States. Abstract Book VIIth Int Conf on AIDS, Florence, June 16–21, 1991, MD 4132.

29 Klee H: The potential for the sexual transmission of HIV: heroin and amphetamine injectors compared. Abstract Book VIIth Int Conf on AIDS, Florence, June 16–21, 1991, TU D 106.

30 Decosas J: The demographic AIDS trap for women in Africa. Implications for health promotion strategies. Abstract Book VIIth Int Conf on AIDS, Florence, June 16–21, 1991, WD 6.

31 D'Eramo JE, Kirschenbaum DZ, McCarthy M, Davis T: Women and minorities have less access to AIDS drug trial. Abstract Book VIIth Int Conf on AIDS, Florence, June 16–21, 1991, WD 4291.

32 Hausermann J, Danziger R: Women and AIDS: A human rights perspective. Abstract Book VIIth Int Conf on AIDS, Florence, June 16–21, 1991, TUD 115.

33 Mahmoud F (moderator): Women and AIDS. Int Conf on AIDS, Florence, June 18, 1991, Communities Challenging AIDS.

34 Ankrah EM, Ssekeboobo A, Asingwire N, Kengeya-Kayondo J: The environment, HIV testing and Ugandan women. Abstract Book VIIth Int Conf on AIDS, Florence, June 16–21, 1991, MD 4246.

35 Rwabukwali C, McGrath JW, Schumann DAM, Mukasa R, Nakayiwa S, Nakyobe L, Namande B, Carroll-Pankhurst C: Socioeconomic determinants of sexual risk behavior among Baganda women in Kampala, Uganda. Abstract Book VIIth Int Conf on AIDS, Florence, June 16–21, 1991, MD 4226.

36 Valdespino JL, Del Rio A, Garcia ML, Morales RA, Basanez RA, Magis C, Sepulveda J: Women and AIDS in Mexico: A socioeconomical approach. Abstract Book VIIth Int Conf on AIDS, Florence, June 16–21, 1991, TUD 113.

37 Del Rio A, Basanex RA, Izazola JA, Valdespino JL, Sepulveda J: Risk perception, sexual practices and AIDS related health service use among women in Mexico. Abstract Book VIIth Int Conf on AIDS, Florence, June 16–21, 1991, MD 4228.

38 Ungchusac K, Thanprasertsuk S, Vichai C, Sriprandh S, Pinichapongsa S, Kunasol P: Trends of HIV spreading in Thailand detected by national sentinel surveillance. Abstract Book VIIth Int Conf on AIDS, Florence, June 16–21, 1991, MC 3246.

39 Amudhamozhi R, Ravinathan R, Meeran M, Sivarajan N, Rosy Vanilla B, Raghu Raman R, Premila M: Oral contraception, IUD, condom use and man to woman heterosexual transmission of STD/HIV infection. Abstract Book VIIth Int Conf on AIDS, Florence, June 16–21, 1991, MD 4232.

40 Dattel BJ, Padian N, Shannon M, Miller J, Crombleholme WR, Sweet RL: HIV serostatus and risk unrelated to pregnancy planning or contraceptive use. Abstract Book VIIth Int Conf on AIDS, Florence, June 16–21, 1991, MD 4233.

41 Rashbaum WK, Lyman WD: Second trimester abortion and HIV: A four year study. Abstract Book VIIth Int Conf on AIDS, Florence, June 16–21, 1991, MD 4737.

42 Armstrong KA, Samost L, Smith HL: HIV related risks of sterilized and non sterilized drug abusing women. Abstract Book VIIth Int Conf on AIDS, Florence, June 16–21, 1991, MD 4241.

43 Gwinn M, Fleming P, Oxtoby M, Green T, Mofenson L, Hannon WH, George JR: HIV seroprevalence in childbearing women and predicted incidence of perinatally acquired AIDS, United States. Abstract Book VIIth Int Conf on AIDS, Florence, June 16–21, 1991, WC 34.

44 Stegagno M, Ippolito G, Costa F, Angeloni P, Aebischer ML, Angeloni U, Guzzanti E: Italian serosurvey of HIV infection in parturients: An update. Abstract Book VIIth Int Conf on AIDS, Florence, June 16–21, 1991, TUD 116.

45 Wise PJ: Effect of perinatal HIV infection on infant mortality: Policy implication. Abstract Book VIIth Int Conf on AIDS, Florence, June 16–21, 1991, TUD 60.

46 Mok J (moderator): Children and AIDS. Int Conf on AIDS, Florence, June 17, 1991, Communities Challenging AIDS.

47 Weslowski V, Andrulis D, Spolarich A, Hintz E: The care of children with AIDS and other HIV diseases in US hospitals. Abstract Book VIIth Int Conf on AIDS, Florence, June 16–21, 1991, MD 113.

48 Samson J, Lapointe N, Contandriopoulos AP, Turpin P: Trends in hospitalization for pediatric HIV-infected patients in Montreal (Sainte-Justine Hospital). Abstract Book VIIth Int Conf on AIDS, Florence, June 16–21, 1991, MD 4204.

49 Principi N, Marchisio P, Tomaghi R, Massironi E, Onorato J, Picco P: Evaluation of medical care needs in HIV positive (HIV+) children. Abstract Book VIIth Int Conf on AIDS, Florence, June 16–21, 1991, MD 4285.

50 Lecroart MC, Levy J, Kenigsman S, Clumeck N, Durieux MP: Familial morbidity and socio-economic background in children with HIV infection in Belgium. Abstract Book VIIth Int Conf on AIDS, Florence, June 16–21, 1991, MD 4258.

51 Haworth A, Kalumba K, Kwapa P, Hamavhwa C, Van Praag E, Nyirenda J: Social consequences of AIDS in 49 Zambian families – a descriptive study. Abstract Book VIIth Int Conf on AIDS, Florence, June 16–21, 1991, MD 4264.

52 Kamenga M, Muniaka N, Batter V, DaSilva M, Ryder RW: Incidence and socioeconomic consequences of AIDS orphans in 422 families with HIV-1(+) mothers in Kinshasa, Zaire. Abstract Book VIIth Int Conf on AIDS, Florence, June 16–21, 1991, TUD 59.

53 Hersh B, Popovici F, Jezek Z, Satteen G, Apetrei R, George R, Hyman D: Risk factors for HIV infection among abandoned Romanian children. Abstract Book VIIth Int Conf on AIDS, Florence, June 16–21, 1991, THD 109.

54 Adams IKB, Andrade Pinto J, Menezes de Paiva JV, Martins RF, da Silva MN, Braganca de Matos MA: Experience of a community based street kids clinic in Belo Horizonte, Brazil: Relevance for HIV intervention. Abstract Book VIIth Int Conf on AIDS, Florence, June 16–21, 1991, WD 4084.

55 Thomson B, Katoof L, Huber J: Utilization of community services by people with AIDS. Abstract Book VIIth Int Conf on AIDS, Florence, June 16–21, 1991, WD 116.

56 Siankanga Z, Chela C, Bailey B, Mwilu R: Home based care for HIV infected individuals in a rural Zambian community. A four year experience of integrated AIDS management. Abstract Book VIIth Int Conf on AIDS, Florence, June 16–21, 1991, WD 5.

57 Chang SW, Arthur J, Hirozawa A, Franks D, Lemp G: Predictors of AIDS rehospitalization in San Francisco. Abstract Book VIIth Int Conf on AIDS, Florence, June 16–21, 1991, MD 4205.

58 Wynn LJ, Katner HP, Amos PE, Beidas SO, Klekamp JW, Smith MU: Effectiveness of outpatient management of HIV infection in central Georgia, USA. Abstract Book VIIth Int Conf on AIDS, Florence, June 16–21, 1991, WD 4090.

59 Van Praag E, Kalumba K, Himonga HB, Chirwa BU, Hartwort A: National AIDS control Programmes, vertical versus horizontal approaches, the Zambian case study. Abstract Book VIIth Int Conf on AIDS, Florence, June 16–21, 1991, WD 59.

60 Defert D (moderator): Community responses to AIDS. Int Conf on AIDS, Florence, June 20, 1991, Communities Challenging AIDS.

61 Andrulis D, Weslowski V, Hintz E: Patients with AIDS and HIV infections: Examining and estimating the total burden of hospital care. Abstract Book VIIth Int Conf on AIDS, Florence, June 16–21, 1991, MD 113.

62 Moien M, Kozak LJ: Hospital utilization for patients with HIV diagnoses. United States, 1984–1989. Abstract Book VIIth Int Conf on AIDS, Florence, June 16–21, 1991, MD 4185.

63 Gejer M: Hospitalization costs, clinical features and epidemiological profile of AIDS patients in a private institution: Hospital Evaldo Foz, Sao Paulo, Brazil, 1985–1990. Abstract Book VIIth Int Conf on AIDS, Florence, June 16–21, 1991, MD 4188.

64 Hellinger FJ: Forecasting the medical care costs of the HIV epidemic in the United States: 1991–1994. Abstract Book VIIth Int Conf on AIDS, Florence, June 16–21, 1991, WD 7.

65 McCollum DP, Awe R, Barden C, Pusanik M, Springer P: Residential hospice care: Alternative to institutional care for AIDS patients. Abstract Book VIIth Int Conf on AIDS, Florence, June 16–21, 1991, WD 4059.

66 Minnefor A, Lazovitx S, Mondrone R, Spector J, Stern A: A nursing home model for care of AIDS patients: A 3 year experience. Abstract Book VIIth Int Conf on AIDS, Florence, June 16–21, 1991, WD 4060.

67 Lieberman B, McCormick WC, Chamberlain D: Development of a new community-based long-term care facility and adult health program for persons with AIDS. Abstract Book VIIth Int Conf on AIDS, Florence, June 16–21, 1991, WD 4081.

68 Martin E, Tolila Y, Mignot F, Matheron S, Bouchaud O, Coulaud JP: 'La Plage' association: A model of psychoaffective care of persons with AIDS. Abstract Book VIIth Int Conf on AIDS, Florence, June 16–21, 1991, WD 113.

69 Roach T, Martin A, Forsythe S: Hospice planning in Barbados. Abstract Book VIIth Int Conf on AIDS, Florence, June 16–21, 1991, MD 4189.

70 Taylor S, Ragsdale D, Kotarba J, Morrow J: Quality of life of hospitalized persons with AIDS: An ethnographic analysis. Abstract Book VIIth Int Conf on AIDS, Florence, June 16–21, 1991, MD 4209.

71 Berman C: The patient as a teacher: People with HIV (PWH) as panelist and in role play/simulation for training health care workers (HCW). Abstract Book VIIth Int Conf on AIDS, Florence, June 16–21, 1991, WD 105.

72 Curran L (moderator): AIDS in prisons. Int Conf on AIDS, Florence, June 17, 1991, Communities Challenging AIDS.

73 Scheib R, McGowan K, Sager S, Florman J, Bradley O, DesRoche A, Morgan B, Gagnon T, Power M: Assessment of health care needs of individuals with AIDS/HIV symptomatic disease released from correctional facilities. Abstract Book VIIth Int Conf on AIDS, Florence, June 16–21, 1991, WD 4234.

74 McCoy RB, Meehan D, Mijch A: Design and implementation of a general practitioner education program for the prevention and treatment of HIV infection. Abstract Book VIIth Int Conf on AIDS, Florence, June 16–21, 1991, WD 4046.

75 Pradier CH, Carles M, Durant J, Bonifassi L, Bernard E, Mondain V, De Salvador F, Dellamonica P: Attitudes of general and specialist practitioner regarding HIV infection in southern France. Abstract Book VIIth Int Conf on AIDS, Florence, June 16–21, 1991, WD 4047.

76 Wenrich MD, Giles LM, Ramsey PG: Primary care patterns of HIV-infected patients in Northwest United States. Abstract Book VIIth Int Conf on AIDS, Florence, June 16–21, 1991, WD 4048.

77 Markson LE, McKee L, Fanning T, Turner BJ: Influence of severity of AIDS complications on health care expenditures. Abstract Book VIIth Int Conf on AIDS, Florence, June 16–21, 1991, MD 115.

78 Bonifassi L, Schneeberger C, Demeulemeester R, Cals M, Brun G: Expanded study of the burden in care related to hospitalization of HIV-infected patients. Abstract Book VIIth Int Conf on AIDS, Florence, June 16–21, 1991, MD 4210.

79 Sullivan C, Campbell S, Henry K: Assimilation of recent knowledge about HIV/AIDS by physicians and nurses at an US ID teaching hospital: A comparative analysis. Abstract Book VIIth Int Conf on AIDS, Florence, June 16–21, 1991, WD 4156.

80 Lauria L, Vieira MAMS, Werneck E, Janini MCR, Carhalho CES, Oliveira JR, Carreira MN, Marques MJO, Lucarevschi BR, Kritski AL, Barroso PF: Knowledge, attitudes and practices (KAP) about HIV infection and AIDS among health care workers (HCWs) in Rio de Janeiro, Brazil. Abstract Book VIIth Int Conf on AIDS, Florence, June 16–21, 1991, WD 4158.

81 Roumeliotou A, Kornarou E, Papaevangelou V, Spiropoulou P, Ktenas E, Stergiou G, Papaevangelou G: Knowledge, attitudes and practices of Greek health care work-

ers in relation to AIDS. Abstract Book VIIth Int Conf on AIDS, Florence, June 16–21, 1991, MD 4157.
82 Gershon R, Curbow B, Vlahov D, Celentano DD, Saarh AJ: Low compliance with universal precautions among hospital employees despite high perceived risk. Abstract Book VIIth Int Conf on AIDS, Florence, June 16–21, 1991, WD 4187.
83 Henry K, Collier P, O'Boyle WC, Campbell S: Observed and self-reported compliance with universal precautions among emergency department personnel at two suburban community hospitals. Abstract Book VIIth Int Conf on AIDS, Florence, June 16–21, 1991, MD 58.
84 Bennett L, Mulhall B: 'Burnout' in nurses in two Australian teaching hospitals. Abstract Book VIIth Int Conf on AIDS, Florence, June 16–21, 1991, WD 4153.
85 Dworkin J: A system approach to AIDS education for medical and psychological care professionals. Abstract Book VIIth Int Conf on AIDS, Florence, June 16–21, 1991, THD 107.
86 Allerton M, Wolfe E, Wibbelsman C, McNalty K, Gordon N, Fuller J, Casal T: Teleconferencing: How 8,000 physicians can gain access to the latest medical education on HIV care and treatment. Abstract Book VIIth Int Conf on AIDS, Florence, June 16–21, 1991, MD 57.
87 Metler R, Ciesielski C, Marcus R, Berkelman R: Occupationally acquired HIV infection, United States. Abstract Book VIIth Int Conf on AIDS, Florence, June 16–21, 1991, WD 4178.
88 Tokars J, Marcus R, Culver DH, McKibben PS, Bell DM: Zidovudine use after occupational exposure to HIV infected blood. Abstract Book VIIth Int Conf on AIDS, Florence, June 16–21, 1991, WD 4184.
89 Ippolito G, Puro V, Italian Collaborative Study Group on Occupational Risk of HIV Infection: Efficacy of HIV transmission after at-risk exposure in health care settings: The Italian multicentric study. Abstract Book VIIth Int Conf on AIDS, Florence, June 16–21, 1991, MD 61.
90 Tokars J, Bell D, Marcus R, Culver D, Chamberland M, McKibben P, Martone W: Percutaneous injuries during surgical procedures. Abstract Book VIIth Int Conf on AIDS, Florence, June 16–21, 1991, THD 108.
91 Papillon M, Houle L, Lebel F: Zidovudine prophylaxis for health care workers following HIV exposure. Abstract Book VIIth Int Conf on AIDS, Florence, June 16–21, 1991, WD 4182.
92 Puro V, Ippolito G, Italian Collaborative Study Group on Occupational Risk of HIV Infection: AZT prophylaxis after occupational exposure to HIV in health care workers. Abstract Book VIIth Int Conf on AIDS, Florence, June 16–21, 1991, WD 4151.
93 Hamel MA, Di Meco P, Lamping DL: HIV seropositive caregivers: Physical and psychological well-being. Abstract Book VIIth Int Conf on AIDS, Florence, June 16-21, 1991, THD 106.
94 Jaffe H, Ou CY, Ciesielski C, Economu N, Furman L, Myers G, Witte J: HIV transmission to patients during dental care. Abstract Book VIIth Int Conf on AIDS, Florence, June 16–21, 1991, THD 110.
95 Bell DM, Martone WJ, Culver DH, Margolis HS, Shapiro CN, Tokars JI, Marcus R, Furman LJ, Jaffe HW, Curran JW, Hughes JM: Risk of endemic HIV and HBV transmission to patients during invasive procedures. Abstract Book VIIth Int Conf on AIDS, Florence, June 16–21, 1991, MD 59.

96 Hammerle C, Robinson P, Winider J, Grassi M: Predictors of dentist willingness to treat people with HIV infection. Abstract Book VIIth Int Conf on AIDS, Florence, June 16–21, 1991, MD 4137.
97 Bertazzoli R, Riengold A, Guerbert B, Bellucci S: Campinas dentists and HIV infection. Abstract Book VIIth Int Conf on AIDS, Florence, June 16–21, 1991, WD 4163.
98 Okanurak K, Sornmani S, Mandel J, Luangjarmekorn L: Dentists and AIDS: Knowledge, attitude and preventive behavior. Abstract Book VIIth Int Conf on AIDS, Florence, June 16–21, 1991, WD 4164.
99 McKusick L, Hilliard R: Multiple loss accounts for worsening distress in a community heavily hit by AIDS. Abstract Book VIIth Int Conf on AIDS, Florence, June 16–21, 1991, WD 2.
100 Stiffman AR, Cunningham R, Earls F, Dore P: Change in AIDS risk behavior from adolescence to adulthood. Abstract Book VIIth Int Conf on AIDS, Florence, June 16–21, 1991, WD 1.
101 Jacobsberg L, Perry S, Fishman B, Frances A: Emotional distress of repeated HIV testing. Abstract Book VIIth Int Conf on AIDS, Florence, June 16–21, 1991, MD 4222.
102 Murphy DA, Kelly JA, Brasfield T, Koob J, Bahr RS, Lawrence J: Predictors of depression among persons with HIV infection. Abstract Book VIIth Int Conf on AIDS, Florence, June 16–21, 1991, WD 4276.
103 Hamblin J, Mallerson SJ: Discrimination and HIV/AIDS: The role of human rights legislation. Abstract Book VIIth Int Conf on AIDS, Florence, June 16–21, 1991, MD 4149.
104 Dab W, Quènel P, Moatti JP, Beltzer N: The role of HIV testing in Paris region. Abstract Book VIIth Int Conf on AIDS, Florence, June 16–21, 1991, MD 4148.
105 Maziero A, Saghafi L, Francioli P: Seroprevalence of HIV: Feasibility of screening with consent among hospitalized patients. Abstract Book VIIth Int Conf on AIDS, Florence, June 16–21, 1991, MD 4158.
106 Christakis NA, Lynn LA, Castelo A: The ethics of clinical research that evaluates cost-effectiveness in the developing world. Abstract Book VIIth Int Conf on AIDS, Florence, June 16–21, 1991, MD 4140.

Prof. Elio Guzzanti, Direzione Scientifica, Ospedale Bambino Gesù,
Piazza S. Onofrio, 4, I–00165 Rome (Italy)

Rossi GB, Beth-Giraldo E, Chieco-Bianchi L, Dianzani F, Giraldo G, Verani P (eds): Science Challenging AIDS. Basel, Karger, 1992, pp 281–292

Communities Challenging AIDS

Manuel Carballo[a], *Giovanni Rezza*[b], *Laura Thomas*[c]

[a]World Health Organization, Geneva, Switzerland;
[b]Istituto Superiore di Sanità, Rome, Italy;
[c]AIDS Treatment News, San Francisco, Calif., USA

The medical and psychosocial complexity of AIDS has necessitated a variety of responses. Some have been planned and implemented as part of the overall development of health and social services. Others have been more spontaneous, driven by perceived or felt needs, and have involved communities and popular action groups. The type and quality of health and social welfare services required to adequately respond to the needs of people with HIV or AIDS has been relatively unprecedented. It has had to be underpinned by ambitious health services research, reallocation of often scarce resources, and carefully targeted initiatives involving individuals and groups that have previously fallen outside traditional service networks. Many of the experiences and benefits now accruing from these actions are already supporting new approaches to health care in domains other than AIDS.

Within the broad range of biomedical and psychosocial problems and challenges facing individuals and families affected by HIV infection and AIDS, however, many necessarily elude the normal capacity and, in some cases, mandate of the formal health and social services system. The past decade has witnessed a growing involvement of community-based groups whose work has effectively complemented the role of the formal sector and has strengthened society's ability to confront AIDS. While this experience has been new and still requires more support and encouragement, community action in the face of AIDS has already made a major contribution to health care principles and practices, particularly in the specific area of AIDS.

Much remains to be done, however, The pandemic continues to affect more and more individuals and to precipitate new demands on them, their social networks, and on health and social service systems. In many countries this pressure is bringing to the forefront the competition for economic and technical resources in the area of health, and is making it all the more important that the role of community action be acknowledged and vitalized. The task will not be an easy one. Community participation in health care has often lacked the official support needed to make maximum use of the potential that grass-roots groups represent. At times it has also lacked the type of commitment required from the community itself. Part of the challenge of the 1990s will be to see how greater community interest and commitment can be mobilized; another will be to determine the best mix of responses and interfaces between non-governmental groups and governmental organizations.

By now a number of areas where the role of the community has been significant have been identified, together with others where more needs to be initiated.

Access to Therapy and Care

Equal access to therapy and care remains a challenge both within and between countries. AIDS has highlighted long-standing, seemingly accepted and institutionalized inequalities in access to and utilization of health and social services. The traditional marginalization of many people at risk of HIV/AIDS because of their personal behaviours has been a contributing factor. Homosexual men, sex workers and people who use illicit drugs have, for a variety of social, political and legal difficulties, often had limited access to health services. In the context of AIDS it has also become ever more apparent that women too, have not always enjoyed a type or quality of health commensurate with their needs. Socio-economic status, always a critical factor, has also become an even more significant concern in light of the high costs of HIV-related treatment.

Access to care and therapy, however, also involves questions of shelter and psychological support. In New York, the Housing Committee of ACT UP has given high priority to funding low-cost, but quality residential space for people affected by AIDS. In so doing, it has demonstrated what can be done through self-help movements and the imaginative, forceful mobilization of resources in the housing sector. It has also served to highlight a problem often overlooked and neglected, namely the physical environmental needs of people with HIV/AIDS, and the fact that many of those affected by life-threatening and chronic diseases are also those who have traditionally lacked decent and secure housing.

In Norway, Italy and Brazil, PLUSS, LILA and the Centro Convival have demonstrated the need for flexibility in the approach to HIV/AIDS and the fact that psychological support is often as necessary as medical care. They have also shown how important it is to reach out to people who would not otherwise initiate health care seeking and might not know what resources are available. In this regard the need to permit individuals to select their own regime of care, support or treatment has also become evident.

AIDS Education in the School

Effective primary prevention of AIDS inevitably lies in reaching and educating young people early in their sexual careers. The importance of the school in this regard remains indisputable. Both with respect to the institutional opportunity it represents for accessing large groups of young people in relatively 'captive audience' conditions, as well as with respect to its technical experience in teaching, the school constitutes an important vehicle for AIDS prevention.

All too often, however, the school remains more a potential resource than a reality. In many cultures it continues to be difficult to use the school for providing sex-related education to young people. AIDS prevention messages have at times suffered as a result, and there is an urgent need for approaches that can be both culturally acceptable and technically sound. International organizations such as WHO, UNESCO, UNFPA, the Commonwealth Secretariat, the US-based Centers for Disease Control, and USAID are among the agencies that have taken up the challenge. Individually and jointly they are supporting research, the development and testing of educational materials, and the training of trainers.

Peer group counselling among young people in school situations is being given special priority. There is growing evidence that the active participation of young people in assessing their own needs, and in designing and implementing interventions, can be particularly productive in terms of improving the acceptability and appropriateness of approaches.

AIDS in Prisons

For a variety of reasons, men and women in prison are often at special risk of having been exposed to HIV prior to entering prison, or being exposed while incarcerated. Many of those who transit through prisons have been arrested for drugs or sex-related offenses, and prostitutes and drug abusers constitute a significant proportion of all those entering and leaving prisons in many countries.

There is also evidence that the use of illicit drugs and same gender sexual relations are relatively common in prison systems and that the lack of education and preventive measures are accentuating the risk of HIV exposure. Because these behaviours are usually considered against the law both inside and outside prisons, the assumption has all too often been that neither occurs in prisons. Perhaps for this reason and the fact that prisoners are often neglected in terms of psychosocial counselling in general, the issue of AIDS and its effect on the lives and well-being of people in prisons has not been given as much attention as it merits either.

The role of non-governmental groups in this domain has become especially important. In Italy, France, USA and Sweden groups such as UDAUH, ANIDFE, ACT UP, Institut Luria and the Noah's Ark Foundation have been active in providing counselling to people entering and leaving prisons as well as working with them while in prison. In the UK the Home Office Prison Department has also been active in exploring ways in which the counselling, care and treatment rights of prisoners affected by or exposed to HIV/AIDS can be met using a mix of formal and non-formal approaches involving community action groups. Policies continue to vary considerably, however, and issues such as the distribution of condoms and access to clean injecting equipment remain controversial in many countries. Much remains to be done in this area and more acknowledgement of the rights of prisoners vis-à-vis HIV/AIDS (as well as other health and welfare needs) is called for. The experiences of those groups already working in this area can serve as an important base for action.

Children and AIDS

The fact that HIV/AIDS was initially observed among men having same gender sex, and that it has consistently been seen as the result of adult behaviour, has tended to deviate attention away from the growing problem of children who are either themselves infected with HIV, or are otherwise affected by HIV in their parents or siblings. Perinatal infection, however, is increasing dramatically in many urban areas, especially in parts of Africa. In the USA and selected cities of Spain and Italy it also constitutes a major health and social concern. In many cases the problems of infected infants is exacerbated by HIV-related illnesses in the parents or, in some areas of the world, by drug-related behaviour which places additional stresses on family life and child care.

The overall problem is made worse by the fact that whether they themselves are infected with HIV or not, increasing numbers of children still dependent on adults for care and protection, are being orphaned as a result of AIDS. Many of the communities, both in developed and developing countries, where this is occurring have been ill-prepared to meet the needs of these children.

HIV/AIDS in the family inevitably places inordinate psychological stress on children. Death of one or both parent even more so. The physical needs of these children also represents a challenge to local authorities particularly since many of the most acute problems are emerging in low-income areas and in economically poor countries. What might otherwise have been traditional ways of coping, including ‘adoption’ by other family members, are sometimes interfered with by a mix of fear, misunderstanding and stigma associated with HIV infection and AIDS. Experience suggests that few communities or authorities have been able to adequately prepare for either the magnitude of the problem or the quality of the care required by children in situations such as these.

Self-help groups and non-governmental organizations have emerged in some instances in collaboration with formal welfare services such as the Lothian Health Board in Scotland, but more often than not, working independently. In the United States of America, innovative approaches have been developed by groups such as 18th Street Services and Housing Works Inc. In Tanzania the UKIMWI Orphans Assistance Programme has also formulated a series of comprehensive strategies designed to meet with the needs of children whose parents have died or are ill with AIDS-related conditions. The magnitude of the problem where the prevalence of heterosexually transmitted HIV/AIDS is high, however, is such that much more concerted and imaginative action is called for. Children have often been neglected in national health services. Their psychosocial needs have traditionally been poorly understood. In the context of AIDS the problem has become all the more complex and its implications all the more far-reaching in terms of the overall development and well-being of children.

Women and AIDS

In many respects the problems confronted by women in the face of the AIDS pandemic have mirrored and been intertwined with those of children, especially with regard to their capacity to negotiate safer sex and protection against HIV. However, women in many parts of the world remain particularly disadvantaged. Their economic and social status continues to limit their autonomy in decision making and imposes an elevated risk of involuntary exposure to the risk of HIV.

As in other areas of health, the specific needs and characteristics of women have tended to escape attention. They have been eclipsed within general approaches to health care which, while sound, have nevertheless been unable to respond adequately to the biological and psychosocial requirements of women and their reproductive health.

Access to care and treatment has also been biased in some communities. In the United States many women's groups have expressed concern at what they have perceived to be unequal involvement in clinical trials of HIV infection therapies. The need for social support networks able to provide guidance help and political expression with respect to AIDS has been highlighted everywhere. In Sudan, SWAA has begun to provide an important source of support for women both from the perspective of prevention as well as care. LILA in Italy, Lifeforce in the USA and Nimehuatzin in Nicaragua are examples of other women's community-based groups that have come into being in order to challenge AIDS and its impact on the health of women. The need for even more networking of women's groups is, however, apparent. Increasingly, action will be required to improve the economic independence of women in such a way that their capacity to take decisions with respect to their health is also enhanced and freed from sexual political constraints.

Alternative and Indigenous Treatments

The advent of alternative and indigenous treatments has occurred in some respects as a result of limited access to care in formal health care systems. In many countries the coverage provided by medical and health services has traditionally been limited and many communities have necessarily recoursed to alternative medical systems. The use and attraction of alternative treatments and medical systems, however, is more complex. Some of the important factors governing the use of these systems is their overall acceptability and the fact that they have often had a much more culture- and user-friendly base than official services.

Alternative treatments can be less costly, involve less-complicated procedures, and can be easier to manage for the patient. Issues of safety and effectiveness, however, continue to pose a dilemma. On the whole there has been a relative neglect by biomedical science of the potential value of alternative and indigenous treatments of AIDS.

What action is occurring in this area is typically being carried out by local groups, indigenous practitioners themselves, and only occasionally by a combination of them with the formal health care system. There is, however, a growing legitimation of these approaches both with respect to what they can offer in terms of palliative treatment, as well as with respect to their role in overall care. Nutritional approaches to health maintenance and the use of traditional medicines and plants has assumed greater importance. In the United States, the Center for Traditional Medicine has taken an important lead in opening up research in this area, and in China the Oriental Healing Arts Institute is experimenting with Compound Q and other naturally derived products. In Brazil the Centro Convival offers a combination of psychosocial support and a wide range of therapies specifically geared to the needs of people with HIV infection and AIDS. The Center for Natural and Traditional Medicine in Tanzania, working with the Muhimbili Medical Centre, has similarly developed a comprehensive programme of research-supported care and treatment.

The complexity of the needs of people affected by HIV and AIDS is such that no single approach is currently feasible or advisable. Individually tailored treatments are likely to be attractive and necessary for some time to come. The use of alternative and indigenous medicines/treatments, especially if they have proven effective in other domains of health care, deserve to be systematically studied and evaluated for their potential contribution in HIV treatment.

Safe Sex Education and Sexual Prevention

Sexual prevention of HIV infection remains the most effective solution to the AIDS pandemic. Although HIV can be transmitted in non-sexual ways, sexual transmission both through heterosexual and same-gender relations is, and will likely remain, the most common. Human sexuality, however, constitutes one of the least well understood areas of human behaviour and, in many sectors of society, one of the most taboo domains.

Ways of overcoming both the lack of information and the resistance to changes in sexual behaviour are urgently needed. It is noteworthy that gay men groups and sex workers have been among the pioneers in this regard. In terms of finding and communicating safe-sex techniques, and in providing the guidance and support that is essential in bringing about modifications in sexual behaviour, groups such as the Colectivo Sol in Mexico, Lucciola in Italy, the San Francisco AIDS Foundation in the USA and Scot-PEP in Scotland, have provided important leadership.

The difficulties inherent in the problem involve reaching out to individuals and groups who have not necessarily come to terms with their sexual preference and identity, or who have been systematically marginalized as a result of dominant social attitudes and values. The role of community-based groups in this regard has proved indispensable.

More applied research in this domain remains a second imperative and the work begun by institutions such as Columbia University in the United States and the Institute of Social Affairs in India and WHO deserves to be built on rapidly.

Hemophilia and AIDS

Among the groups most systematically affected by HIV/AIDS, people with hemophilia have been particularly touched. In many countries, the treatment procedures followed in treating hemophilia placed many individuals and their families at a high risk of exposure. In all cases, the lifestyles and responses of people with hemophilia to their personal health and health care has been profoundly affected.

The World Federation of Hemophilia, working closely with its affiliated groups and institutions such as the Centro Emofilia e Trombosi and the Sclavo S.P.A. in Italy, the Rehabilitation Hospital and Hemophilia

Centre in Germany, and the Royal Free Hospital, Royal Victoria Infirmary and the UK Hemophilia Society in the UK, have developed counselling programmes, care initiatives and family support actions that have helped confront many of the problems faced by people with hemophilia and HIV infection. They have also helped meet many of the needs experienced by people with hemophilia in terms of new life styles and maintenance of individual autonomy in health care.

The response of people with hemophilia to the AIDS pandemic has served as an important model for the health care of all individuals affected by HIV, and has highlighted the role of comprehensive care that encompasses family dynamics as well as the health and welfare of the individual.

Substance Use and HIV

Substance use, especially where sharing of injecting equipment is concerned, has become the second most important source of HIV transmission. In selected countries and cities it has become the most important vehicle of transmission. Responses to substance use have traditionally been impaired by social, legal and cultural attitudes. Interventions have primarily been of a repressive nature and, on the whole, have failed to significantly alter the course of drug use.

The human damage subsequent to drug abuse calls for more imaginative and community-friendly approaches that build on the interests and capacities of drug users to change and adopt less risky behaviours. Policies and philosophies among action groups vary considerably, however. On the one hand, needle-exchange schemes have proved effective in involving drug injectors who would otherwise have not come into contact with helping agencies and groups. They have also been effective in reducing the amount of needle-sharing. On the other hand schemes of this kind have nevertheless aroused antipathy from some sectors of the public who see these initiatives primarily as legitimation of drug abuse, not prevention of disease.

Within the community of ex-drug users, schemes that seek to reduce HIV exposure through risk reduction have also been heavily criticized. Groups such as IDRET in Italy and ADDEPOS in France call for much more attention to abstinence and the more open discussion of HIV antibody status. They have consistently argued against the need to protect anonymity. Other groups such as Gruppo Abele and Co. R.A. in Italy, as well as the Bangkok Metropolitan Administration in Thailand, Narcotic, and Drug Research Inc. in the USA, and the University of Glasgow in Scotland, have meanwhile assumed a more user-tailored approach that stresses attention to the sociocultural and psychological context in which the drug user functions, and have supported confidentiality.

Clinical Trials of AIDS Treatments

The demand for, and expectations associated with AIDS treatments has placed new and often difficult-to-meet pressures on biomedical research. These have at times been associated with criticism of the procedures followed for testing and approval of new drugs. Many of these criticisms have reflected on the one hand a desire by individuals and groups to participate in trials and, in so doing, have access to new drugs of sufficient promise to justify testing, and, on the other, a desire to influence the nature and procedure of trials.

In many respects the advent of HIV/AIDS has often highlighted gaps and problems in the routine planning and management of drug trials. It has given new significance to the concept of informed consent and to the need for the anticipated implications of drugs to be explained in detail. In part, this has been necessitated by the at-times complicated regimens involved, and at other times by the relative benefits and discomfort associated with certain therapies. The fact that AIDS has emerged at a time when many educated patient population are increasingly well informed about the technical properties of drugs and the nature and implications of HIV and its course, has meant a different quality of dialogue between researchers and patients.

The perceived urgency for new developments has also given rise to the involvement of communities in organizing their own clinical trials, following protocols they have been instrumental in formulating, and involving candidate drugs that have not necessarily been approved for testing by national authorities. As more candidate AIDS treatments emerge and the complexity of clinical trials increases so as to ensure multi-population representativeness and meet scientific and ethical criteria, greater community participation in determining what, how, when and on whom drugs are tested, can be anticipated and should be encouraged. Groups such as Project Inform and ACT UP in the United States, and Solidarité Plus in France are among the many groups already active in this area.

Community Responses to AIDS

The AIDS pandemic has been relatively unique in mobilizing community action around not only the medical but also the social, legal and economic aspects of the problem. In part because many of the communities affected by AIDS have felt neglected or inadequately understood by health and welfare systems, there have been fairly strong movements to set up complementary networks of people at risk. While some of these have focussed on providing psychological and social support using self-help and 'buddy'-type systems, many others have become more comprehensive and today constitute major political pressure groups both in their countries and internationally.

Community participation in health and social welfare has rarely become so well developed or so expressive of the overall needs of its constituents as in the case of AIDS. The organization of this participation, and of the groups in question, however, has necessarily been time and energy consuming. At times there has been resistance from official health and welfare sectors to their involvement, and there have been instances of resistance from within communities themselves to potential politicalization of their interests.

How to maintain the role and function of these groups into the second decade of the pandemic and how to sustain their vitality has become an issue of singular importance. Groups such as AIDES in France, Commune di Roma in Italy, Riksforbundet for HIV Positive in Sweden, Black and White Men Together in the USA, as well as others, have already gone far in demonstrating what can be achieved. Ensuring the continued contribution of these groups in consonance with that of formal health and social service system will depend on the constituencies they represent, their ability to maintain a high and legitimate visibility, and their capacity to maintain a working relationship with the formal sector.

Conclusions

Community involvement in the fight against AIDS has now attained unprecedented proportions. Community-based groups have sprung up in developed countries everywhere and are increasingly active in developing countries too. Their contribution in confronting AIDS is now indisputable and major lessons have to be learned from them. Application of their experiences to other areas of health and health care merits consideration. As we move into the second decade of the pandemic, the work of these groups will need to be strengthened. To do so will require concerted action by the communities in question, national authorities and the health and social services that ultimately stand to gain from the work of these community action groups.

These 'parallel' discussions benefitted from, and would have been impossible without the contributions of the moderators and all the speakers who brought to the sessions a wealth of personal experience and commitment.

Dr. Manuel Carballo, Programme on Substance Abuse,
World Health Organization (WHO), CH–1211 Geneva 27 (Switzerland)

Subject Index